T0393196

Applications of Nanomaterials for Energy Storage Devices

Electrochemical energy storage devices are the prime interest of researchers and students. This book provides a comprehensive introduction to nanomaterials and their potential applications specifically for electrochemical devices (rechargeable batteries, supercapacitors, and so forth) in a coherent and simple manner. It covers fundamental concepts of nanomaterials, chemical and physical methods of synthesis, properties, characterization methods, and related applications.

Features:

- Introduces the evolution of nanoparticles in electrochemical energy storage devices.
- Provides the detailed information on step-by-step synthesis of nanoparticles.
- Discusses different characterization methods (structural, electrical, optical, and thermal).
- Includes the use of nanoparticles in various electrochemical devices.
- Aims to bridge the gap between the material synthesis and the real application.

This book aims at Senior Undergraduate/Graduate students in Material Chemistry, Electrochemistry and Chemical Engineering, and Energy Storage.

Applications of Nanomaterials for Energy Storage Devices

Edited by

Amit Saxena, Bhaskar Bhattacharya and
Felipe Caballero-Briones

CRC Press is an imprint of the
Taylor & Francis Group, an **informa** business

First edition published 2023
by CRC Press
6000 Broken Sound Parkway NW, Suite 300, Boca Raton, FL 33487-2742

and by CRC Press
4 Park Square, Milton Park, Abingdon, Oxon, OX14 4RN

CRC Press is an imprint of Taylor & Francis Group, LLC

ISBN: 9781032106311 (hbk)
ISBN: 9781032106328 (pbk)
ISBN: 9781003216308 (ebk)

DOI: 10.1201/9781003216308

Typeset in Times New Roman
by codeMantra

Contents

Preface

Three years heretofore, suddenly the "Beauty of Nature" enthralled me, while I was sitting in my garden. That experience was related to the origin of nature, the world, and the universe. This set me on the path to drafting a book that could talk about nature and its creations. Initially, I asked myself a question about how such a wonderful and attractive world was created. According to Hindu mythology, Lord Brahma wove this entire universe. However, being a person of physics, the answer to this question must be more scientific. Hence, this book is a scientific way of explaining creation with the help of nanoscience. While planning for this book, it was arduous to get the appropriate text and chapters. Therefore, I bellied up to my other editors/contributors (Prof. Felipe and Prof. Bhattacharya). Finally, because of my mentors' support, we got a group of authors who are experts in nanoscience/nanotechnology. This book is a collection of important topics, concepts, and chapters of all expertise, which would show the pathway between nanoscience and nature.

This book is a compilation of ten important chapters. Chapter 1 is dedicated to understanding nature in the view of nanoscience. In this chapter, the author has taken much effort to explain the importance of nanoscience in every corner of this universe. Furthermore, Chapters 2 and 3 explain the fundamentals of nanoscience and nanotechnology. Chapters 4 and 5 discuss the various physics and chemical methods of the preparation of nanoparticles. Appropriate experimental techniques with diagrams are included in these chapters. Furthermore, Chapter 6 focuses on the important properties of nanoparticles, like electrical, mechanical, optical, etc. Chapter 7 includes the various characterization techniques which are important to analyze the nanoparticles. Furthermore, the next three chapters discuss the application part of nanoparticles for various devices, rechargeable batteries in particular. This collection of chapters would be a single-point knowledge center for all readers. We hope the reader enjoys going through the book and will get the best out of this.

This book is dedicated to our parents and teachers. We are indebted to the contributors/authors of this book. Because of their efforts and in-time completion of their chapters, this book was completed in a time span. We also thank all the publishers and authors for their permission to use the materials like figures/tables from their publications. We are also thankful to Shri. Vaishnav Vidyapeeth Vishwavidyalaya, Indore; MMV, Banaras Hindu University, Varanasi; and National Polytechnic Institute, Materials, and Technologies for Energy, Health and Environment (GESMAT), CICATA Altamira, for supporting us to complete the edition of this book. Finally, we are indebted to CRC Publication for accepting our proposal and publishing the work.

We end with the words of Prof Richard Feynman: "There's Plenty of Room at the Bottom: An Invitation to Enter a New Field of Physics"

Editors

Dr. Amit Saxena

I am a regular faculty from the Department of Physics, Shri. Vaishnav Vidyapeeth Vishwavidyalaya, Indore, India. I am also a lifetime member of IAPT, Kanpur (Indian Association of Physics Teachers), and NMS (Nano and Molecular Society), New Delhi. My area of research interest is the application of carbon-based materials, nanoparticles, and biomaterials for electrochemical applications. My research group is extensively working on rechargeable batteries, DSSC, OSC, quantum dots, supercapacitors, etc. Very recently, we are being actively involved in the extraction of biopolymers from various biomaterials for biodegradable electrochemical devices. Also, we filed a patent in this direction.

My formation is B.Sc. in Electronics, M.Sc. in Materials Science, and Ph.D. in Mixed Conducting Polymer Electrolytes. I received a gold medal for my PG studies. I have two patents, two books, three chapters, and eight research papers in reputed international journals and publications. My passion is teaching, reading mythological text, and collaborative research communications. I am so thankful to Prof. Felipe Caballero-Briones and Prof. Bhaskar Bhattacharya for accepting my invitation for being coeditor of this book.

Dr. Bhaskar Bhattacharya

Dr. Bhattacharya is a Professor of Physics at Mahila Mahavidyalaya, Banaras Hindu University, Varanasi, India. Dr. Bhattacharya has teaching and research experience of more than 30 years and has more than 100 research papers, 4 book chapters, and 2 patents to his credits. His field of research includes polymer electrolytes, dye-sensitized solar cells, quantum dots, carbon nanostructures, and ion beam interaction.

Dr. Bhattacharya has obtained his B.Sc., M.Sc., and Ph.D. degrees in Physics from Banaras Hindu University and has worked at different academic institutions. In addition to teaching and research, Dr. Bhattacharya has been involved in academic administration in different capacities. He has worked as a research professor at KAIST, Korea, and has intensively worked on nanostructures and quantum dots.

Dr. Felipe Caballero-Briones

I am a member of the Mexican Academy of Science and Group Leader of GESMAT, Materials and Technologies for Energy, Health and Environment Group at Instituto Politecnico Nacional-CICATA Altamira close to Tampico, at the Gulf of Mexico Northeast Coast.

The strength of GESMAT is to develop approaches to prepare and modify graphene-based semiconductors and oxide materials by simple chemical/ electrochemical routes. Within our collaborations, our materials are being applied in solar cells (TFSCs, DSSCs, QDSSCs, and BHJSCs), microbial and polymer fuel cells, thermoelectric, supercapacitors, batteries, and photocatalytic among others. Also, water remediation, artificial photosynthesis, lubrication, coating for surface protection, and materials for dosimetry and for cancer treatment lie in our fields of interest.

My formation is B.Sc. in Industrial Chemistry, M.Sc. in Advanced Technology, and Ph.D. in Materials Science and Technology. Traveling, overseas collaboration, as well as science communication are among my passions. I am grateful to Dr. Saxena for the invitation as a coeditor of this book and for his incredible interest in teaching, science, and culture.

Contributors

Bhaskar Bhattacharya
Department of Physics, MMV
Banaras Hindu University
Varanasi, India

Felipe Caballero-Briones
National Polytechnic Institute, Materials and Technologies for Energy, Health and Environment (GESMAT)
CICATA Altamira
Altamira, México

Yarazett Hernández Castillo
IPN Research Center
Applied Science and Advanced Technology
Altamira, Mexico

N. Cruz-González
National Polytechnic Institute, Laboratory of Functional Materials
CICATA Legaria
Miguel Hidalgo, Mexico City

Nelson Y. Dzade
Department of Energy and Mineral Engineering
Pennsylvania State University
University Park, Pennsylvania

F.J. Espinosa-Faller
Escuela de Ingeniería, Universidad Marista de Mérida
Mérida, México

N.E. García-Martínez
National Polytechnic Institute, Materials and Technologies for Energy, Health and Environment (GESMAT)
CICATA Altamira
Altamira, Mexico

N. Gnanaseelan
Instituto Politécnico Nacional, Materiales para Energía, Salud y Medioambiente (GESMAT)
CICATA Altamira
Altamira Tamaulipas, México

Meenal Gupta
Department of Physics, School of Basic Sciences and Research
Sharda University
Greater Noida, India

Ashok Jadhavar
Department of Physics
New Arts, Commerce and Science College
Ahmednagar, India

Sandesh R. Jadkar
Department of Physics
Savitribai Phule Pune University
Pune, India

Hareesh K.
Department of Physics
RV College of Engineering
Bengaluru
India
and
Center of Excellence on Macro-Electronics, Interdisciplinary Research Center
RV College of Engineering
Bengaluru, India

S.K. Kamaraj
Tecnológico Nacional de México
Instituto Tecnológico El Llano
El Llano, México

Mahesh M. Kamble
Department of Physics
DEA'S Anantrao Pawar College
Pune, Indoa

Ashwani Kumar
Nanoscience Laboratory
Indian Institute of Technology
Roorkee, India

Yogesh Kumar
Department of Physics
Govt. College Sec-12 Palwal
Haryana, India

Yogesh Kumar
Department of Physics, A R S D College
University of Delhi
New Delhi, India

D.K. Kushvaha
Department of Physics
Birla Institute of Technology
RanchiJharkhand, India

R. M. Mehra
School of Engineering and Technology
Sharda University
Greater Noida, India

S. J. Montiel-Perales
National Polytechnic Institute, Materials and Technologies for Energy, Health and Environment (GESMAT)
CICATA Altamira
Altamira, Mexico

M.S. Ovando-Rocha
Universidad Tecnológica de Altamira, Boulevard de los Ríos km 3+100, 89608 Puerto Industrial, Altamira, México

Puneeth Kumar P.
Centre for Advanced Materials
REVA University
Bengaluru, India

Ganesh K. Rahane
Department of Materials Engineering
Indian Institute of Science (IISc)
Bangalore, India

Sachin R. Rondiya
Department of Materials Engineering
Indian Institute of Science (IISc)
Bengaluru, India

S.K. Rout
Department of Physics
Birla Institute of Technology
Ranchi, India

Anurag Roy
Environment and Sustainability Institute
University of Exeter, Penryn Campus
Cornwall, United Kingdom

F. Ruiz-Perez
Instituto Politécnico Nacional, Materiales y Tecnologías para Energía, Salud y Medio Ambiente (GESMAT)
CICATA Altamira
Altamira, México

Amit Saxena
Department of Physics
Shri Vaishnav Vidyapeeth Vishwavidyalaya
Indore, India

Sweta Sharma
Department of Physics
GSCW Kolhan University
Jharkhand, India

Pushpa Singh
Department of Zoology, SSN College
University of Delhi
New Delhi, India

Rahul Singh
Department of Chemical and Biomolecular Engineering
Sogang University
Seoul, Republic of Korea

Mahesh P. Suryawanshi
School of Photovoltaic and Renewable Energy Engineering
University of New South Wales
Sydney, Australia

R.V. Tolentino-Hernandez
Instituto Politécnico Nacional, Materiales y Tecnologías para Energía, Salud y Medio Ambiente (GESMAT)
CICATA Altamira
Altamira, México

Santosh J. Uke
Department of Physics
J.D.P.S. College
Amravati, India

1 Motivation
Nature to Nano

Amit Saxena, Bhaskar Bhattacharya, and Felipe Caballero-Briones

CONTENTS

DOI: 10.1201/9781003216308-1

1.1 INTRODUCTION

This is the first and most fundamental chapter of this book. It will allow the reader to understand and deep into the concepts of nanoscience and nano entities, first through the observation of nature examples and the explanation of the properties that made the entities at the nanoscale so interesting. Out of these, a few examples from nature are structuring at the nanoscale of the lotus leaf which makes it superhydrophobic, the iridescence of colors of butterfly wings, fireflies that glow in the night, many flower pigments, minerals, human skin, feathers, horns, hair, bones, and much more. Here, the term natural to nano means the materials, which offer properties like nanomaterials without any processing or human interference. The atomic/molecular arrangement defines the chemical, physical, or electrical properties of the substance, wherein the biological properties of the objects are due to their supramolecular structures. The interaction of water, light, and other materials with these biological substances provides freak properties, which can be appreciated [1].

These all are a point of fascination among scientists for a long time due to their structural and physical properties. All special properties offered by these are due to nanosized particles as per the known scientific facts.

1.2 HISTORICAL DEVELOPMENT OF NANOPARTICLES

Humans have developed and used ceramic materials from around 7500 years back [2]. Recent studies have shown that many of the encountered objects' properties such as some colors and opalescence of iridescence are in fact due to nanomaterials. For example, close to 5000 BC, in Cyprus, clothes were bleached with the help of clay; thereof, the size of these clay particles was of few nanometers in dimension [3]. Some 4000 years back, Egyptians used synthetic chemical-based materials for hair dyes, wherein the size of these dyes was close to 5 nm in diameter [4]. Also, the first artificial pigment, popularly known as "Egyptian Blue", was synthesized by the Egyptians with sintered, nanosized glass and quartz; thereof, around the third century BC, it was used for the purpose of coloring objects [5]. Some more examples of nanomaterials from the Bronze Age around 1200–100 BC are the objects for which red color was created by the plasmonic effect of Cu nanoparticles [6]. From 801 BC to 900 BC, the use of glazed ceramics was popular among Mesopotamians for decoration purposes [1]; due to the presence of silver and/or copper nanoparticles in between the glaze layers, an astounding optical effect was seen in this type of decoration, i.e., the glittering green and blue color due to the reflection phenomenon. When transmission electron microscopy (TEM) analysis of these decorative materials was performed, it was revealed that silver nanoparticles were deposited on the surfaces separated 430 nm apart from each other, with the smaller particles (5–10 nm) deposited on the outer layer and the large particles (5–20 nm) deposited on the inner layers. This construction gave an effect of interferences. Hence whenever light is scattered due to these inner and outer layers, there is a phase shift between the reflected light rays, wherein due to phase shift during scattering, a different wavelength appears. Other examples are the Cu- and cuprous oxide-based Celtic red enamels, from around 400 to 100 BC [7]. Apart from all these, the Roman glass (Lycurgus) cup is also a popular product, wherein the color of the cup changes with the incident light direction, looking red when light is incident from the back, while it looks green in color when light is incident from the front [8]. These Lycurgus cups consist of alloys of silver–gold nanoparticles, wherein their ratio is 7:3; thereof, also 10% copper is present[9]. Furthermore, stained glass, which is red and yellow, was also found in the churches constructed during the 11–14th centuries, wherein the colloidal gold and silver nanoparticles are the main contents [10]. Based on the same technique, the popular Satsuma glass was produced in Japan in the mid of the 19th century. Here, the ruby color of the Satsuma glasses is due to the absorption effect of copper nanoparticles [11].

In modern times, Michael Faraday was the first who described the synthesis of colloidal gold in 1857 and revealed the differences between the bulk and colloidal counterparts. Furthermore, in 1908, Mie explained the cause of certain colors of metal colloids [12]. The use of silicon dioxide nanoparticles in place of carbon black was started in the 1940s; thereof, the initial purpose was for rubber augmentation [13]. Nowadays, the synthesis of nanoparticles can offer remarkable improvement in the mechanical, electrical, optical, and thermal properties of bulk materials; thereof, these improvements can be further used for various applications and device fabrications. Even the use of nanoparticles in the field of biology can offer various self-healing, antibacterial, and anti-freezing properties as well. Regardless of all the abovementioned important properties of nanoparticles, there are huge possibilities for the future of nanoparticles for several technologies. Many manufacturing companies use the idea of nanoparticles for their electronic, electrical, and mechanical products; for example, in 2003, Samsung used silver nanoparticles for antibacterial technology [14]. Also, nanoparticles and nanosized materials are broadly in use for the improvement of tires to increase the harshness of the vehicle surface, increase the efficiency of the engine, make the car body lighter and stronger, etc. [15]. Nanoparticles have also been used for metallic and non-metal paints, which can make scratch-resistant glasses/surfaces [16]. In series of all these development and enhancement of nanoparticles, nowadays more than 1800 products based on nanoparticles are available for consumer products commercially in over 20 countries [17].

1.3 NATURE AND NANO

Nature is awash in nanomaterials and scientists are studying these nanostructures to determine their possible applications and properties through a research field called biomimicry. There are many distinct types of nanostructures in nature. It includes some organics structures like feathers and skin strands. Furthermore, it shows many natural inorganic materials like clays and carbonous soot. Also, proteins and chitins are common examples of organic nanostructures. The structure of the planetary atmosphere plays a role in several natural phenomena, such as the wetness of surfaces, the shimmering of butterfly wings, and the adhesion of gecko feet. Nanostructured materials, which have great mechanical strength and toughness, have been fabricated by nature and evolved sophisticated bottom-up processes. Nacre is one of the hardest materials which is the shimmering shell of mollusks. Mollusks deposit $CaCO_3$ over porous polysaccharide chitin layers to produce nacre. In time, the mineral crystallizes, producing layers of organic material separated by stacks of $CaCO_3$. Here, the rigidity of nacre is because of the inter-locked assembly of molecules. Nowadays, the research focus is on the synthesis of strong materials by biomimetic nanocomposites. These materials can be used for lightweight defense systems, transportation, electric devices, durable electronic devices, aviation, and many others. We know that the colors are because of the pigments, but nature creates them due to structural color. For example, the wings of some insects contain chitin which is hexagonally packed. The order of these wings is between 200 and 100 nm, which enables the wings to act as self-cleaning and anti-reflective layers. It also imparts mechanical durability and enhances aerodynamics, etc. This stunning blue color of the morpho butterflies is due to crystals of natural photonic, which are between the wing ribs. Interestingly, there is no involvement of pigments in wing color. One more example of natural nanomaterials is the lotus leaf. Here, the concept of self-cleaning of the lotus leaf is associated with chemical and physical properties at the nanoscale level. In 1994, a famous botanist, Wilhelm Barthlott, filed a patent for the lotus effect. This patent was for the micro- and nano-properties of the lotus leaf surface. Here, it is important to mention that the size of the protrusion is about 100 nm in height. Due to the surface tension of the water droplets and the surface morphology of the lotus leaf, while rolling, it would clear the surface of the leaf. This can be seen in Figure 1.1.

However, still, it is primitive in many areas of development and manufacturing. Researchers/scientists need to do a lot to get close to natural accuracy. To date, no one has succeeded in achieving the capacity of photosynthesis to energy storage. We are even not close to the efficiency of biomolecules for energy transfer. More interestingly, although we have many varieties of **reverse osmosis** (RO) technologies

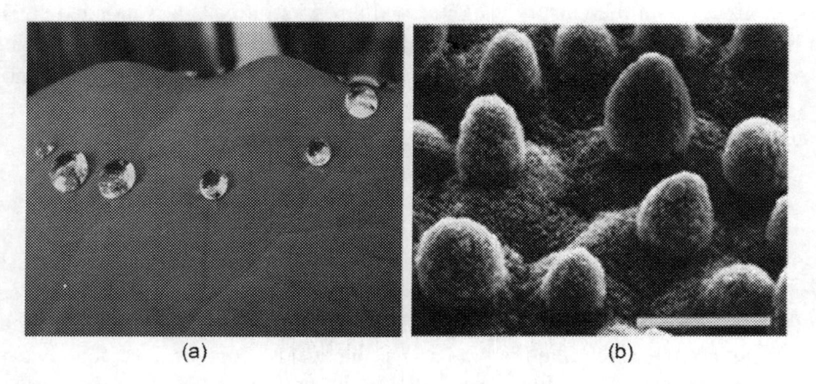

FIGURE 1.1 A classic example of a macroscopic property in nature. (a) A self-cleaning concept in a lotus leaf. (Adopted originally from "*Natural necklace*" by Tanaka who is licensed under CC BY 2.0). (b) Electron micrograph of a lotus leaf (surface); size appears to be 20 μm [18]. (Adopted with permission from Barthlott, W., Neinhuis, C. *Purity of the sacred lotus, or escape from contamination in biological surfaces. Planta* 202, 1–8 (1997), Copyright: Springer (1997), https://doi.org/10.1007/s004250050096.)

to date, we are still no closer to the purity and efficiency of storage as coconut water and watermelons. We are in the era of the digital world, where we store our day-to-day details in digital memory cards/drives, but still, the data which can be stored in a human brain is not comparable to date. Images and moving objects, which can be seen by our eyes, cannot be compared with any digital cameras. Even the sensitivity of olfactory receptors in a dog is much more than any latest sensors which we have developed. The conventional energy transformation has not ensured the appropriate conversion of energy. To date, the highest conversion efficiency of conventional photovoltaic panels is not more than 25%. Even the best engine in the world has shown only 52% of efficiency, and only 38% of energy is used for cooking, which is produced by gas. In a comparison of all the above examples, our body can utilize whole energy as produced; therein, some other examples are plants, bacteria, etc. That is why we can say that if we want to achieve our highest technology, then we must walk toward nature and learn from nature. In the coming sections, we will learn the way of coordination, sources, and development between nature and nanoparticles.

1.4 OVERVIEW OF NATURAL NANOPARTICLES AND SOURCES

There are many natural materials in the universe that own the properties of nanostructures due to their compositions, like spider silk, flower colors, clay, paper, etc. [19,20]. Usually, based on the origin of the sources, nanomaterials can be classified into three major categories; these are shown below in Figure 1.2.

Now, let us discuss these sources of nanoparticles to understand the evaluation of nanoscience from nature. It is not complete about every detail, but a focused reader can find further details at the end of the chapter; thereof, detailed biography has been provided there.

1.4.1 NATURAL SOURCES OF NANOPARTICLES

There are numerous ways of producing natural nanoparticles, where some of them are fires in forests, an explosion of volcanoes, reactions during the photochemical process, etc. This all-natural way of producing nanoparticles affects the air quality worldwide. Even shedding of animal skin and hair and the shedding of leaves or skin of plants also contribute toward the natural nanoparticle's compositions. In addition to this, a huge quantity of synthetic nanoparticles is generated due to industrialization, fuel burning, vehicle use for transportation, etc., which is an emergency condition for the present world. However, it was reported that in the atmosphere, only 10% of aerosols are due to humans, while the remaining 90% are due to nature [21].

1.4.1.1 Sandstorm and Cosmic Dust

Sandstorm or dust storm is a common phenomenon in the domain of meteorology, which usually takes place in dry or semi-dry regions. These phenomena arise due to strong hurricanes. These hurricanes take the fine granules of dust from the dry or semi-dry areas to another location and deposited them.

In this series of dust storms, a star (Eagle Nebula) which is about 6500 light years distant from the earth, has dust particles, which are disk-shaped and have the

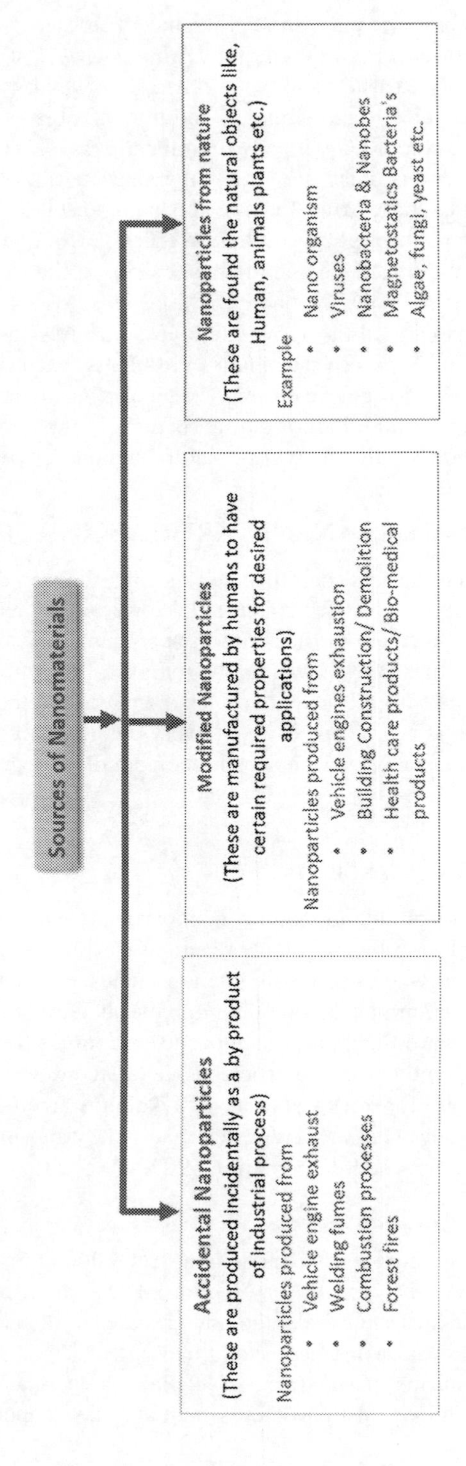

FIGURE 1.2 Classification of sources of nanomaterials.

ability to develop a solar system [22]. While analyzing the celestial activities and stardust during various space missions, it was concluded that the major components of this stardust or meteorites are in the dimension of nanoparticles, wherein the main components are carbon, nitride, silicate, etc. [22]. Here, it would be interesting to mention that in the Murchison meteorite, the existence of diamond was found, which is the best example of the existence of nanoparticles in space [23]. This example eventually proves that there is an existence of various nanoparticles all around the universe, wherein mixed, classified, and modified nanoparticles are important. These nanoparticles in space are due to sudden changes in temperature, electromagnetic radiations, physical crashes, and shock waves [22]. A similar analysis of the existence of nanoparticles in the atmosphere was reported by Prashant and Al-Dabbous, wherein they mentioned that in the summertime and on roadsides of Kuwait, there is airborne pollution of nanoparticles [22]. Even this type of air pollution due to airborne nanoparticles can be found in many places in the world, and these are the primary reasons behind many health problems. Figure 1.3 shows the presence of nanosized dust particles in the air.

1.4.1.2 Nanoparticles through Natural Decay and Volcanic Activity

Nanoparticles are part of our metallurgic world because they commonly occur during weathering and volcanic explosions. During the venting of volcanoes, particles ranging from a few micrometers to a few nanometers in size spread into the atmosphere, wherein the number of nanoparticles may be about 30×10^6 tons [25–28]. These released nanoparticles flow throughout the atmosphere and deposit on the lowest layer of the troposphere and stratosphere. However, this effect is more prominent in the regions, which are close to the volcanoes. The volcanic explosion majorly affects sunlight by scattering and blocking it, which further affects plant, human, and animal activities. Most of the time, these volcano explosions release heavy metals which are harmful to humans, causing skin problems, headaches, nose and eye infections, asthma, and Kaposi's sarcoma [29,30].

FIGURE 1.3 Field emission SEM: (scale 5 μm). (a) Dust in the course of and (b) dust after the storm. 19/03/2002 [24]. (Adopted with permission from Senlin Lu, T. P. Jones, Longyi Shao, et al. *Microscopy and mineralogy of airborne particles collected during severe dust storm episodes in Beijing, China. Journal of Geophysical Research: Atmospheres*, 110 (2005) doi:10.1029/2004JD005073. Copyright: 2005 by the Amer. Geophys. Union.)

1.4.1.3 Volcanic Ashes

Usually, the origin of natural nanoparticles is considered due to erosion and the explosion of volcanoes. The temperature of the cinder may reach the limit of 1400°C upon release during the explosion of volcanoes. When these cinders reach the available sources of water or soil near the volcano, they cause chemical reactions and create different ranges of nanoparticles that may be deposited over theirs. This could be an environmental hazard. Many times, it is been found that the places close to such volcanoes have a negative impact on health as well. For example, the size of these cinder particles may vary between 100 and 200 nm in diameter, and these particles may be suspended easily in the air. Whenever these are inhaled by someone, they may create serious respiratory problems due to deposition in the respiratory tract.

1.4.1.4 Jungle Fire and Sea Water Evaporation

Forest fires are due to tremendous flashes of lightning and human interference in jungle areas, worldwide. As a result of a fire in the jungle, a huge quantity of fumes, cinders, and small nanoparticles are produced and expand over a long distance, which affects the overall ambient quality of air [31]. Figure 1.4 shows the existence of nanoparticles in smoke. It has been reported by many researchers that due to the deposition of fumes, cinders, and small carbon-based nanoparticles on the surface of the Himalayan region, glaciers are melting faster [32,33].

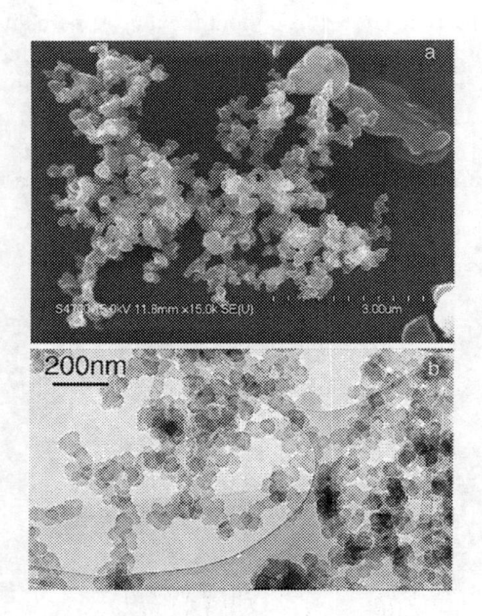

FIGURE 1.4 (a) SEM image of aggregated particles of carbon of smoke, 20/08/2000, South Africa. (b) TEM image of aggregated particles of carbon, 05/09/2000, Zambia [34]. (Adopted by permission from Jia Li, Mihály Pósfai, Peter V. Hobbs, Peter R. Buseck. *Individual aerosol particles from biomass burning in Southern Africa: 2, Compositions and aging of inorganic particles. Journal of Geophysical Research: Atmospheres*, 108 (2003) https://doi.org/10.1029/2002JD002310, Copyright 2003: the Amer. Geophy. Union.)

Here, it would be interesting to mention that even due to the evaporation of seawater and sea waves, various types of nanosized materials are produced [35]. Usually, the size of these nanosized materials is in the range of ~100 nm to a few micrometers; thereof, these are produced due to vapor and natural precipitation during sea vaporization [36]. Even in many places, it has been reported that these nanosized materials near the sea area can also be produced due to temperature changes. Here, the spraying of these nanosized materials due to sea salt carries many microorganisms and other pollutants to nearby locations and causes many health issues.

1.4.2 Engineered Nanomaterials

Apart from all-natural ways of production of nanoparticles, as discussed in the previous section, there are many more ways of production of nanoparticles, wherein the engineered nanoparticles are one of the important classifications, where the production of nanoparticles take place due to various reasons such as mechanical/chemical/biomedical industrialization, automobile fuel combustion, wood combustion during cooking, fuel combustion for power generation, welding, and much more. These all-anthropogenic activities produce nanoparticles such as carbon nanoparticles, silica nanoparticles, TiO_2, Al_2O_3, carbon black, Co, Ni nanoparticles, etc., due to which health problem is a big issue.

1.4.2.1 Nanoparticles from Fuel Combustion

Worldwide, the major source of natural/atmospheric nanoparticles is fuel combustion, wherein the burning of diesel produces nanoparticles of the order of 20–130 nm, whereas the burning of gasoline products, like petrol, benzine, etc., produces nanoparticles of the order of 20–60 nm [37–39]. It is also found that carbon nanotubes (CNTs) and other fibers were also produced as byproducts [40]. It is especially important to mention here that 90% of carbon nanoparticles are due to vehicles and other devices which combust fuel for various purposes [41]. Here, many studies show the existence of CNTs in human cells can produce granulomatous reactions, neoplasia in the lungs, stress, swelling, and skin cancer [42,43].

1.4.2.2 Demolition of Building and Cigarette Smoke

Demolishment of structures and smoking are the anthropogenic processes that produce and spread nanoparticles in the atmosphere. Smoke is one of the complicated compositions wherein approximately 1 lakh chemical compounds are observed [44]. Likewise, during the demolishment of buildings, nanoparticles of the order of 10 nm are produced [45]. Even during building demolishment, some other nanoparticles are also produced, like asbestos fibers, lead, building waste, glass, etc.

Smoking is one of the major causes of respiratory problems, heart problems, pancreatic cancer, genetic disorder, asthma, etc. [46]. Here, it is important to mention that the chance of many health problems can be reversed by stopping smoking [47]. To date, the effect of various dust particles and nanoparticles on humans is still not disclosed completely. However, after the rescue process at the World Trade Centre (11/09/2001), a severe problem related to cough and bronchial was reported among the firefighters [48]. This shows that there is a requirement for detailed studies on the ill effects of nanoparticles on humans and nature.

1.4.2.3 Nanoparticles from Healthcare Products

In most beauty products, there is a use of nanoparticles; for example, the anti-reflective and antioxidant properties in sunscreen lotion are because of nanoparticles [49]. Interestingly, most of the nanoparticles which are used for cosmetic applications are engineered nanoparticles that are prepared with various biological, physical, and chemical processes [50–52]. Furthermore, the nanoparticles are extensively in use for other commercial products, wherein paints, personal care, and clothes are a few examples [53]. The white color in most beauty products is due to TiO_2 nanoparticles which are larger than 100 nm in size [54]. Similarly, in some other cosmetic and food products, like toothpaste, wipes, food packets, and shampoos, there is a use of silver nanoparticles [55]. There are many nanoparticles that are under research for their potential uses in biomedical/cosmetic applications. Despite the potential application and growth of nanoparticle-based products, the harmful effect of these products on humans is still not known clearly.

1.4.3 Natural Sources of Nanomaterials

In the sequence of the nanoparticles in nature, as discussed in the above two segments, there are some more examples of nanoparticles that are available naturally. These examples are fundamentally from plants, microorganisms, algae, animals, insects, bacteria, sea species, viruses, and humans. The latest enhancement in optical instruments helps us a lot to analyze the size, structural, and morphological properties of nanoparticles. This study can lead to a greater understanding of the presence of nanosized organisms in nature, whereof this can enhance the biomedical applications of these microorganisms. The nanostructures in insects are formed through a regular process of development due to which they can live in difficult living conditions. Even the presence of nano-biominerals in the plants is due to water and soil, which they develop during their growth. Interestingly, the floaty/lightweight wings and attractive colors of the animals and insects are due to nano-wax and nanoparticles.

In this series, humans are one of the interesting examples where the existence of nanomaterials is very much reported like in bones, enzymes, antibodies, DNA, RNA, etc., nanostructured in dimension. DNA and RNA, the genetic materials, are the key components for cell development and even for the proper functioning of the living cell/tissues, whereof these are also nanostructured. Hence, finally, it can be concluded that nanostructures are the basic units for all types of life on earth. Now, in the undermentioned text, we will get some more details about the presence of nanostructures in living organisms.

1.4.3.1 Nanoscale Organisms

Nanoscale organisms are also known as nano-organisms which are found everywhere in this universe and in our bodies as well. Fundamentally, these nano-organisms are a class of nanomaterials that incorporates a huge count of organisms. For example, yeast, nanobacteria, fungi, and algae, can produce nanoparticles.

1.4.3.2 Viruses

Viruses, Figure 1.5, which may be living or non-living, are the biggest structurally identified molecular assemblies to date. They are usually found inside a host cell.

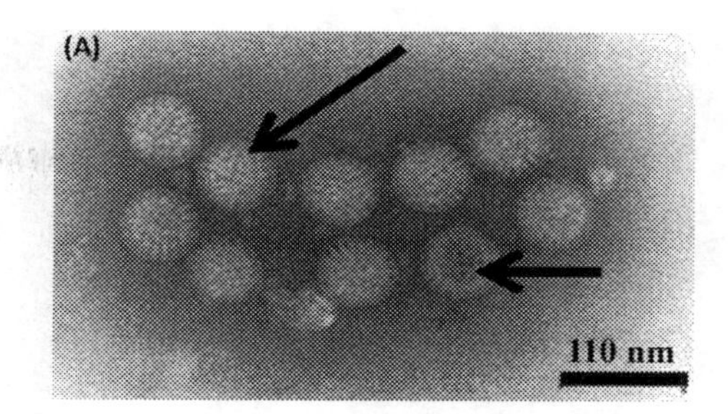

FIGURE 1.5 *Rotavirus* (Negative Strain) with complete and empty particles in swine excrement [66]. (Adopted originally from, M.H.B. Catroxo and A.M.C.R.P.F. Martins (02/09/2015). *Veterinary diagnostic using transmission electron microscopy. The Transmission Electron Microscope–Theory and Applications*, Khan Maaz, IntechOpen, DOI: 10.5772/61125. Available from: https://www.intechopen.com/chapters/48878, Copyright 2015, Cartaxo & Martin. Under http://creativecommons.org/licenses/by/3.0.)

Often, they are injurious to living organisms, humans, plants, and animals [56–59]. A very recent example of the harmful effects of a virus is the COVID-19 disease.

Nowadays, the recent development in molecular biology has enhanced the probability of tailoring viruses genetically, which can be further used for various biomedical applications. Some specific properties of viruses make them special in the category of nanoparticles, wherein the composition, ability to control the genome manipulation, various shapes, rapid growth, vulnerability to smaller molecules, and stability for temperature are important [60,61]. The use of viruses for targeted drug delivery through nanocargoes is popular in the biomedical field; these nanocargoes are prepared by removing the genetic medium from viruses [62]. The latest research on compound encapsulation in nanostructured cages has shown the potential use of viral nanoparticles [58,63]. In this area of drug delivery, the use of plant viruses is more potential due to their nontoxic behavior [64,65].

1.4.3.3 Bacterial Spores, Fungi, Algae and Yeast

It is very interesting, as shown in Figure 1.6, that silver nanoparticles can be formed by *Chlorella vulgaris* algae [67], cadmium sulfide nanoparticles by *Phaeodactylum tricornutum* [68], and many more (for more details, kindly refer bibliography). However, less research work has been done in this area, so the process of formation of nanoparticles through algae is still not identified [69]. Similarly, a research study shows that fungi have various enzymes, which can be handled easily, and this gives them a chance to synthesize metal and metal sulfide-based nanoparticles [70]. More information is summarized in Table 1.1.

We have seen how we can analyze the existence of nanoparticles in nature, and now we will discuss some special properties of these nanoparticles in the coming sections.

1.5 EFFECTS AT NANOSCALE

Nanoscience is also known as the "science of tiny". It will be remarkably interesting to know how size affects the properties of materials. In this chapter, we will discuss

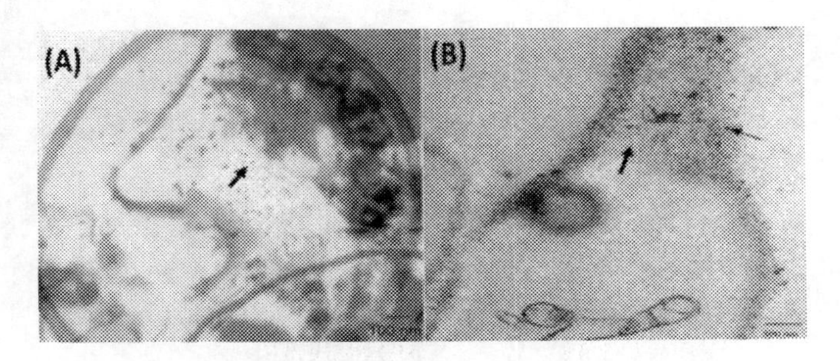

FIGURE 1.6 Synthesis of nanoparticles in algae and fungi, intracellularly. (a) Micrograph-TEM image showing Cu nanoparticles [71]. (From Salvadori, M. R., Ando, R. A., Oller do Nascimento, C. A., Corrêa, B. *Intracellular biosynthesis and removal of copper nanoparticles by dead biomass of yeast isolated from the wastewater of a mine in the Brazilian Amazonia. PLoS One* 9 (1), e87968 (2014). https://doi.org/10.1371/journal.pone.0087968, Copyright 2014 Salvadori et al. under http://creativecommons.org/licenses/by/3.0.) (b) Photomicrograph-TEM image showing intracellular nickel oxide nanoparticles [72]. (From Salvadori, M. R., Ando, R. A., Oller do Nascimento, C. A., Corrêa, B. *Extra and intracellular synthesis of nickel oxide nanoparticles mediated by dead fungal biomass. PLoS One* 10 (6), e0129799 (2015). https://doi.org/10.1371/journal.pone.0129799, Copyright 2015, Salvadori et al. under http://creativecommons.org/licenses/by/3.0.)

TABLE 1.1
List of Various Fungai/Algae and Yeast with Nanoparticles Synthesis

Fungi/Algae/Yeast	Type of Nanoparticles Syntheses	Refs.
Fungi–*Fusarium oxysporum* and *Verticillium* sp.	Au, Ag, and Au–Ag alloy NP synthesis	[73–75]
Fungi–*Fusarium oxysporum*	CdS quantum dots, zirconium particles	[76,77]
Algae–*Chlorella vulgaris*	Silver nanoparticles	[67]
Algae–*Phaeodactylum tricornutum*	Cadmium sulfide nanoparticles	[68]
Yeasts–*Candida glabrata*	CdS quantum dots	[78,79]
Yeasts–*Torulopsis* sp.	PbS nanocrystals	[80]
Yeasts–*Schizosaccharomyces pombe*	Silver nanoparticles	[81]

Source: L. *Filipponi*, iNANO, Aarhus University, Creative Commons ShareAlike 3.0

only the introductory part of these properties. Further details about the various properties and applications of nanoparticles are discussed in the upcoming chapters of this book.

1.5.1 Materials at Nanoscale

Now let us discuss the effect of nanoscale on material properties. Generally, the physical properties of any substance, mainly the conductivity, melting point, etc., can be measured by the study of samples, which can be easily done in the laboratory.

Fundamentally, 1 mole of any substance has 6.022×10^{23} entities (atoms and molecules), wherein the weight of 1 mole of the same substance corresponds to its atomic, molecular, or formula mass expressed in grams per mole. This is because the particle at the nanoscale follows quantum mechanics rather than Newtonian physics. Therefore, it is more accurate to say that the properties of the materials are size-dependent. It would be a fantastic way of learning in a classroom that the property of any material (gas, liquid, or solid) is due to how the molecules or atoms that build the material and how they are attached to each other (chemical bonding). Usually, the size of a particle is not referred to as a crucial element. Maybe students expect the color of gold as golden, whatever size it has, big or small. It may be true at the macroscopic level or even at the microscopic level, but it will certainly not be correct at the nanoscale because of the quantum effect. Here the example of gold is more precise for students because at the nanoscale the color of gold becomes ruby red. As shown in Figure 1.7.

In a more simplified manner, we can understand the size effect of nanosized particles with the example of gold nanoparticles. As shown in Figure 1.7.

1.5.2 Physics at Nanoscale

In practice, nanoscale means a group of molecules or atoms, for example, a group of 8 H-atoms or a group of 3.5 atoms of Au. Hence, if we try to compare a nanosize particle, then it would be found close to a single atom, rather than bulk materials. As we all know that quantum mechanics is a branch of science that deals with the properties (energy or motion) of microscopic particles, so to study the behavior of nanosized particles, we need to understand and use quantum mechanics. However, we are not discussing the detailed idea of quantum mechanics here since it is out of the area of the chapter. Still, the reader can get the details somewhere else. Here, some details of quantum effects would be discussed that would be appropriate to understand nanosized particles.

- As nanosized particles have small mass and size, so to understand the behavior of nanoparticles, electromagnetic forces are prominent.
- **Wave-Particle Duality:** As per the statement of de Broglie, every tiny moving particle is associated with a wave, so nanosize particles can exhibit a wavelike behavior, and their position can be explained by the wave function.
- **Quantum Confinement:** Nanoparticles would always be confined to boundary conditions, rather than wander freely in bulk materials.

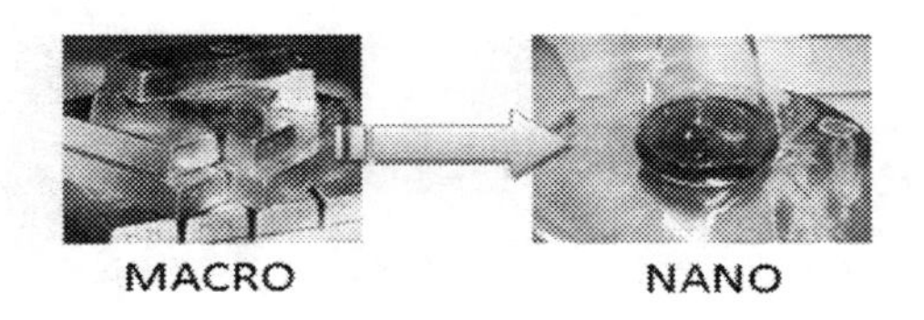

FIGURE 1.7 Size-dependent color of gold.

- **Quantized Energy:** Nanosized particles always show the effect of quantized energy, rather than discrete energy levels.
- **Irregular Motion of Molecular:** In a bulk material, the particles/molecules always move because of kinetic energy and this irregular molecular motion is present always. This irregular motion is not appreciable in the case of macroscale in comparison to small size. However, in the case of nanoscale particles, this irregular motion shows a piece of valuable information about the behavior of nanoparticles.
- **Enhanced Ratio of Surface-to-Volume:** This is one of the most important properties of nanosized particles, where the surface area of a nanomaterial is higher in comparison to bulk materials.

1.5.3 Chemistry at Nanoscale

As we have already discussed in Section 1.4.2, nanosized particles are due to a collection of few atoms (a group of 8 H-atoms or a group of 3.5 atoms of Au), and these all atoms/molecular are bound chemically with each other. Therefore, it would be important to understand these chemical bindings between the nanoparticles as well. Usually, these can be classified as follows.

- **Chemical Interaction (Intra-Molecular Bonds):** In these types of bonding, the molecular structure is found to change chemically. These comprise metallic, ionic, and covalent bonding.
- **Physical Interaction (Inter-Molecular Bonds):** In these types of bonding, there is no change in the molecular structure chemically. These comprise ion–ion and dipole–ion, hydrophobic, van der Waals, and hydrogen bonds.

Usually, the van der Waals and hydrogen bonds are weak bonding, but the bonding force would be large once a greater number of molecules bond together. For example, DNA (with 2 nm in cross-section) is bound in two helixes by an enormous number of hydrogen bonds. This concept becomes more appreciable in the area of nanoscience because of the large surface-to-volume ratio and strong force.

With a simple example, we can explain to the students inside the classroom how the force between molecules increases as the molecules grow. To illustrate this example, we can use two books. Initially, a few pages of these two books are stuck to each other and now if these two books are pulled then both of them will be separated from each other easily. However, if we stick all the pages of these books to each other and then if we try to bring them apart, then it would be not possible. This is one of the simplest examples which can explain how bonding forces increase between molecules once they are large in count and with a large surface area.

1.6 DISTINCTIVE PROPERTIES OF NANOSCALE MATERIALS

As discussed in the previous few sections, the size of particles is incredibly significant for various applications. Hence, in the further sections, we will have a look at the different important properties of nanoparticles in detail. These are as follows.

1.6.1 Surface Properties

"Surface science" is a branch in which we discuss the biological, chemical, and physical properties of any surface. The chemical and physical feature of any bulk or nanosized materials is due to of surface property of that material. The surface of any material offers various properties, like floating of substances on water, energy at interfaces of the materials, speed of chemical reactions (like catalysts), etc. Normally, in place of surface, the interface is a more common term to express the boundary between the materials and the nearby environment (like gas, liquid, or solid).

For example, the surface area of the material appreciably increases if the material is divided into a small segment, wherein there is no change in the total volume of the material. This can be shown below in Figure 1.8.

Here in this figure, it can be concluded that the surface-to-volume if the small-sized particle increases in comparison to the bulk parent material.

Now to understand the concept of change in the surface area, let us take an example, where we start with the material of size 1 m^3 cube. Gradually, these materials are cut into a smaller cube until it reaches the size of 1 nm^3. Even this can be understood with the following Table 1.2.

In the further chapters, we will find more details about the effect of this surface area on various properties of nanosized particles.

1.6.2 Electrical Properties

We can classify the materials into three classes, based on their energy-level diagram. These are insulators, conductors, and semiconductors. These are because of the band gap between the valence band and conduction band. We all are aware of this definition from your earlier classes. If a material has a bigger band gap between the valence

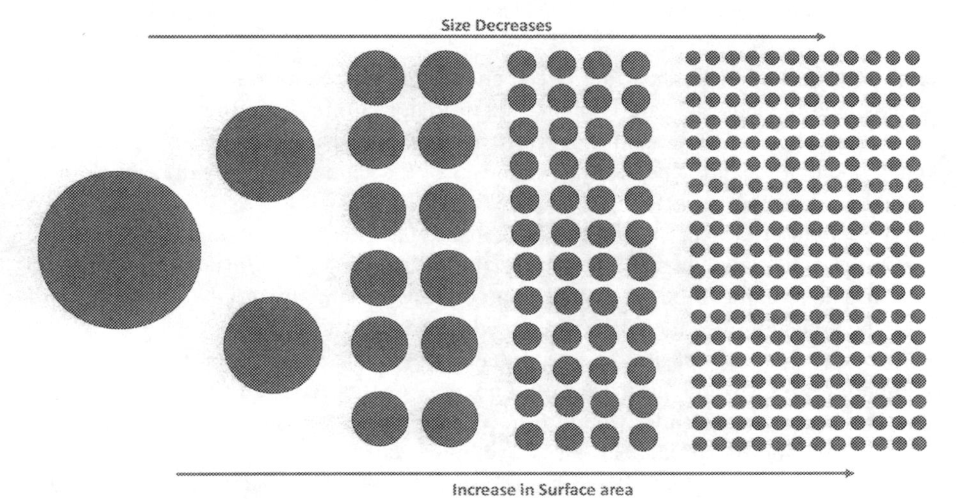

FIGURE 1.8 Diagrammatic showing the increase in surface-to-volume ratio with a decrease in size.

TABLE 1.2
Size vs. Surface Property of a Material

Size	Number of Cubes	Surface Area
$1\,m^3$	1	$6\,m^2$
$0.1\,m^3$	1000	$60\,m^2$
$0.01\,m^3$	$10^6 = 1$ million	$600\,m^2$
$0.001\,m^3$	$10^9 = 1$ billion	$6000\,m^2$
$10^{-9}\,m^3$	10^{27}	$6 \times 10^9 = 6000\,km^2$

Source: *L. Filipponi*, iNANO, Aarhus University, Creative Commons ShareAlike 3.0.

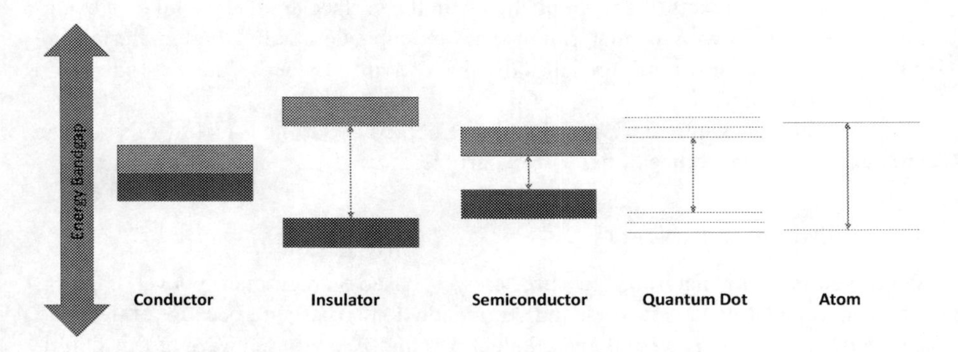

FIGURE 1.9 The band gap between the band (arrow) in a conductor, insulators, semiconductor, quantum dot, and atom. The band gap increases with the loss of energy states.

and conduction bands, then there is no chance of electron movement between these bands, and the materials are known as an insulator. Similarly, in the case of conductors, these bands overlapped, and electrons can easily move between the valence and conduction band, wherein the properties of the semiconductor can be explained based on the quantum effect.

The quantum effect can increase the band gap energy, as shown in Figure 1.9. Even in small-size/nanosized materials, the overlapping of energy bands vanishes, and the band gap can be seen. This can define why some materials offer semiconducting behavior once their size is reduced.

Here, the concept of an increase in band gap is due to the quantum effect. Due to quantum imprisonment, there would be more energy required to get absorbed by the band gap of materials, wherein more energy means a small wavelength (blue shift). Even nanosized particles also have higher energy (due to surface effect), due to which they radiate fluorescent light and show the blue shift. Here, this concept can explain the properties of light absorption and emission by controlling the crystallite size of the semiconductor nanoparticles. More details of optical properties would be discussed in the upcoming chapters.

Few nanomaterials offer some exceptional electrical properties due to their distinctive structures. The examples are CNTs and fullerenes. Here, CNTs show the properties of semiconductors or conductors, based on their structure. Further details on electrical properties would be discussed in Chapter 6 of this book.

1.6.3 Optical Properties

We all know how we see the color of an object. Whenever light is incident on the surface of any object, then a particular wavelength is absorbed by the object, and the remaining range of wavelength gets reflected from the surface, wherein the color of that object is due to the wavelength which was absorbed. This is one of the simple color-appearing phenomena, which we are aware of. However, some nanomaterials show quite different properties in comparison to their bulk counterparts, like the color of the particle, transparency, etc.

In the further text, we will discuss the cause of these optical changes with some examples. Initially, we will discuss how the nanosized particles generate the color, wherein we will discuss how light interacts with the nanosized objects in detail. In common, nanosized objects offer unusual optical properties when light interacts with the nanosized object. Table 1.3 shows an overview of a few of them.

Now to explain the optical properties in more detail, let us discuss the plasmon effect first. A quantum or quasiparticle associated with a local collective oscillation of charge density, popularly known as plasmon, is responsible for the various properties of metal nanosized particles. In this mechanism, whenever light falls on the surface of metals, which may be any size, some percentage of light waves grow along the surface of the metals, wherein they produce plasmon at the surface. Here, this plasmon is nothing but a set of electrons that propagate parallel to the interface

TABLE 1.3
Colors from Nanomaterials while Interacting with Light

Color Effect Due To...	Example of Nanosized Materials...
Interference. This is an optical phenomenon of constructive and destructive interference (superposition) of light waves, after interacting with nanomaterials.	Colors of wings of a butterfly, color from soap surface (liquid crystal)
Scattering. The dissimilar particle scatters a different wavelength, which causes colors.	Colloids (milk)
Surface plasmons. This phenomenon, is in charge of the bright color of the nanoparticles and metal colloids.	Metal colloids (Example: nano-gold)
Quantum fluorescence. Once the semiconductor absorbed energy then due to quantum imprisonment, it leads to distinct energy levels through which emission of energy takes place, known as fluorescence.	Semiconductor quantum dots (QDs)

of the metal/dielectric. It is interesting to mention that if this plasmon is generated in a bulk material, then no appreciable effect is observed, but the plasmon phenomenon in nanoparticles is visible as a resonance of the surface plasmon of nanoparticles, also known as LSPR (localized surface plasmon resonance). Resonance means the synchronization between the frequency of light and the oscillation of nanoparticles.

In LSPR, due to of resonance of plasmons, nanosized particles offer high visible absorption, wherein this absorption affects the color of colloids of nanoparticles, which is not there in the bulk counterpart of the materials. We have already discussed this effect in Figure 1.7, where the color of gold nanoparticles is shown. Even this property of nanoparticles can be used to produce optical sensors. A similar property can be observed in quantum dots as well, which shall be discussed in detail in further chapters of this book.

1.6.4 Magnetic Properties

For various applications, the magnetic property of a material is one of the important parameters. These magnetic properties can be analyzed by the magnetic curve of the material. Popularly, this magnetic curve is known as the B–H curve or hysteresis loop of the materials. We all are aware of the B–H curve in detail in our previous classes. In the B–H curve, the coercive field is one of the important parameters to identify the properties of the material. For example, in soft-magnetic materials, the required coercive field must be extremely low or zero to reduce the energy loss (heat loss) once it is placed in an alternating magnetic field. While for hard magnets (useful for permanent magnets), the coercive field should be large and high saturation of the magnetic loop.

Here, it is worthwhile to mention that the properties of soft and hard magnets can be improved by reducing the size of bulk material in the nanostructure. In common, these magnetic properties depend on the materials' temperature and structure. Usually, the expected classical size of the magnetic domain is of the order of 1 mm, and once the size of the domain reduces to the order of a nanometer, then due to the surface-to-volume ratio, the quantum effect becomes appreciable and new properties would be shown by magnetic materials. The giant magnetoresistance effect (GME) is one of the examples of phenomena in these types of nanosized magnetic materials; this is a basic nano-effect and is in use in memory devices in computers/mobiles.

1.6.5 Mechanical Properties

It is well known that the mechanical properties of the materials are fundamentally due to their structure. Similarly, a nanosized particle would also offer some intrinsic unusual mechanical properties due to its structure. For example, CNTs are small honeycomb-like tubes that have the same structures as graphite, wherein these CNTs offer distinctive characteristics as compared to graphite. It is important to mention that these CNTs are lighter (6 times) but stronger (100 times) than steel. The details about the properties and applications of CNTs would be discussed in Chapter 6 of this book.

REFERENCES

[1] D. Schaming and H. Remita, "Nanotechnology: from the ancient time to nowadays," *Found. Chem.*, vol. 17, no. 3, pp. 187–205, Jul. 2015, doi: 10.1007/s10698-015-9235-y.

[2] F. J. Heiligtag and M. Niederberger, "The fascinating world of nanoparticle research," *Mater. Today*, vol. 16, no. 7–8. pp. 262–271, Jul. 2013, doi: 10.1016/j.mattod.2013.07.004.

[3] G. Rytwo, "Clay minerals as an ancient nanotechnology: historical uses of clay organic interactions, and future possible perspectives," *La Rev. Macla*, vol. 9, pp. 15–17, 2008.

[4] P. Walter et al., "Early use of PbS nanotechnology for an ancient hair dyeing formula," *Nano Lett.*, vol. 6, no. 10, pp. 2215–2219, Oct. 2006, doi: 10.1021/nl061493u.

[5] D. Johnson-Mcdaniel, C. A. Barrett, A. Sharafi, and T. T. Salguero, "Nanoscience of an ancient pigment," *J. Am. Chem. Soc.*, vol. 135, no. 5, pp. 1677–1679, Feb. 2013, doi: 10.1021/ja310587c.

[6] G. Artioli, I. Angelini, and A. Polla, "Crystals and phase transitions in protohistoric glass materials," *Ph. Transit.*, vol. 81, no. 2–3, pp. 233–252, Feb. 2008, doi: 10.1080/01411590701514409.

[7] N. Brun, L. Mazerolles, and M. Pernot, "Microstructure of opaque red glass containing copper," *J. Mater. Sci. Lett. 1991 1023*, vol. 10, no. 23, pp. 1418–1420, Jan. 1991, doi: 10.1007/BF00735696.

[8] U. Leonhardt, "Invisibility cup," *Nat. Photon.*, vol. 1, no. 4, pp. 207–208, 2007, doi: 10.1038/nphoton.2007.38.

[9] I. Freestone, N. Meeks, M. Sax, and C. Higgitt, "The Lycurgus Cup—A Roman nanotechnology," *Gold Bull.*, vol. 40, no. 4, pp. 270–277, 2007, doi: 10.1007/BF03215599.

[10] J. Jeevanandam, A. Barhoum, Y. S. Chan, A. Dufresne, and M. K. Danquah, "Review on nanoparticles and nanostructured materials: history, sources, toxicity and regulations," *Beilstein J. Nanotechnol.*, vol. 9, no. 1. pp. 1050–1074, Apr. 03, 2018, doi: 10.3762/bjnano.9.98.

[11] I. Nakai, C. Numako, H. Hosono, and K. Yamasaki, "Origin of the red color of satsuma copper–ruby glass as determined by EXAFS and optical absorption spectroscopy," *J. Am. Ceram. Soc.*, vol. 82, no. 3, pp. 689–695, 1999, doi: 10.1111/J.1151-2916.1999.TB01818.X.

[12] G. Mie, "Beiträge zur optik trüber medien, speziell kolloidaler metallösungen," *Ann. Phys.*, vol. 330, no. 3, pp. 377–445, Jan. 1908, doi: 10.1002/ANDP.19083300302.

[13] M. N. Rittner and T. Abraham, "Nanostructured materials: an overview and commercial analysis," *JOM*, vol. 50, no. 1, pp. 36–37, 1998, doi: 10.1007/S11837-998-0065-4.

[14] "Samsung and its attractions—Asia's new model company," 2011. http://www.economist.com/node/21530984 (Accessed: June 20, 2021).

[15] I. Malsch, et al., "Benefits, risks, ethical, legal and social aspects of nanotechnology," www.nanoforum.org, no. June 2004.

[16] C. Alexiou et al., "Therapeutic efficacy of ferrofluid bound anticancer agent," *Magnetohydrodynamics*, vol. 37, no. 3, pp. 318–322, 2001, doi: 10.22364/mhd.37.3.16.

[17] M. E. Vance, T. Kuiken, E. P. Vejerano, S. P. McGinnis, M. F. Hochella, and D. R. Hull, "Nanotechnology in the real world: redeveloping the nanomaterial consumer products inventory," *Beilstein J. Nanotechnol.*, vol. 6, no. 1, pp. 1769–1780, 2015, doi: 10.3762/BJNANO.6.181.

[18] M. Ghasemlou, F. Daver, E. P. Ivanova, and B. Adhikari, "Bio-inspired sustainable and durable superhydrophobic materials: from nature to market," *J. Mater. Chem. A*, vol. 7, no. 28, pp. 16643–16670, Jul. 2019, doi: 10.1039/C9TA05185F.

[19] "Top 3 things you never knew about nanotechnology > promethean particles." https://prometheanparticles.co.uk/top-3-things-you-never-knew-about-nanotechnology/ (Accessed: September 04, 2021).

[20] S. Griffin et al., "Natural nanoparticles: a particular matter inspired by nature," *Antioxidants*, vol. 7, no. 1, p. 3, Dec. 2017, doi: 10.3390/ANTIOX7010003.
[21] D. A. Taylor, "Dust in the wind," *Environ. Health Perspect.*, vol. 110, no. 2, 2002, doi: 10.1289/EHP.110-A80.
[22] A. N. Al-Dabbous and P. Kumar, "Number and size distribution of airborne nanoparticles during summertime in Kuwait: first observations from the middle east," *Environ. Sci. Technol.*, vol. 48, no. 23, pp. 13634–13643, Dec. 2014, doi: 10.1021/es505175u.
[23] Z. R. Dai, J. P. Bradley, D. J. Joswiak, D. E. Brownlee, H. G. M. Hill, and M. J. Genge, "Possible in situ formation of meteoritic nanodiamonds in the early Solar System," *Nat. 2002 4186894*, vol. 418, no. 6894, pp. 157–159, Jul. 2002, doi: 10.1038/nature00897.
[24] Z. Shi, L. Shao, T. P. Jones, and S. Lu, "Microscopy and mineralogy of airborne particles collected during severe dust storm episodes in Beijing, China," *J. Geophys. Res. Atmos.*, vol. 110, no. D1, pp. 1–10, Jan. 2005, doi: 10.1029/2004JD005073.
[25] C. Buzea, I. I. Pacheco, and K. Robbie, "Nanomaterials and nanoparticles: sources and toxicity," *Biointerphases*, vol. 2, no. 4, pp. MR17–MR71, Dec. 2007, doi: 10.1116/1.2815690.
[26] M. Ammann and H. Burtscher, "Characterization of ultrafine aerosol particles in Mt. Etna emissions," *Bull. Volcanol.*, vol. 52, no. 8, pp. 577–583, Nov. 1990, doi: 10.1007/BF00301209.
[27] K. A. Meeker, R. L. Chuan, P. R. Kyle, and J. M. Palais, "Emission of elemental gold particles from Mount Erebus, Ross Island, Antarctica," *Geophys. Res. Lett.*, vol. 18, no. 8, pp. 1405–1408, 1991, doi: 10.1029/91GL01928.
[28] S. Guo, G. J. S. Bluth, W. I. Rose, I. M. Watson, and A. J. Prata, "Re-evaluation of SO_2 release of the 15 June 1991 Pinatubo eruption using ultraviolet and infrared satellite sensors," *Geochemistry, Geophys. Geosystems*, vol. 5, no. 4, Apr. 2004, doi: 10.1029/2003GC000654.
[29] E. Yano, K. Maeda, Y. Yokoyama, H. Higashi, S. Nishii, and A. Koizumi, "Health effects of volcanic ash: a repeat study," *Arch. Environ. Health*, vol. 45, no. 6, pp. 367–373, 1990, doi: 10.1080/00039896.1990.10118757.
[30] J. A. Mott et al., "Wildland forest fire smoke: health effects and intervention evaluation, Hoopa, California, 1999," *West. J. Med.*, vol. 176, no. 3, p. 157, May 2002, doi: 10.1136/EWJM.176.3.157.
[31] Amir Sapkota et al., "Impact of the 2002 Canadian Forest Fires on Particulate Matter Air Quality in Baltimore City," *Environ. Sci. Technol.*, vol. 39, no. 1, pp. 24–32, Jan. 2004, doi: 10.1021/ES035311Z.
[32] Ö. Gustafsson et al., "Brown clouds over South Asia: biomass or fossil fuel combustion?," *Science (80-.)*., vol. 323, no. 5913, pp. 495–498, Jan. 2009, doi: 10.1126/SCIENCE.1164857.
[33] S. Smita, S. K. Gupta, A. Bartonova, M. Dusinska, A. C. Gutleb, and Q. Rahman, "Nanoparticles in the environment: assessment using the causal diagram approach," *Environ. Heal. 2012 111*, vol. 11, no. 1, pp. 1–11, Jun. 2012, doi: 10.1186/1476-069X-11-S1-S13.
[34] M. Pósfai, R. Simonics, J. Li, P. V. Hobbs, and P. R. Buseck, "Individual aerosol particles from biomass burning in Southern Africa: 1. compositions and size distributions of carbonaceous particles," *J. Geophys. Res. Atmos.*, vol. 108, no. 13, 2003, doi: 10.1029/2002JD002291.
[35] P. R. Buseck and M. Pósfai, "Airborne minerals and related aerosol particles: effects on climate and the environment," *Proc. Natl. Acad. Sci. U. S. A.*, vol. 96, no. 7, pp. 3372–3379, 1999, doi: 10.1073/pnas.96.7.3372.
[36] "SeaWiFS: lake Michigan brightens again," September 2001. Accessed: June 25, 2021. [Online]. Available: https://visibleearth.nasa.gov/view.php?id=56765.

[37] J. Kagawa, "Health effects of diesel exhaust emissions–a mixture of air pollutants of worldwide concern," *Toxicology*, vol. 181–182, pp. 349–353, Dec. 2002, doi: 10.1016/S0300-483X(02)00461-4.

[38] D. Westerdahl, S. Fruin, T. Sax, P. M. Fine, and C. Sioutas, "Mobile platform measurements of ultrafine particles and associated pollutant concentrations on freeways and residential streets in Los Angeles," *Atmos. Environ.*, vol. 39, no. 20, pp. 3597–3610, Jun. 2005, doi: 10.1016/j.atmosenv.2005.02.034.

[39] C. Sioutas, R. J. Delfino, and M. Singh, "Exposure assessment for atmospheric ultrafine particles (UFPs) and implications in epidemiologic research," *Environ. Health Perspect.*, vol. 113, no. 8, pp. 947–955, Aug. 2005, doi: 10.1289/EHP.7939.

[40] K. F. Soto, A. Carrasco, T. G. Powell, K. M. Garza, and L. E. Murr, "Comparative in vitro cytotoxicity assessment of some manufacture dnanoparticulate materials characterized by transmissionelectron microscopy," *J. Nanoparticle Res.*, vol. 7, no. 2–3, pp. 145–169, Jun. 2005, doi: 10.1007/s11051-005-3473-1.

[41] M. Singh, H. C. Phuleria, K. Bowers, and C. Sioutas, "Seasonal and spatial trends in particle number concentrations and size distributions at the children's health study sites in Southern California," *J. Expo. Sci. Environ. Epidemiol.*, vol. 16, no. 1, pp. 3–18, Jan. 2006, doi: 10.1038/SJ.JEA.7500432.

[42] J. Vermylen, A. Nemmar, B. Nemery, and M. F. Hoylaerts, "Ambient air pollution and acute myocardial infarction," *J. Thromb. Haemost.*, vol. 3, no. 9, pp. 1955–1961, 2005, doi: 10.1111/J.1538-7836.2005.01471.X.

[43] G. Hoek, B. Brunekreef, S. Goldbohm, P. Fischer, and P. A. Van Den Brandt, "Association between mortality and indicators of traffic-related air pollution in the Netherlands: a cohort study," *Lancet*, vol. 360, no. 9341, pp. 1203–1209, Oct. 2002, doi: 10.1016/S0140-6736(02)11280-3.

[44] Z. Ning, C. S. Cheung, J. Fu, M. A. Liu, and M. A. Schnell, "Experimental study of environmental tobacco smoke particles under actual indoor environment," *Sci. Total Environ.*, vol. 367, no. 2–3, pp. 822–830, Aug. 2006, doi: 10.1016/J.SCITOTENV.2006.02.017.

[45] D. Stefani, D. Wardman, and T. Lambert, "The implosion of the calgary general hospital: ambient air quality issues," *J. Air Waste Manag. Assoc.*, vol. 55, no. 1, pp. 52–59, 2005, doi: 10.1080/10473289.2005.10464605.

[46] R. L, "Health impact of environmental tobacco smoke in the home," *Rev. Environ. Health*, vol. 19, no. 3–4, pp. 291–309, Jul. 2004. Accessed: September 04, 2021. [Online]. Available: https://pubmed.ncbi.nlm.nih.gov/15742675/.

[47] N. S. Godtfredsen, M. Osler, J. Vestbo, I. Andersen, and E. Prescott, "Smoking reduction, smoking cessation, and incidence of fatal and non-fatal myocardial infarction in Denmark 1976–1998: a pooled cohort study," *J. Epidemiol. Community Health*, vol. 57, no. 6, pp. 412–416, Jun. 2003, doi: 10.1136/JECH.57.6.412.

[48] E. M. Fireman et al., "Induced sputum assessment in New York city firefighters exposed to World Trade Center dust," *Environ. Health Perspect.*, vol. 112, no. 15, pp. 1564–1569, Nov. 2004, doi: 10.1289/ehp.7233.

[49] N. A. Monteiro-Riviere, K. Wiench, R. Landsiedel, S. Schulte, A. O. Inman, and J. E. Riviere, "Safety evaluation of sunscreen formulations containing titanium dioxide and zinc oxide nanoparticles in UVB sunburned skin: an in vitro and in vivo study," *Toxicol. Sci.*, vol. 123, no. 1, pp. 264–280, Sep. 2011, doi: 10.1093/TOXSCI/KFR148.

[50] G. Cao and Y. Wang, "Nanostructures fabricated by physical techniques," *Nanostructures Nanomater.*, pp. 369–432, Jan. 2011, doi: 10.1142/9789814340571_0007.

[51] T. Hyeon, "Chemical synthesis of magnetic nanoparticles," *Chem. Commun.*, vol. 3, no. 8, pp. 927–934, 2003, doi: 10.1039/B207789B.

[52] P. Mohanpuria, N. K. Rana, and S. K. Yadav, "Biosynthesis of nanoparticles: technological concepts and future applications," *Journal of Nanoparticle Research*, vol. 10, no. 3. pp. 507–517, Mar. 2008, doi: 10.1007/s11051-007-9275-x.

[53] P.Westerhoff,I.A.Fulton,andT.Woodruff,"Nano-particlesinbabyformula:tinynewingredientsareabigconcern,"*JeremyTager,FriendsEarthAust.*,2016,[Online].Available:http://webiva-downton.s3.amazonaws.com/877/eb/2/8482/FOE_Nano BabyFormulaReport_13.pdf (Accessed June 28, 2021).

[54] K. Donaldson, V. Stone, C. L. Tran, W. Kreyling, and P. J. A. Borm, "Nanotoxicology," *Occup. Environ. Med.*, vol. 61, no. 9, pp. 727–728, Sep. 2004, doi: 10.1136/OEM.2004.013243.

[55] International Center for Scholars, "Inventory finds increase in consumer products containing nanoscale materials • news archive • nanotechnology project," *AnalysisdPEN Consumer Products Inventory*, 2019. http://www.nanotechproject.tech/news/archive/9242/. (Accessed: June 26, 2021).

[56] D. H. Duckworth and P. A. Gulig, "Bacteriophages: potential treatment for bacterial infections," *BioDrugs*, vol. 16, no. 1. Adis International Ltd, pp. 57–62, 2002, doi: 10.2165/00063030-200216010-00006.

[57] B. Picó, M. J. Díez, and F. Nuez, "Viral diseases causing the greatest economic losses to the tomato crop. II. the tomato yellow leaf curl virus–a review," *Scientia Horticulturae*, vol. 67, no. 3–4. pp. 151–196, Dec. 1996, doi: 10.1016/S0304-4238(96)00945-4.

[58] T. P. D. Van Den Berg, "Acute infectious bursal disease in poultry: a review," *Avian Pathol.*, vol. 29, no. 3, pp. 175–194, 2000, doi: 10.1080/03079450050045431.

[59] T. Uchiyama, "Human T cell leukemia virus type I (HTLV-I) and human diseases," *Annu. Rev. Immunol.*, vol. 15, pp. 15–37, 1997, doi: 10.1146/ANNUREV.IMMUNOL.15.1.15.

[60] M. L. Flenniken, D. A. Willits, S. Brumfield, M. J. Young, and T. Douglas, "The small heat shock protein cage from methanococcus jannaschii is a versatile nanoscale platform for genetic and chemical modification," *Nano Lett.*, vol. 3, no. 11, pp. 1573–1576, Nov. 2003, doi: 10.1021/nl034786l.

[61] M. L. Flenniken et al., "Melanoma and lymphocyte cell-specific targeting incorporated into a heat shock protein cage architecture," *Chem. Biol.*, vol. 13, no. 2, pp. 161–170, Feb. 2006, doi: 10.1016/J.CHEMBIOL.2005.11.007.

[62] K. Saunders, F. Sainsbury, and G. P. Lomonossoff, "Efficient generation of cowpea mosaic virus empty virus-like particles by the proteolytic processing of precursors in insect cells and plants," *Virology*, vol. 393, no. 2, pp. 329–337, Oct. 2009, doi: 10.1016/J.VIROL.2009.08.023.

[63] Q. Wang, T. Lin, L. Tang, J. E. Johnson, and M. G. Finn, "Icosahedral virus particles as addressable nanoscale building blocks," *Angew. Chemie*, vol. 114, no. 3, pp. 477–480, 2002, doi: 10.1002/1521-3757(20020201)114:3<477::aid-ange477>3.0.co;2-2.

[64] P. Singh et al., "Bio-distribution, toxicity and pathology of cowpea mosaic virus nanoparticles in vivo," *J. Control. Release*, vol. 120, no. 1–2, pp. 41–50, Jul. 2007, doi: 10.1016/J.JCONREL.2007.04.003.

[65] R. Yupeng, "Application of hcrsv protein cage for anticancer drug delivery ren yupeng a thesis submitted for the degree of doctor of philosophy department of pharmacy," pp. 1–92, 2007. Accessed: July 05, 2021. [Online]. Available: https://123docz.net/document/3053002-application-of-hcrsv-protein-cage-for-anticancer-drug-delivery.htm.

[66] M. H. B. Catroxo and A. M. C. R. P. F. Martins, "Veterinary diagnostic using transmission electron microscopy," *Transm. Electron Microsc.–Theory Appl.*, Sep. 2015, doi: 10.5772/61125.

[67] M. Hosea, B. Greene, R. Mcpherson, M. Henzl, M. Dale Alexander, and D. W. Darnall, "Accumulation of elemental gold on the alga Chlorella vulgaris," *Inorganica Chim. Acta*, vol. 123, no. 3, pp. 161–165, Mar. 1986, doi: 10.1016/S0020-1693(00)86339-2.

[68] G. Scarano and E. Morelli, "Properties of phytochelatin-coated CdS nanocrystallites formed in a marine phytoplanktonic alga (phaeodactylum tricornutum, Bohlin) in response to Cd," *Plant Sci.*, vol. 165, no. 4, pp. 803–810, Oct. 2003, doi: 10.1016/S0168-9452(03)00274-7.

[69] N. Krumov, I. Perner-Nochta, S. Oder, V. Gotcheva, A. Angelov, and C. Posten, "Production of inorganic nanoparticles by microorganisms," *Chem. Eng. Technol.*, vol. 32, no. 7, pp. 1026–1035, Jul. 2009, doi: 10.1002/CEAT.200900046.

[70] M. Sastry, A. Ahmad, M. I. Khan, and R. Kumar, "Biosynthesis of metal nanoparticles using fungi and actinomycete," *Curr. Sci.*, vol. 85, no. 2, pp. 162–170, Sep. 2003, [Online]. Available: http://www.jstor.org/stable/24108579.

[71] M. R. Salvadori, R. A. Ando, C. A. O. do Nascimento, and B. Corrêa, "Intracellular biosynthesis and removal of copper nanoparticles by dead biomass of yeast isolated from the wastewater of a mine in the brazilian amazonia," *PLoS One*, vol. 9, no. 1, p. e87968, Jan. 2014, doi: 10.1371/JOURNAL.PONE.0087968.

[72] M. R. Salvadori, R. A. Ando, C. A. O. Nascimento, and B. Corrêa, "Extra and Intracellular Synthesis of Nickel Oxide Nanoparticles Mediated by Dead Fungal Biomass," *PLoS One*, vol. 10, no. 6, p. e0129799, Jun. 2015, doi: 10.1371/JOURNAL.PONE.0129799.

[73] S. Senapati, A. Ahmad, M. I. Khan, M. Sastry, and R. Kumar, "Extracellular biosynthesis of bimetallic Au–Ag alloy nanoparticles," *Small*, vol. 1, no. 5, pp. 517–520, May 2005, doi: 10.1002/smll.200400053.

[74] A. Mukherjee P., Senapati, S., Manda, D., Ahmad, "Extracellular synthesis of gold nanoparticles by the fungus fusarium oxysporum," *Chem Bio Chem*, vol. 3, no. 5, pp. 461–463, 2002, doi: https://doi.org/10.1002/1439-7633(20020503)3:5<461::AID-CBIC461>3.0.CO;2-X.

[75] A. Ahmad et al., "Extracellular biosynthesis of silver nanoparticles using the fungus fusarium oxysporum," *Colloids Surf. B Biointerfaces*, vol. 28, no. 4, pp. 313–318, May 2003, doi: 10.1016/S0927-7765(02)00174-1.

[76] R. Sastry, M., Ahmad, A., Khan, M., and Kumar, "Biosynthesis of metal nanoparticles using fungi and actinomycete," *Curr. Sci.*, vol. 85, no. 2, pp. 162–170, 2003. Accessed: July 04, 2021. [Online]. Available: https://www.jstor.org/stable/24108579.

[77] Absar Ahmad et al., "Enzyme mediated extracellular synthesis of cds nanoparticles by the fungus, fusarium oxysporum," *J. Am. Chem. Soc.*, vol. 124, no. 41, pp. 12108–12109, Oct. 2002, doi: 10.1021/JA027296O.

[78] R. N. Reese and D. R. Winge, "Sulfide stabilization of the cadmium-gamma-glutamyl peptide complex of schizosaccharomyces pombe.," *J. Biol. Chem.*, vol. 263, no. 26, pp. 12832–12835, 1988, doi: 10.1016/s0021-9258(18)37635-x.

[79] C. T. Dameron et al., "Biosynthesis of cadmium sulphide quantum semiconductor crystallites," *Nature*, vol. 338, no. 6216, pp. 596–597, 1989, doi: 10.1038/338596a0.

[80] "Microbial synthesis of semiconductor pbs nanocrystallites–kowshik–2002–advanced materials–Wiley Online Library." Accessed: September 04, 2021. [Online]. Available: https://onlinelibrary.wiley.com/doi/10.1002/1521-4095(20020605)14:11%3C815::AID-ADMA815%3E3.0.CO;2–K.

[81] M. Kowshik et al., "Extracellular synthesis of silver nanoparticles by a silver-tolerant yeast strain MKY3," *Nanotechnology*, vol. 14, no. 1, pp. 95–100, Dec. 2003, doi: 10.1088/0957-4484/14/1/321.

2 Introduction to Nanoscience

D.K. Kushvaha and S.K. Rout

CONTENTS

2.1 WHAT ARE NANOSCIENCE AND NANOTECHNOLOGY?

The rapid development in domestic as well as industrial demands compelled researchers to produce nanoscale (one-millionth of a millimeter) materials with enhanced properties for more sophisticated technology. Therefore, this demand motivated researchers over the globe to dedicate their research to synthesize and characterize nanoscale materials for distinct applications. In this regard, the role of nanoscience and nanotechnology becomes vital to fulfill the objective of future technology. The branch of science that deals with the study of physicochemical property within nanoscale matter or particles are referred to as nanoscience. Nanoscience is a multidisciplinary field that engages researchers and scientists from a wide field of physics, chemistry, biology, medicine, materials science, engineering, etc. to understand cost-effective, efficient, ecofriendly routes of synthesis and unique behavior of nanoscale

DOI: 10.1201/9781003216308-2

materials. Nanoscience enables us to understand and predict the possible changes in the properties of nanoscale materials for suitable utilization in devices. Moreover, the study of properties of nanoscale particles is always a very fascinating task due to remarkable changes with respect to their bulk counterpart.

Literally, nanotechnology (or nanotech) means reliable technology based on nanomaterials for the real world. Thus, the word "nanotechnology" accommodates the manipulation, synthesis, and application of nanoscale structured materials based on suitable electrical, optical, thermal, chemical, or mechanical behavior of the materials [1]. Several chemical or physical methods are being used to produce nanomaterials which include the sol-gel process, high-energy ball milling, coprecipitation method, etc. Nanotechnology uses two different approaches to produce or fabricate nanomaterials: (a) top-down approach and (b) bottom-up approach. In top-down approaches, nanomaterials are produced by physically or chemically breaking down larger particles. On the other hand, the bottom-up (or sometimes called self-assembly) approaches utilize physical or chemical forces for assembling basic atomic or molecular units to produce nanoparticles. Nanotechnology is a widely accepted technology for industrial as well as domestic purposes and has the potential to bring revolution in areas such as energy, biotechnology, nanoelectronics, medicine and healthcare, information technology, national security, etc. Thus, nanotechnology is engaged in a broader area of science, engineering, and technology which includes nano-robotics, nano-fluids, nanobiology, etc. Further miniaturization of the present technology is directly associated with the advancement of nanoscience and nanotechnology with better understanding and continuous effort to revolutionize future technology.

2.2 CLASSIFICATION OF NANOSTRUCTURES – NANOSCALE ARCHITECTURE

It has been observed that the size of a material particle impacts its properties including physical, chemical, optical, thermal, electrical, or mechanical properties, etc. This means that the property of a nanoscale material is different from that of bulk materials. For example, the bulk gold particles appear yellowish, but on breaking the bulk gold into nano-size particles, their color changes from yellow to red, orange, purple, etc. depending upon the particle size. The details have been explained in a later section of this chapter. Based on the dimension, the nanostructured materials (NSMs) may be classified as shown below [2,3].

- **Zero-dimensional (0D) Nanomaterials:** If all three dimensions of the material in 3D space are in the nanoscale (<100 nm), then the material is referred to as a zero-dimensional (0D) nanomaterial. Quantum dots, nanodots, fullerenes, etc. are common examples of 0D nanomaterials.
- **One-dimensional (1D) Nanomaterials:** If any single dimension of material in 3D space is beyond nanoscale, i.e., >100 nm, then the material is referred to as a 1D nanomaterial. Nanowires, nanorods, nanotubes, nanopillars, carbon nanotubes, etc. are common examples of 1D nanomaterials.

- **Two-dimensional (2D) Nanomaterials:** If any two dimensions of the material in 3D space are beyond the nanoscale, i.e., >100 nm, then the material is referred to as a 2D nanomaterial. Graphene, nanofilms, nanolayers, nanocoating, etc. are common examples of 2D nanomaterials.
- **Three-dimensional (3D) Nanomaterials:** The dimensions of bulk materials are not confined to the nanoscale. Each dimension of three-dimensional nanomaterials is greater than 100 nm. Thus, the 3D nanomaterial is technically excluded from the category of nanomaterials.

The following Figure 2.1 shows the length scale with 0.1–10^8 nm along with the approximate range size of objects to visualize the dimension of nanomaterials and to employ the technique to measure its dimension. From the figure, it may be observed that the nanomaterial lies within the range of 1–100 nm. In order for visualization, the dimension of nanomaterials may be compared with viruses whose approximate size lies within the range of 20–200 nm and hereditary DNA molecules with a dimensional range of 5–10 nm [4,5]. The complete dimension of protein and part of virus species dimension lies within the nanoscale region. The approximate diameter of human hair is 20–200 μm. Thus, 1 nm is approximately the 10,000th part of the thinnest human hair.

Particle size measurement is one of the challenging tasks for researchers. Several techniques which include dynamic image analysis, static image analysis, dynamic light scattering (DLS), laser diffraction analysis, sieve analysis, etc. have been employed to overcome the challenges. Out of these, the basic overview of the laser diffraction technique and DLS has been focused on in the present chapter. The laser diffraction technique (also known as static light scattering) is the most commonly used technique for the analysis of particle size distribution. This technique works on the principle of diffraction of the laser beam by dispersed particles in liquid or air. Within this technique, the diffraction angle and intensity are the characteristics of particle size and have the capability to measure particle size of range 0.01–2.8 μm (sometimes, the range may vary with the instrument). The larger particle size is more preferred for laser diffraction analysis due to the relatively high diffraction intensity

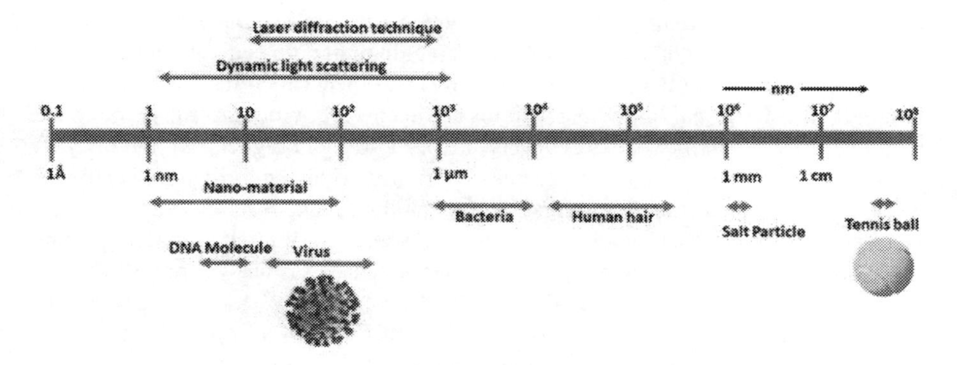

FIGURE 2.1 Length scale from 1 Å to 10 cm with the corresponding object and measurement technique of particle size.

and low diffraction angle. DLS is another technique to measure particle size within the approximate range of 0.6 nm to 6 μm. The technique is based on the Brownian motion, i.e., random thermal motion of suspended particles. The particle size is evaluated using the Stokes–Einstein equation as shown below

$$D_h = \frac{k_B T}{3\pi\eta D_t} \tag{2.1}$$

where D_h, D_t, k_B, T and η represent the hydrodynamic diameter of the particle, translational diffusion coefficient, Boltzmann's constant, thermodynamic temperature, and dynamic viscosity, respectively.

2.3 SUMMARY OF THE ELECTRONIC PROPERTIES OF ATOMS AND SOLIDS

2.3.1 The Isolated Atom and Giant Molecular Solids

An atom consists of three basic elements: the proton, electron, and nucleus. The core components of the atom called the nucleus contain positively charged protons and neutral neutrons, while the negatively charged electrons cloud surround the core nucleus. Despite charge being present in individual elements, atoms are electrically neutral but may gain positive or negative ionic charge by losing their electrons to or accepting electrons from other atoms. The charges on protons and electrons are entirely responsible for the interaction of one atom with other atoms in the solid or molecule. The physical interaction between the same or different atoms in molecules or solids is referred to as a chemical bond. The atoms may interact by different physical forces such as magnetic force, electrostatic force, gravitational force, and van der Waals force. However, the magnetic force has a very weak effect and gravitational force has a negligible effect on the cohesion of atoms. The details of atomic bonding have been explained in Section 2.3.3 of this chapter.

An isolated atom has discrete energies corresponding to the orbital electrons of that atom. Thus, when two atoms A and B are placed at an infinite distance from each other, then both the atoms will have their own energy level, as shown in Figure 2.2. When atom B is brought near atom A to form a diatomic molecule, then the orbital electrons of each atom begin to interact with each other, which results in modification of their original energies. When atoms are in close proximity to each other, then their energy splits into two energy levels slightly lower and higher than the original level for the system of diatomic molecules. Similarly, when the third atom is brought in close proximity to the diatomic system of atoms A and B, then their energy is modified into three discrete levels. Eventually, when a very large number of atoms (say N) are brought closer to each other, then the energy of the system will have N discrete levels very close to each other. These close energies of N atoms appear continuously distributed like a band and are called the band of allowed energy. There is a gap between two such allowed energy bands called the forbidden gap of energy.

Depending upon the size of the band of the forbidden gap of energy, the materials may be classified into three categories: metal, semiconductor, and insulator.

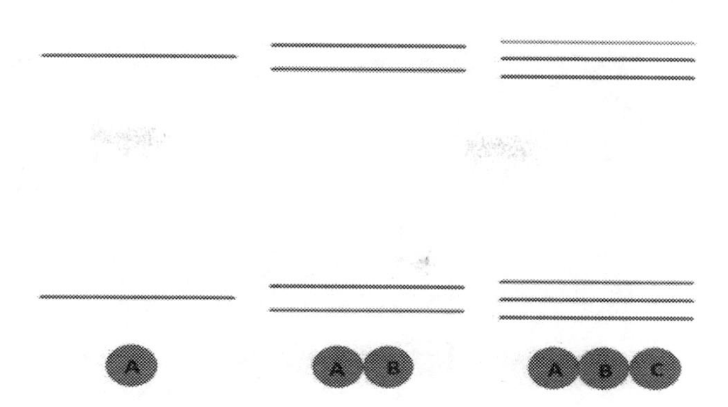

FIGURE 2.2 Band formation mechanism for three atoms A, B and C.

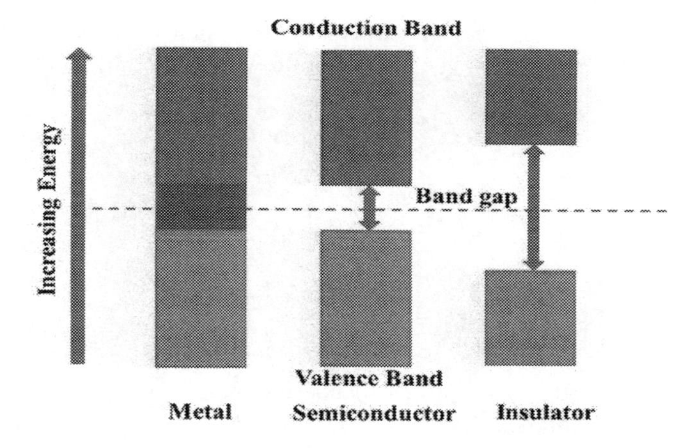

FIGURE 2.3 Comparison between the band gap for metal, semiconductor and insulator.

The forbidden gap of the insulator is very large, whereas the valence band and conduction band of a conductor overlap each other. The forbidden gap of a semiconductor lies between the insulator and conductor, as shown in Figure 2.3. The details have been given in Section 2.3.2.

The energy band gap of nanomaterials is particle shape- and size-dependent. Figure 2.4 shows the variation of the bandgap with the size of a nanomaterial. The figure shows that the band gap of the nanomaterial is larger than its bulk counterpart. Furthermore, the band gap width increases and bandwidth narrows with a decrease in particle size. Due to a decrease in the particle size, the number of overlapping orbitals or energy levels reduces and results in a narrowing bandwidth and increase in the band gap.

Several researchers have proposed a theoretical model to predict the variation of shape- and size-dependent band gap within the nanoscale dimension of particles. The proposed models are in good agreement with the experimentally observed

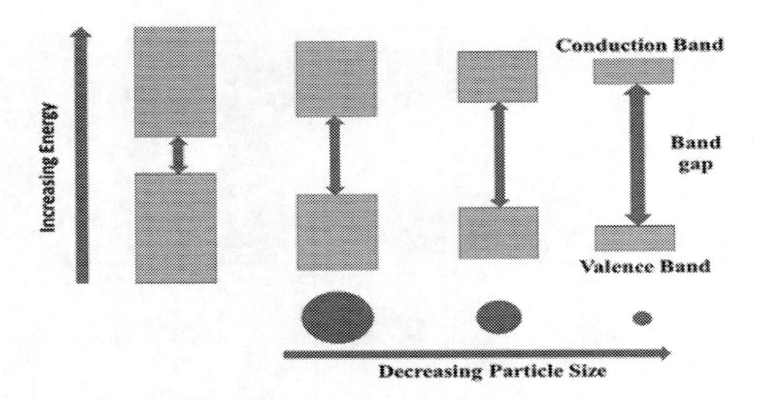

FIGURE 2.4 Variation of the band gap with particle size in nanoscale.

results. One of the models for spherical nanosolids, nanowires, and nanofilm gives the band gap energy by following expression [6]

$$E(D) = E_g(bulk)\left(1 + \frac{2d}{D}\right) \tag{2.2a}$$

$$E(l) = E_g(bulk)\left(1 + \frac{4d}{3l}\right) \tag{2.2b}$$

$$E(h) = E_g(bulk)\left(1 + \frac{2d}{3h}\right) \tag{2.2c}$$

where $E(D).E(l)$ and $E(h)$ represent band gap energy for spherical nanosolids, nanowires, and nanofilms, respectively. Moreover, $E_g(bulk), d, D, l$ and h represent the band gap corresponding to the bulk counterpart, diameter of an atom, diameter of spherical nanosolid, diameter of nanowire, and width of nanofilm, respectively. Figure 2.5 represents the variation of band gap energy within a range of 1–50 nm particle size for a spherical nanosolid with the atomic diameter $d = 0.268$ nm and band gap energy of the bulk counterpart $E_g(bulk) = 1.74eV$. The figure shows that the band gap of spherical nanosolid increases with decreasing particle size from 50 to 1 nm.

2.3.2 Electronic Conduction

The conductivity of the solid materials is the measure of ease of charge (electron or hole) flowing through it. The electrons which reside in the conduction band of a band structure are free to move within the solid. Since most solid forms bond to completely fill the valence band, the electrons need a sufficient amount of energy to cross the forbidden gap by jumping from the valence band to the conduction band to initiate the conduction process. The conduction band and valence band of a conductor overlap with each other, which results in a sufficiently large number of free electrons

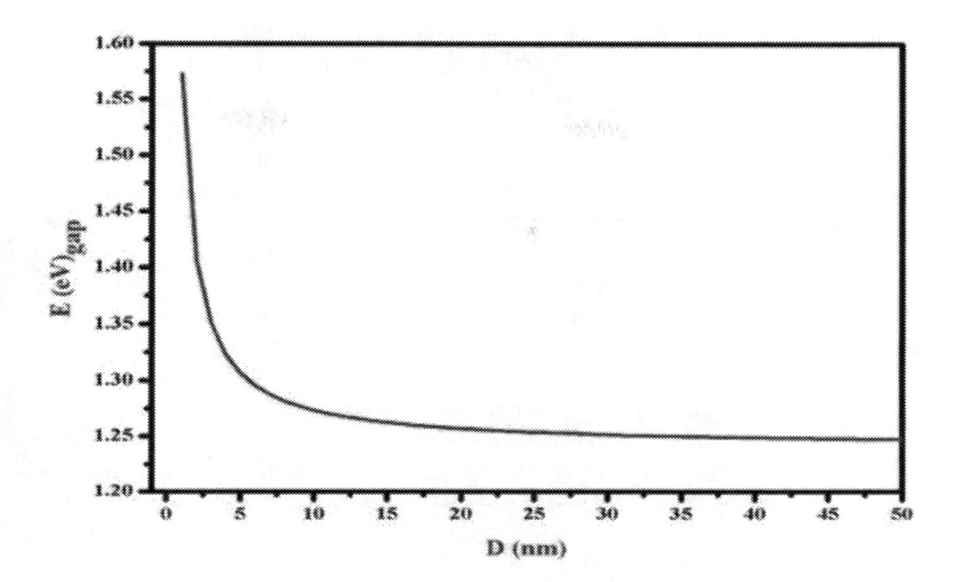

FIGURE 2.5 Variation of the band gap with particle size for spherical nanosolids.

that could ease their migration to increase conductivity, while, the insulators are characterized by a very large band gap or forbidden gap between the completely filled valence band and completely empty conduction band. The gap is large enough to cross the electrons even though at large energy. Thus, the free electrons within the conduction band are very small as compared to the conductor on the application of electric field across it, which results in extremely low conductivity. Within a semiconductor, the forbidden gap is relatively much smaller than in the insulators. For example, the forbidden gap of Si and Ge semiconductors are 1.1 and 0.7 eV, respectively. However, the band gap of diamond is 6 eV, which is significantly larger than Si or Ge. The semiconductor has a completely filled valence band and a completely empty conduction band at absolute temperature and thus behaves like an insulator. This is because the electron cannot jump from the valence band to the conduction band at absolute temperature, but with increasing temperature, the electron begins to reach the conduction band crossing the forbidden gap. At room temperature, sufficient amounts of electrons fill the conduction band by acquiring an adequate amount of energy showing good conductivity.

2.3.3 Bonding between Atoms

The solids are an aggregation of billions of closely packed atoms and their arrangement within materials. These solids are characterized by their strength, directionality, and character of binding force. The binding force which holds the adjacent atoms, ions, or molecules together is called a bond. The physical, chemical, electrical, thermal, or mechanical properties of materials are highly dependent upon the type of bonding between atoms. The bonding between atoms may be categorized into the following five types: (a) ionic bond, (b) covalent bond, (c) metallic bond, (d) van der Waals bond, and (e) hydrogen bond. The van der Waals bonding is a relatively weak binding force between inert gas atoms because of completely filled shells. van der Waals suggested that the binding between inert gas atoms may be due to instantaneous dipole moment which arises from the instant separation of the nucleus and electrons. On the other side, the hydrogen bond refers to the binding force between the hydrogen atom of a molecule and the electronegative atom of another molecule. In this chapter, only ionic, covalent, and metallic bond has been focused on due to their deep engagement in nanomaterial solids. Depending upon the nature of the binding force between atoms, ions, or molecules, the solid materials are categorized as ionic, covalent, and metallic solids.

a. **Ionic Solid:** Ionic solids are composed of electropositive and electronegative species. The bonds between ions of such solids are chemical bonds that form due to the complete transfer of one or more electrons from an atom to the nearest neighbor to attain an inert configuration. The ionic bond is a Coulombic or electrostatic attractive intermolecular force between two oppositely charged ions. The ionic species in simple ionic crystals lose or gain electrons to acquire a closed electronic shell of inert gas. NaCl is a typical example of an ionic solid that consists of positively charged Na^+ ions and negatively charged Cl^- ions. The Na ($Z=11$) and Cl ($Z=17$) atom has electronic configuration $1s^2.2s^2.2p^6.3s^1$ or $[Ne].3s^1$ and $1s^2.2s^2.2p^6.3s^2.3p^5$ or $[Ne].\ 3s^2.3p^5$. The Cl atom gains one electron in the $3p^5$ orbital from the $3s^1$ orbital of the Na atom to attain a stable closed electronic shell of inert Ar and Ne gas configuration, respectively.
b. **Covalent Solid:** The binding force within covalent solids is a type of chemical binding due to sharing of unpaired electrons between two atoms. The covalent bond is a strong bond and forms by the sharing of one, two, or three unpaired electrons from each atom to attain a stable configuration. The simplest example of a covalent bond is the sharing of one unpaired electron between two hydrogen atoms to form a hydrogen molecule. Si, Ge, graphite, diamond, etc. are some examples of solids that form a covalent bond. The binding between unpaired electrons is the strongest when their spins are antiparallel. The relative spin orientation-dependent binding force between unpaired electrons of atoms is due to the Pauli exclusion principle in quantum mechanics, which states that two or more electrons within atoms or molecules can never have identical states among all four quantum states.

   The saturation capacity and directional characteristics are special features of a covalent bond. The saturation refers to the definite number of covalent bonds with neighboring atoms. Within the H_2 molecule, the interaction between two antiparallel (+1/2 and −1/2 spin orientation) electrons in the outermost shell of two H-atom forms a covalent bond in the ground state. Now, it is not possible to form a second covalent bond between an H_2 molecule and another H-atom to achieve H_3 molecule due to saturation of the covalent bond. For N electrons in the valence shell of an electrically neutral atom, the number of unpaired electrons or number of covalent bonds is given by (8-N). Since there are four electrons in the valance shell of a carbon atom with electronic configuration $1s^2.2s^2.2p^2$. Thus, a carbon atom has $(8-4)=4$ unpaired electrons that form a covalent bond with other atoms by sharing with another atom. The carbon atom in diamond forms four covalent bonds with four different neighboring carbon atoms in a tetragonal configuration.

Within quantum mechanics, the electronic states are represented by the solution of the Schrodinger wave equation. The solution is described in terms of wavefunction ψ_{n,l,m_l}, which depends upon three constants n (principal quantum number with integer value 1 to ∞), l (angular momentum quantum number with value 0 to $(n-1)$), and m_l (magnetic quantum number with value $m_l=-l$ to $m_l=+l$). The solution ψ_{n,l,m_l} depicts a three-dimensional space with a 95% probability of finding electrons and terms as orbital. The atomic orbitals are generally designated by alphanumeric characters such as 1s, 2s, 2p, 3s, etc. The directionality characteristic in the covalent bond appears because of the nature of atomic orbitals such as s, p, d, …. orbitals. The covalent bond forms due to head-on overlapping of atomic orbitals in different directions as shown in Figure 2.6. Therefore, the bond should also form in the direction of overlapping orbitals.

Polar Covalent Bond: When two atoms with different electronegativity characters form a covalent bond by sharing their unpaired electrons, the covalent bond is called the polar covalent bond. The electronegativity is a tendency of atoms to attract the shared paired electron toward itself. Thus, due to the relatively higher electronegativity of one atom between shared electrons of covalent atoms, the electron cloud partially shifts toward highly electronegative atoms. This partial shift of electron cloud induces a partial charge on the covalent atom and is denoted with a lower Greek delta (δ) as shown below. The hydrogen chloride (HCl) molecule consisting of hydrogen and chlorine atoms forms a polar covalent bond [7]. The chlorine atom in HCl has higher electronegativity relative to the hydrogen atom, which

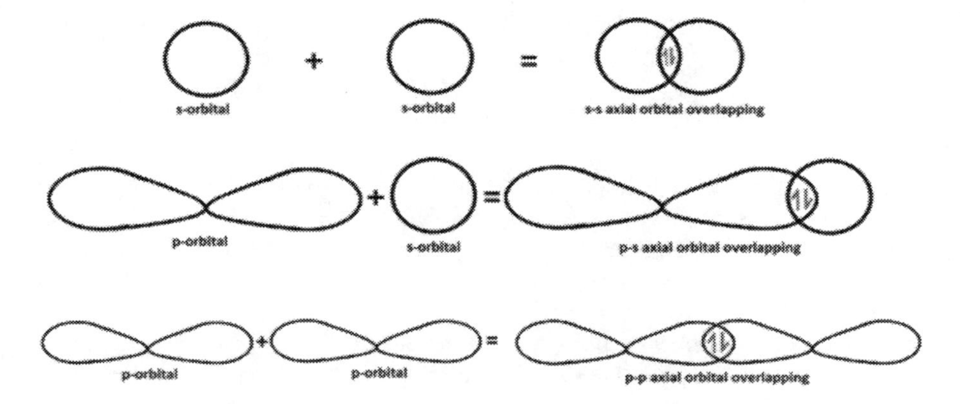

FIGURE 2.6 Head-on overlapping of s-s, p-s and p-p orbitals.

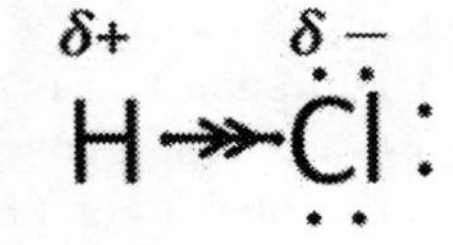

$$\overset{\delta+}{H} \twoheadrightarrow \overset{\delta-}{\ddot{\underset{\cdot\cdot}{Cl}}}:$$

FIGURE 2.7 Lewis dot structure of HCl.

results in a partial shift of shared pair electrons toward chlorine atoms as shown below in Figure 2.7.

Due to the partial shift of shared paired electron partial positive and negative charge appears on hydrogen and chlorine atom, respectively. The partial charge leads to develop the dipole moment in the HCl atom.

c. **Metallic Solid:** Most of the metallic atoms contain one or two valence electrons in the outermost atomic shell at the ground state. The outer atomic shell electrons for metallic atoms are bounded by very weak force such that they could easily be detached from the parent atoms and become free to move within the material. Thus, the core positively charged ions hold free electrons via very weak electrostatic or Coulombic force. The motion of the free electrons of metallic solid materials is responsible for electrical and thermal conductivity. Since these free electrons participate in the conduction process of materials, they are called conduction electrons. The larger interatomic distance of metallic solid reduces the kinetic energy of conduction electrons [8]. The metallic solid is generally a good conductor of heat and electricity due to the presence of high concentrations of free electrons. The metallic bonds are influenced by factors such as core cationic radii, the strength of cationic charge, and a number of delocalized electrons.

2.4 THE FREE ELECTRON MODEL AND ENERGY BANDS

The density of states (DOS) may be defined as an electronically allowed state or number of allowed energy levels per unit volume per unit energy such that

$$D(E) = \frac{\partial N(k)}{\partial E} \tag{2.3}$$

where $N(k)$ represents the total number of states [9]. The DOS calculation allows us to find the distribution of state as a function of energy or momentum. This function is very important in describing the bulk behavior of materials such as specific heat, paramagnetic susceptibility and other transport phenomena of conductive solids.

When a large number of particles such as electrons are confined in the three-dimensional periodic potential for a crystalline solid, then from the Bloch theorem (explained in the next section), we know that the wave function is periodic in nature within a periodic crystalline structure.

$$\psi(x,y,z) = \psi(x+L_x,\ y+L_y,\ z+L_z)$$
$$= \exp\left[i\left\{k_x(x+L_x)+k_y(\ y+L_y)+k_z(\ z+L_z)\right\}\right]$$
$$= \exp\left[i\left\{k_x x+k_y y+k_z z\right\}\right].\exp\left[i\left\{k_x L_x+k_y L_y+k_z L_z\right\}\right]$$
$$= \psi(x,y,z).\exp\left[i\left\{k_x L_x+k_y L_y+k_z L_z\right\}\right]$$

where $L_x = L_y = L_z = L$ is the period of the cubic crystal. This equation is only valid when

$$Lk_x = 2\pi n_x,\ Lk_y = 2\pi n_y,\ Lk_z = 2\pi n_z \tag{2.4}$$

where n_x, n_y and n_z are integers.

The number of electrons within the energy range dE is represented by $N(E)dE$, which can be expressed in terms of the number of available states $g(E)$ and the probability of filling states $f(E)$ as given bellow

$$N(E)dE = f(E)g(E)dE$$

where $g(E)dE$ is the number of states at dE and $f(E)dE$ is the probability of filling states $g(E)dE$ by electrons.

In order to calculate the DOS function, let us consider a sphere in 3D space, as shown in Figure 2.8 of radius r and $r+dr$, which accommodates all states less than

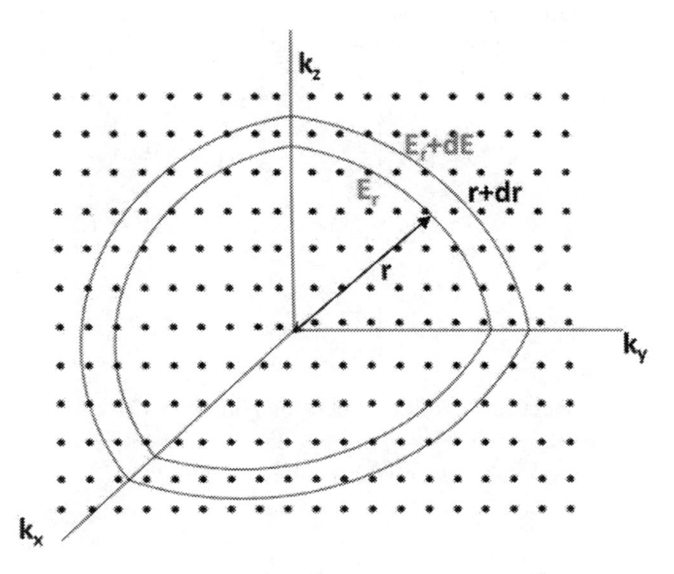

FIGURE 2.8 Lattice point in the first quadrant of k-space and two concentric spheres with radius r and $r+dr$.

E_r and $E_r + dE$, respectively. The number of states $g(E_r)$ in the first quadrant of 3D space within radius r is

$$g(E_r) = \frac{1}{8} \cdot \frac{4\pi}{3} \cdot n^3 \tag{2.5}$$

Also, the energy of a free particle confined in 3D periodic potential is given as (using the value of k from equation (2.4))

$$E_r = \frac{\hbar^2 k^2}{2m} = \frac{h^2 n^2}{8mL^2} \tag{2.6}$$

$$n^2 = \frac{8mL^2}{h^2} \cdot E_r$$

$$n^3 = \left(\frac{8mL^2}{h^2} \cdot E_r \right)^{3/2} \tag{2.7}$$

Using equation (2.7) in (2.5) and dropping subscript r,

$$g(E) = \frac{4\pi V}{3h^3} \cdot (2m)^{3/2} \cdot E^{3/2} \tag{2.8}$$

$$g(E)d(E) = \left[\frac{2\pi V}{h^3} \cdot (2m)^{3/2} \right] \cdot E^{1/2} dE$$

According to the Pauli principle, two electrons may be present in a single state; therefore, multiplying by a factor of 2 for +1/2 and −1/2 spin of electrons in the above equation, we get

$$g(E)d(E) = \left[\frac{4\pi V}{h^3} \cdot (2m)^{3/2} \right] \cdot E^{1/2} dE$$

Now, the number of states per unit volume per unit energy or DOS $D(E)$ in 3D is given as

$$D(E) = \left[\frac{4\pi}{h^3} \cdot (2m)^{3/2} \right] \cdot E^{1/2} \tag{2.9}$$

The DOS $D(E)$ for electrons in three-dimensional materials has been plotted against energy E in Figure 2.9.

The total number of electrons $N(E)$ per unit volume up to the Fermi energy E_{Fo} is given as

$$N(E) = \int_0^{E_{Fo}} D(E) f(E) dE$$

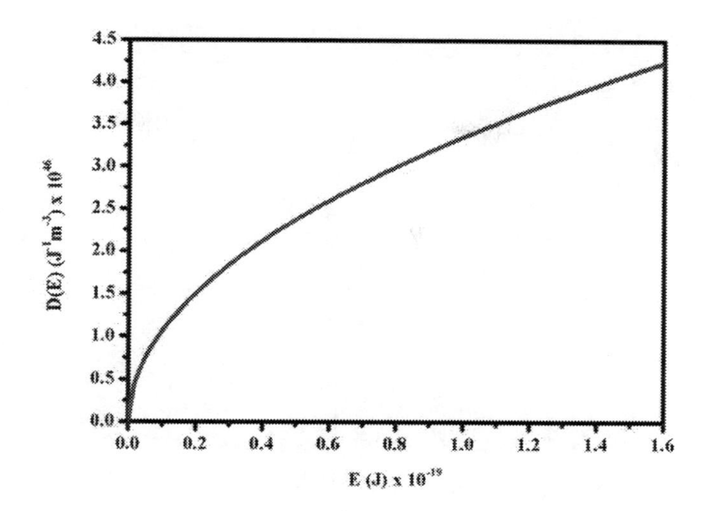

FIGURE 2.9 Plot of the DOS $D(E)$ as a function of electron energy E in 3D materials.

We know that $f(E)=1$ at absolute temperature; thus

$$N(E)=\int_{0}^{E_{Fo}} D(E)dE$$

$$N(E)=\int_{0}^{E_{Fo}} \left[\frac{4\pi}{h^3}\cdot(2m)^{3/2}\right]\cdot E^{1/2}dE$$

$$N(E)=\left[\frac{4\pi}{h^3}\cdot(2m)^{3/2}\right]\cdot E_{Fo}^{3/2} \tag{2.10}$$

Quantum Well (Two-Dimensional Nanomaterials)

The motion of a particle in quantum well systems is confined within the $k_x - k_y$ plane and restricted in the k_z plane of k-space. Thus, the DOS for such systems is derived by considering the planar confinement of particles. The total number of states per unit area $N(k)$ up to the Fermi energy is calculated by the ratio of a circular area of radius k and area of primitive cell $\left(2\pi/L\right)^2$ per unit area A. Hence

$$N(k)=2\cdot\frac{\pi k^2}{A}\cdot\frac{1}{\left(2\pi/L\right)^2}$$

$$=2\cdot\frac{\pi k^2}{(2\pi)^2}\cdot\frac{L^2}{A}$$

$$=2\cdot\frac{\pi k^2}{(2\pi)^2} \tag{2.11}$$

where factor 2 is the spin degeneracy of the electrons and accounts due to the Pauli principle, L is the length of a 2D square, $A = L^2$, and $\frac{2\pi}{L}$ is the distance between two lattice points in the k-space.

From equation (2.6)

$$E = \frac{\hbar^2 k^2}{2m}$$

$$k = \sqrt{\frac{2mE}{\hbar^2}}$$

$$\frac{\partial k}{\partial E} = \left(\frac{2m}{\hbar^2}\right)^{1/2} \cdot \frac{1}{2\sqrt{E}} \tag{2.12}$$

Therefore, the DOS for 2D quantum wells is given as

$$D(E) = \frac{\partial N(k)}{\partial E} = \frac{\partial N(k)}{\partial k} \cdot \frac{\partial k}{\partial E}$$

Using equations (2.11) and (2.12), we get

$$D(E) = \left(\frac{k}{\pi}\right)\left(\frac{2m^*}{\hbar^2}\right)^{1/2} \cdot \frac{1}{2\sqrt{E}}$$

$$= \frac{m^*}{\pi\hbar^2} \tag{2.13}$$

This equation shows that the total number of allowed states per unit area per unit energy or DOSs in 2D quantum wells is independent of energy. For n number of confined states, the DOS becomes

$$= \sum_{i=1}^{n} \frac{m^*}{\pi\hbar^2} \cdot Y(E - E_i) \tag{2.14}$$

where Y represents a step function such that

$$Y(E - E_i) = \begin{cases} 1 & for\ E < E_i \\ 0 & for\ E > E_i \end{cases}$$

Figure 2.10 represents the DOS for three confined sub-bands corresponding to $i = 1$, 2, and 3 of quantum well energy levels.

Quantum Wire (One-Dimensional Nanomaterial)

The quantum wire or nanowire is a quasi-one-dimensional nanostructure that possesses a unique behavior due to the constraints of the charge carrier to transport in a

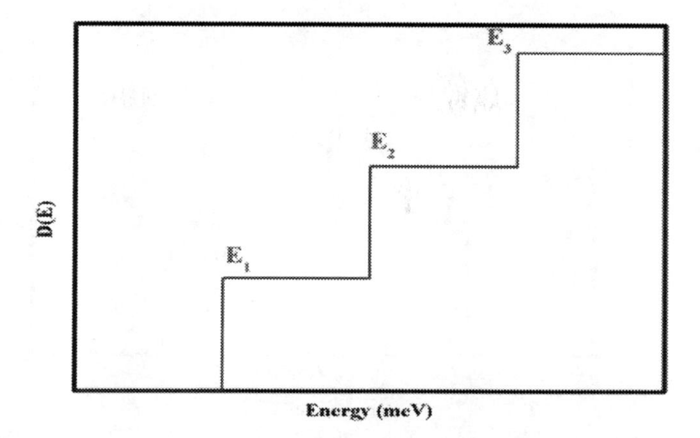

FIGURE 2.10 Plot of the DOS $D(E)$ as a function energy E for three confined sub-bands in 2D nanomaterials.

single direction. Thus, a free charge carrier consists of a single degree of freedom due to the reduction of dimensionality with two confined directions and one unconfined direction for the charge carrier. Assume that the free electron is confined from the y- and z-directions but can move freely in the x-direction (unconfined direction). In this case, the total number of states per unit length $N(k)$ up to the Fermi energy is given by

$$N(k)=2\cdot\frac{2k}{L}\cdot\frac{1}{\left(2\pi/L\right)}=\frac{2k}{\pi} \tag{2.15}$$

where factor 2 appears because of electronic spin degeneracy and accounts due to the Pauli principle. Therefore, the DOS for quantum wires is given as

$$D(E)=\frac{\partial N(k)}{\partial E}=\frac{\partial N(k)}{\partial k}\cdot\frac{\partial k}{\partial E}$$

$$=\left(\frac{2m^*}{\hbar^2}\right)^{1/2}\cdot\frac{1}{\pi\sqrt{E}}. \tag{2.16}$$

For n number of confined states, the DOS becomes

$$=\left(\frac{2m^*}{\hbar^2}\right)^{1/2}\sum_{i=1}^{n}\frac{1}{\pi\sqrt{E-E_i}}\cdot Y(E-E_i) \tag{2.17}$$

where Y represents the step function such that

$$Y(E-E_i)=\begin{cases} 1 & for\ E<E_i \\ 0 & for\ E>E_i \end{cases}$$

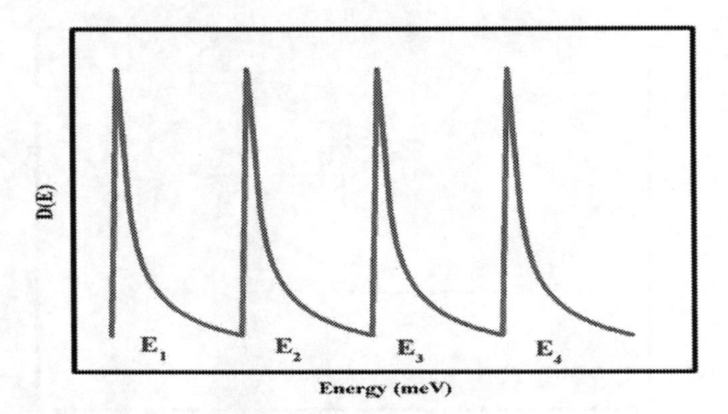

FIGURE 2.11 Plot of the DOS $D(E)$ as a function energy E for four confined sub-bands in 1D nanomaterials.

Figure 2.11 represents the DOS for four confined sub-bands corresponding to $i = 1, 2, 3$, and 4 of the quantum wire energy level.

Quantum Dots (Zero-Dimensional Nanomaterials)

A quantum dot is one of the simplest practical examples of zero-dimensional (0D) materials. Quantum dots are nm-sized tiny semiconductor particles or nanocrystals that confine electrons, holes or electron pairs to a small space called the quantum box. They show a unique shape-, size- and composition-dependent electronic and optical properties which can be used in broader fields of electronic and biological industries. The quantum dots have a zero degree of freedom due to the confinement of all three-dimensional directions in the k-space. Thus, the energy of quantum dots may be expected to be discrete, and their DOS may be evaluated by dividing the number of confined states with the energy level. The DOS is given in terms of the Dirac delta function when the energy interval tends to be zero as shown below

$$D(E) = 2 \cdot \sum_{i} \delta(E - E_i), \tag{2.18}$$

where δ represents the Dirac delta function and the factor 2 is introduced due to two possible spins +1/2 and −1/2 at the same energy level E_i. Figure 2.12 represents the DOS for three confined sub-bands corresponding to $i = 1, 2$, and 3 of quantum wire energy levels.

2.5 BLOCH THEOREM

We know that the general Schrodinger equation for a wave function $\psi(r)$ in crystal potential $V(r)$ of a crystal is given as

$$\left[-\frac{\hbar^2}{2m}\nabla^2 + V(\vec{r})\right]\psi(\vec{r}) = E\psi(\vec{r}), \tag{2.19}$$

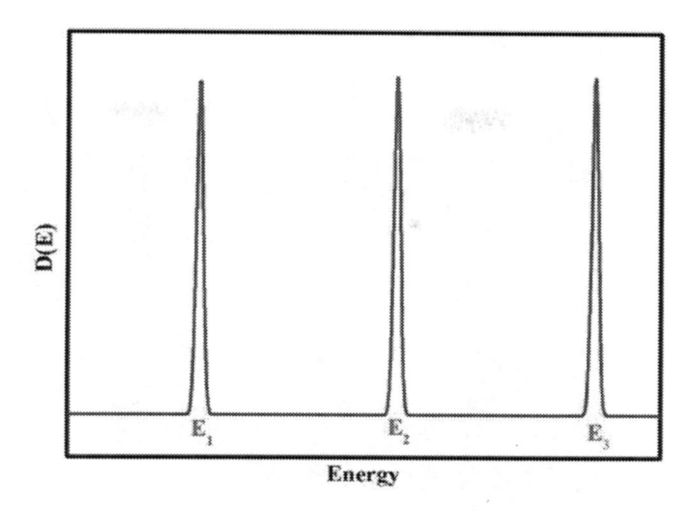

FIGURE 2.12 Plot of the DOS $D(E)$ as a function energy E for three confined sub-bands in 0D nanomaterials.

where E is the energy of electrons. Bloch (1928) assumed the nature of Coulomb potential $V(r)$ as a periodic inside a crystal due to the periodic arrangement of the lattice [10]. Hence,

$$V(\vec{r}) = V(\vec{r} + \vec{T}),$$

where $\vec{T} = n_1\vec{a} + n_2\vec{b} + n_3\vec{c}$ (n_1, n_2 and n_3 are integers 0, ±1, ±2, ±3, ….. and $\vec{a}$, $\vec{b}$ and $\vec{c}$ are lattice constants) represents an arbitrary translational vector in a crystallographic lattice. Figure 2.13 shows the periodic 1D simplified Coulombic potential, which is of the same height within the lattice but the real crystal potential is comparatively high at the boundary, as shown in Figure 2.14.

The Bloch theorem is valid for the potential of the same height throughout the crystal. In this regard, Bloch ignored the abrupt change in potential at the boundary as shown in Figure 2.14 and considered that the potential is of the same height.

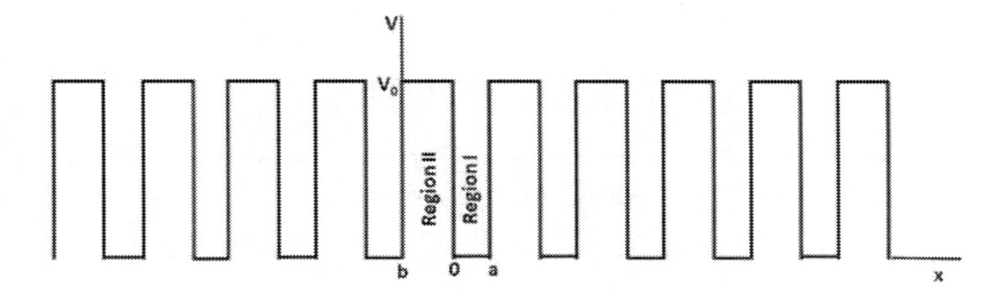

FIGURE 2.13 Ideal Bloch potential for a periodic lattice.

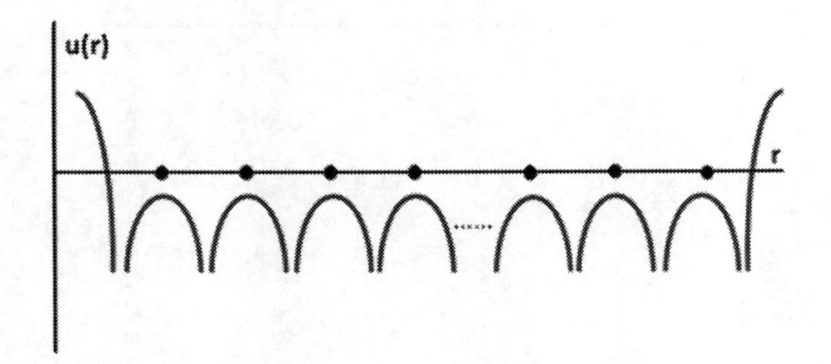

FIGURE 2.14 Real potential for a periodic crystal lattice with higher potential on the surface.

Moreover, Bloch also considered an ideal crystal by ignoring the crystal imperfection and impurities in the lattice points.

Bloch predicted the wave function of crystals in the periodic potential or solution of equation (2.19), which is expressed as the product of the plane-wave equation e^{iKr} and periodic $u(r)$ with the same periodicity of lattice, i.e.,

$$\psi(\vec{r}) = e^{i\vec{k}\cdot\vec{r}} \cdot u(\vec{r}), \tag{2.20}$$

where

$$u(\vec{r}) = u(\vec{r} + \vec{T}) \tag{2.21}$$

Here, $\vec{T}$ represents the translational vector. When $\vec{r} \rightarrow \vec{r} + \vec{T}$

$$\psi(\vec{r} + \vec{T}) = e^{i\vec{k}\cdot(\vec{r}+\vec{T})} \cdot u(\vec{r} + \vec{T}),$$

Now using periodic potential in equation (2.21) and rearranging the above one, we get

$$\psi(\vec{r} + \vec{T}) = e^{i\vec{k}\cdot\vec{T}} \cdot e^{i\vec{k}\cdot\vec{r}} \cdot u(\vec{r}),$$

$$\psi(\vec{r} + \vec{T}) = e^{i\vec{k}\cdot\vec{T}} \cdot \psi(\vec{r})$$

The above equation shows that the wave function in periodic potential will be periodic when it satisfies the condition

$$e^{i\vec{k}\cdot\vec{T}} = 1 = e^{i.2\pi n} \tag{2.22}$$

where $n = 0, 1, 2, 3, \ldots$

Now, since the wave function of periodic potential $u(\vec{r})$ satisfying the above condition is periodic in nature, all other associated observable physical quantities such as electron probability density $|\psi(\vec{r})|^2$ are periodic.

$$|\psi(\vec{r})|^2 = |\psi(\vec{r}+\vec{T})|^2 \tag{2.23}$$

2.6 CRYSTALLINE SOLIDS – PERIODICITY OF CRYSTAL LATTICES

During the explanation of the free electron theory of solids, we had assumed that conduction electron or charge carrier experiences zero or constant potential that makes them move freely within the material (about the crystal). The free electron theory successfully explains several physical phenomena including specific heat, electrical conductivity, thermal conductivity, thermionic emission, paramagnetism, etc. However, this model is not suitable for explaining the band gap of the material that distinguishes insulators, conductors and semiconductors from each other. Hence, there is a need for some other model that could explain the band gap of the material accurately.

The Kronig Penney model utilizes the Schrodinger equation to explain the nature of electrons within the periodic potential of a crystal lattice. In order to solve the problem, Kronig Penney considered a repetition of a periodic potential well and potential barrier as shown in Figure 2.13. This arrangement of the potential well and barrier is very close to real crystal potential as shown in Figure 2.14 and is suitable for simplified calculation. Despite the artificial chosen potential, this model derives many important characteristics of electronic behavior in periodic crystalline solids. Regions I and II in Figure 2.14a show the potential well ($0<x<a$) and barrier ($-b<x<0$), respectively. Thus, the Schrodinger equation for regions I and II is as follows

$$\frac{d^2\psi}{dx^2} + \frac{8\pi^2 m}{h^2} E\psi = 0 \quad \text{for } 0<x<a \tag{2.24}$$

$$\frac{d^2\psi}{dx^2} + \frac{8\pi^2 m}{h^2}(E - V_o)\psi = 0 \quad \text{for } -b<x<0 \tag{2.25}$$

where V_o is the potential barrier height. Since the electron is confined within the potential well and cannot cross the potential barrier, E is always less than V_o, i.e., $E<V_o$

$$\frac{d^2\psi}{dx^2} + \alpha^2\psi = 0 \quad \text{for } 0<x<a \tag{2.26}$$

$$\frac{d^2\psi}{dx^2} - \beta^2\psi = 0 \quad \text{for } -b<x<0 \tag{2.27}$$

where $\alpha^2 = \frac{8\pi^2 mE}{h^2}$ and $\beta^2 = \frac{8\pi^2 m}{h^2}(V_o - E)$

Using the Bloch theorem for periodic potential, the suitable solution of equations (2.26) and (2.27) may be given as

$$\psi(x) = u_k(x)e^{ikx} \tag{2.28}$$

On differentiating equation (2.28) twice with respect to x, we get

$$\frac{d\psi}{dx} = u_k iKe^{ikx} + e^{ikx}\frac{du_k}{dx}$$

and
$$\frac{d^2\psi}{dx^2} = -K^2 e^{iKx}u_k + 2iKe^{iKx}\cdot\frac{du_k}{dx} + e^{iKx}\frac{d^2u_k}{dx^2}$$

Using this equation in equations (2.26) and (2.27), we get

$$\frac{d^2u_1}{dx^2} + 2ik\frac{du_1}{dx} + \left(\alpha^2 - k^2\right)u_1 = 0 \quad \text{for } 0 < x < a \tag{2.29}$$

$$\frac{d^2u_2}{dx^2} + 2ik\frac{du_2}{dx} - \left(\beta^2 + k^2\right)u_2 = 0 \quad \text{for } -b < x < 0 \tag{2.30}$$

The solution of equations (2.29) and (2.30) may be expressed in the following form

$$u_1 = Ae^{i(\alpha-k)x} + Be^{-i(\alpha+k)x} \tag{2.31}$$

$$u_2 = Ce^{(\beta-ik)x} + De^{-(\beta+ik)x} \tag{2.32}$$

Now, the wave function and first derivative of the wave function of regions I and II will be continuous at $x=0$ due to the finite height of the potential barrier. Thus, applying boundary condition,

$$(u_1)_{x=0} = (u_2)_{x=0}$$

and,

$$\left(\frac{du_1}{dx}\right)_{x=0} = \left(\frac{du_2}{dx}\right)_{x=0}$$

Using equations (2.31) and (2.32) in the above boundary conditions, we get

$$A + B = C + D \tag{2.33}$$

and,

$$A\left[i(\alpha-k)\right] - B\left[i(\alpha+k)\right] = C\left[(\beta-ik)\right] - D\left[(\beta+ik)\right] \tag{2.34}$$

Furthermore, due to the periodic nature of $u_k(x)$, the following conditions may be imposed on u_1 and u_2

$$(u_1)_{x=a} = (u_2)_{x=-b}$$

and,

$$\left(\frac{du_1}{dx}\right)_{x=a} = \left(\frac{du_2}{dx}\right)_{x=-b}$$

Using equations (2.31) and (2.32) in the above conditions, we get

$$Ae^{i(\alpha-k)a} + Be^{-i(\alpha+k)a} = Ce^{-(\beta-ik)b} + De^{(\beta+ik)b} \tag{2.35}$$

and,

$$Ai(\alpha-k)e^{i(\alpha-k)a} - Bi(\alpha+k)e^{-i(\alpha+k)a} = C(\beta-ik)e^{-(\beta-ik)b} - D(\beta+ik)e^{(\beta+ik)b} \tag{2.36}$$

For the nonvanishing solution of equations (2.33), (2.34), (2.35), and (2.36), the determinant of coefficients A, B, C, and D must vanish. Thus

$$\begin{vmatrix} 1 & 1 & 1 & 1 \\ i(\alpha-k) & -i(\alpha+k) & (\beta-ik) & -(\beta+ik) \\ e^{i(\alpha-k)a} & e^{-i(\alpha+k)a} & e^{-(\beta-ik)b} & e^{(\beta+ik)b} \\ i(\alpha-k)e^{i(\alpha-k)a} & -i(\alpha+k)e^{-i(\alpha+k)a} & (\beta-ik)e^{-(\beta-ik)b} & -(\beta+ik)e^{(\beta+ik)b} \end{vmatrix} = 0$$

By solving the above determinant, we get

$$\frac{(\beta^2-\alpha^2)}{2\alpha\beta}\sin\alpha a \sin h\beta b + \cos h\beta b \cos\alpha a = \cos k(a+b) \tag{2.37}$$

The solution of equation (2.37) is complicated to solve. Thus, Kronig Penney assumed $V_o \to \infty$ and $b \to \infty$ in such a way that $V_o b$ remains finite to simplify the solution, and the term $V_o b$ is referred to as the barrier strength [11]. This assumption models a potential with a very deep potential well and a very small width potential barrier. Thus, $\sin h\beta b \to \beta b$ and $\cos h\beta b \to 1$ as $b \to 0$, which shapes Eq. (2.15) as follows

$$\frac{(\beta^2-\alpha^2)}{2\alpha} b \sin\alpha a + \cos\alpha a = \cos ka$$

Using the value of α and β in the above equation, we get

$$\left(\frac{mV_o ab}{\hbar^2}\right)\frac{\sin\alpha a}{\alpha a} + \cos\alpha a = \cos ka$$

$$P\frac{\sin\alpha a}{\alpha a}+\cos\alpha a=\cos ka \tag{2.38}$$

where the dimensionless quantity $P=\frac{mV_oab}{\hbar^2}$ is a measure of electron attraction strength toward the ion in the crystal lattice site and is referred to as the scattering power of the potential barrier. The above equation represents the relation between energy $\alpha^2=\frac{8\pi^2mE}{h^2}$ and wave vector k and gives the condition for the existence of wave function. The RHS term $\cos ka$ of the above equation is bound to have a value between +1 and −1. The curve of $P\frac{\sin\alpha a}{\alpha a}+\cos\alpha a$ against αa has been plotted in Figure 2.15. Thus, the region of the curve between +1 and −1 corresponding to the LHS term which satisfies the RHS is the only allowed value, which represents the allowed range of energy or allowed band and the rest represents the forbidden energy range of the forbidden energy band. The shaded region in the figure shows the allowed energy band and the unshaded one shows the forbidden energy band. Thus, there exists a forbidden energy band in between allowed energy bands for electrons in a periodic potential of a crystal. The figure also shows that the forbidden band narrows, whereas the allowed bands broaden toward higher energy.

Now, when we use $\alpha a=\pm n\pi$, then equation (2.38) becomes

$$\cos n\pi=\cos ka$$

$$\text{or, } \pm n\pi=ka$$

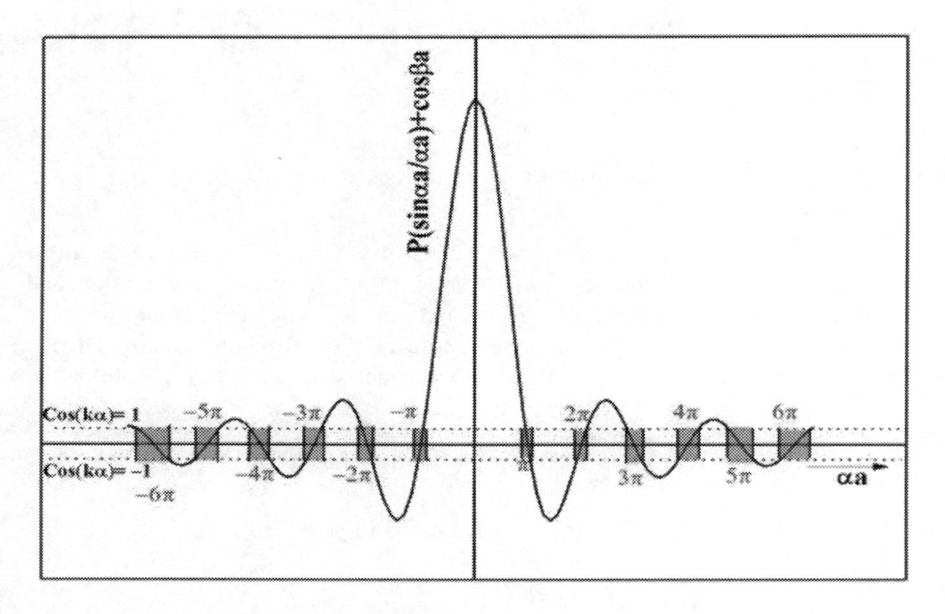

FIGURE 2.15 The curve of $.p\frac{sin\ aa}{aa}+cos\ aa$. against aa...

$$k = \pm\frac{n\pi}{a}$$

Thus, a discontinuity is found at $k = \pm\frac{n\pi}{a}$ in the E–k curve for electrons in a crystal. Figure 2.16 shows the E–k diagram for free electrons shown by the parabolic dotted line with energy $E = \frac{\hbar^2 k^2}{2m} = \frac{h^2 k^2}{8\pi^2 m}$ and solid lines for electrons in the periodic potential. The allowed region between $k = +\frac{\pi}{a}$ and $k = -\frac{\pi}{a}$ is called the first Brillouin zone. Due to discontinuity at $k = \pm\frac{\pi}{a}$, a forbidden gap appears between two allowed bands, as shown in Figure 2.16. The region followed by the first Brillouin zone, i.e., $k = +\frac{\pi}{a}$ to $k = +\frac{2\pi}{a}$ and $k = -\frac{\pi}{a}$ to $k = -\frac{2\pi}{a}$ is called the second Brillouin zone. Similarly, a higher-order Brillouin zone occurs with an increasing value of n.

Moreover, it may be shown that by varying $P = 0$ to $P \to \infty$, electrons may cover the entire range of energy from a completely free state to a completely bound state. To show this, let us consider the case with infinite scattering power of the potential barrier, i.e., $P \to \infty$. In this case, $\sin \alpha a = 0 = \sin n\pi$

Thus, $\alpha a = \pm n\pi$

$$\alpha^2 = \frac{n^2\pi^2}{a^2} = \frac{8\pi^2 mE}{h^2}$$

$$E = \frac{n^2 h^2}{8ma^2} \tag{2.39}$$

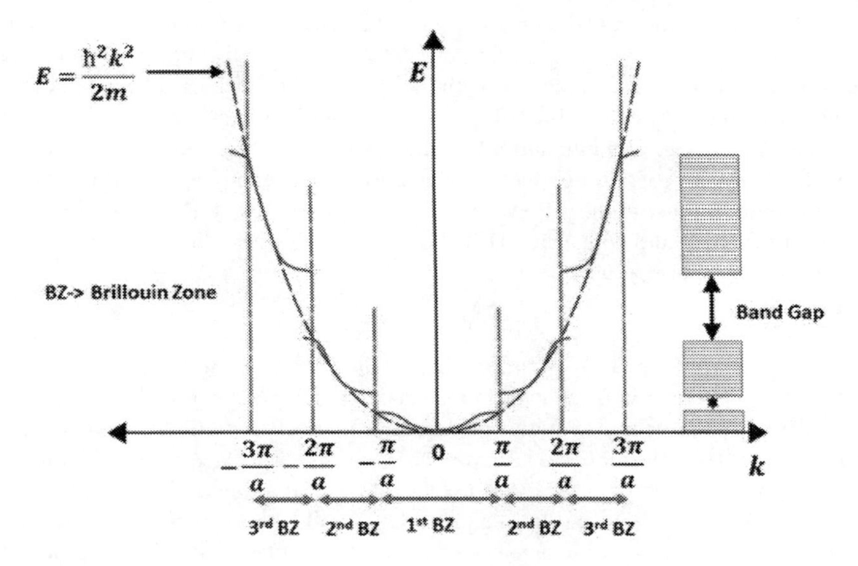

FIGURE 2.16 *E* versus *k* curve for electrons in periodic potential in a 1D lattice and Brillouin zone.

This equation is similar to the energy of the particles in a box, and hence, we may infer that the condition $P \rightarrow \infty$ represents the completely bound state of particles. The energy value in the above equation is independent of k. Now, in order to show the completely free state of electrons, we use the condition $P = 0$ in equation (2.38) and we get

$$\cos \alpha a = \cos ka$$

$$\alpha^2 = k^2 = \frac{8\pi^2 mE}{h^2} = \frac{2mE}{\hbar^2}$$

$$E = \frac{\hbar^2}{2m} \cdot k^2$$

$$E = \frac{\hbar^2}{2m} \cdot \frac{p^2}{\hbar^2} = \frac{1}{2} mv^2 \tag{2.40}$$

This energy value is equivalent to the energy of the free particles moving with velocity v. In such a situation, all possible energies are allowed for electrons. Thus, two extreme conditions of P, i.e., $P = 0$ and $P \rightarrow \infty$, represent the completely free and bounded state of electrons, respectively.

2.7 EFFECTS OF THE NANOMETER LENGTH SCALE

Very few changes in the behavior of the solid material are observed with the change in particle size from a visible length scale to the regular optical microscope scale. However, the changes become significant with the reduction of the particle size of the material to a nanoscale dimension. Thus, the behavior of a solid material such as fluorescence, melting point, magnetic permeability, electrical conductivity, chemical reactivity, etc. changes significantly with the change of particle size from bulk to the nanoscale, which has been reported in several scientific research articles. The origin of a special behavior of nanomaterials is highly influenced by the high surface area-to-volume ratio and quantum effect at the nanoscale. For example, metallic gold (Au) is often found in a shining yellow color, but gold particles of 10–20 nm dimension absorb green light and appear red [12]. These properties have been explained in the later section of this chapter.

Surface Effect

A larger surface area-to-volume ratio (or simply, surface-to-volume ratio) is an important characteristic of nanoscale materials, which creates possibilities for new materials with significantly different physical or chemical properties. Nanotechnology enables us to fabricate a layered structure, where a larger surface area is required for many applications such as to serve as potent catalysis, to improve wear resistance, to serve as a thermal barrier in thin films, etc. To visualize the variation in the surface-to-volume ratio with the dimension, let us consider a cube of length $a = 10$ cm. The surface area of one face of the cube is $a \times a = 100\,\text{cm}^2$. Since there are six faces of a

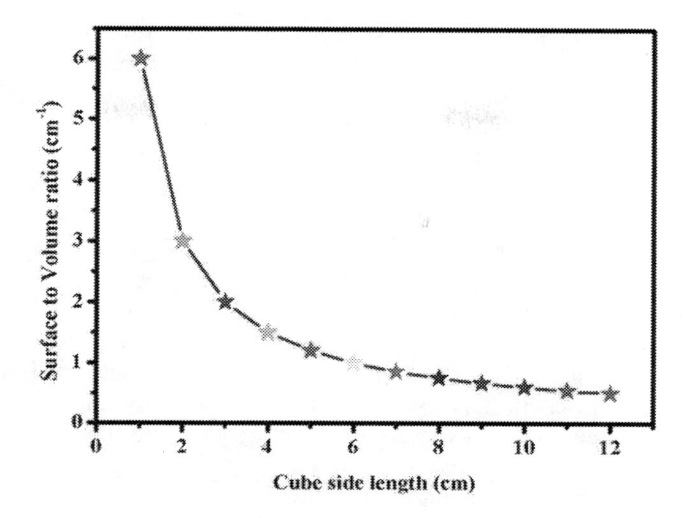

FIGURE 2.17 Plot of surface-to-volume ratio as a function of cube length.

cube, the total surface area of the cube is equal to 6 × the surface area of one face. Hence, the total surface area of the cube $= 6 \times 100\,cm^2 = 600\,cm^2$ and the volume of cube will become $a^3 = 1000\,cm^3$. Thus, the surface area-to-volume ratio becomes $600/1000\,cm^{-1} = 0.6\,cm^{-1}$. Now, if we replace this cube with another smaller cube of length $a = 1$ cm, then the total number of smaller cubes required to fill the bigger cube is equal to the ratio of the volume of the bigger cube to the volume of the smaller cube, which is equal to $10^3/1^3 = 1000$ cube. The surface area-to-volume ratio of the smaller cube will become $6 \times 1^2 cm^2/1^3 cm^3 = 6$. Thus, the surface area-to-volume ratio increases upon lowering the dimension of the particle. Figure 2.17 shows the variation of the surface-to-volume ratio against the dimension of a cube. Alternatively, the surface-to-volume ratio for a cubical particle may be given as $6a^2/a^3 = 6/a$.

2.7.1 Changes to the System Total Energy

We have already seen in Section 2.3 that the band gap of a material is size-dependent and widens with the decrease in particle size. Thus, with the reduction in particle size, a higher amount of energy is absorbed or released because of the transition between the conduction band to the valence band and vice versa. The widening of the energy band gap with particle size and shape may be interpreted from equation (2.2). At the same time, with decreasing system particle size, the allowed energy band considerably narrows due to the decrease in the number of atoms within smaller particles. Moreover, the surface energy of materials is dependent on their surface-to-volume ratio. The surface energy is always a positive quantity that influences its physical and chemical properties including melting point, vapor pressure and reactivity. The surface atoms are unstable and unsaturated because of dangling bond formation. The unstable dangling bond increases the surface energy, which promotes the atom to react with other species in order to achieve stability. We have seen in the preceding section that the surface-to-volume ratio increases with decreasing particle size. Thus, nanoscale particles participate in surface reactions more promptly as compared with their bulk counterpart. Thus, nanoscale particles are more suitable for catalytic reactions due to their larger surface-to-volume ratio.

In recent decades, researchers rely on the application of nanomaterials for electrochemical, dipolar energy storage and energy conversion in systems such as batteries, capacitors and solar panels due to refined ionic transport and electronic conduction observation [13]. The electrochemical storage system is a method to store electrical energy in a chemical form and is capable of performing reversible conversion of energy [14]. Nanostructured transition-metal oxides such as oxides of iron, manganese, cobalt, zinc, chromium, etc. are attractive candidates for high energy-density electrodes in lithium-ion batteries to fulfill the growing demand of electronic and automobile industries [15]. Chalcogenides, carbides and lithium alloys are also being investigated outside oxides with versatile shape and chemical compositions. On the other hand, a capacitor is also a type of dipolar electrical energy storage device based on interfacial charging and discharging due to a change in dipole moment. Batteries are characterized by high energy density and low power density, but the reverse is true for capacitors [16]. However, electrochemical capacitors also referred to as supercapacitors, are bridges between batteries and capacitors in terms of energy density and power density. Nanomaterials have also been investigated for improving the electrical energy storage capacity of existing capacitors or electrochemical capacitors [16–19].

2.7.2 Changes to the System Structure

Nanocrystal structures of some materials change due to their reduced size despite the same chemical composition and formula as that of their bulk counterpart. However, in practice, the structure remains the same for most of the solid materials after the reduction of particle size up to a few nanometers. The electronic and optical properties of crystalline solids have been observed closely related to the crystal structure. Thus, critical information regarding the crystal structure and distortion in the crystal structure may be collected with the study of these properties via different diffraction and spectroscopic techniques. The collective vibrational mode of atoms associated with solids is called a phonon, which reflects the important character of crystal structures in spectroscopic studies [20].

The X-ray powder diffraction technique has been widely accepted for phase identification of crystalline solids and provides crucial information on the lattice parameter, atomic position, symmetry, etc. The technique solely is based on Bragg's diffraction law

$$2d\sin\theta = n\lambda \tag{2.41}$$

where d is the spacing between the lattice of two consecutive planes, λ is the wavelength of incident X-ray light, θ is the glancing angle, and n is an integer 0, 1, 2, 3….. representing the order of diffraction. The powder diffraction pattern peaks for a nanoparticle are broader than their bulk counterpart despite the same chemical formula and composition [21]. The broadness in peaks is due to the smaller crystallite size. The crystallite size D may be calculated from the Scherrer formula applicable to X-ray diffraction peaks [22]. The Scherrer formula is given as

$$D = \frac{K\lambda}{\beta \cos\theta} \tag{2.42}$$

where $\lambda = 1.5405$ Å, K, β and θ represent the wavelength of X-rays, a crystallite shape-dependent dimensionless quantity called the shape factor with a typical value of 0.9, full width half-maximum (FWHM) of the diffraction peak, and Bragg's diffraction angle, respectively.

Metal nanoclusters are agglomerates of few to hundreds of atoms or molecules, which represent a bridge between a single atom and plasmonic nanoparticles. The particle size of metal nanoclusters is generally less than 2 nm and may be considered ultrasmall. The study of small clusters of noble metal materials is of profound interest for researchers over the globe due to their unique optical, electronic, magnetic, etc. properties. Only some specific numbers of atoms are allowed in stable nanoclusters due to minimized energy. The specific numbers are called structural magic numbers. For example, one of the combinations of 55 metallic gold atoms shows a stable structure of nanoclusters; thus, 55 is a structural magic number for gold nanoclusters. Most of the metallic atoms such as Ag, Au, Co, Al, Cu, etc. are forms of face-centered cubic (FCC) close-packed structures, and the number of atoms in a cluster N is given as

$$N = \frac{1}{3}\left[10n^3 - 15n^2 + 4n - 3\right] \tag{2.43}$$

whereas the number of atoms on the surface of a cluster is given as

$$N_{surf} = 10n^2 - 20n + 12 \tag{2.44}$$

where n represents the number of layers.

2.7.3 How Nanoscale Dimensions Affect Properties

2.7.3.1 Effect on Chemical Property

It is well established that during the reaction of any material, its surface area exposed to other reactants plays a vital role in the chemical reaction. Therefore, a higher surface area of the material is required for better chemical reactivity as well as catalytic activity and is expressed in terms of the surface-to-volume ratio. The surface-to-volume ratio has been already explained in the surface effect section of this chapter, which shows that nanoscale materials have a significantly high surface-to-volume ratio relative to bulk materials. Thus, a nanocatalyst has the potential to increase the rate of reaction, selectivity and chemical reaction significantly due to its nanodimension. The nanogold particle deposited on partially reactive oxides such as Fe_2O_3, MnO, and titanium is found to have an excellent catalytic behavior.

2.7.3.2 Effect on Mechanical Property

The size distribution, agglomeration, defect, material synthesis process, crystal structure, grain size, etc. are some factors that influence the mechanical behavior of materials. It is well established that the toughness of a nanomaterial is inversely proportional to the fabrication technique or defect of the materials. Polycrystalline nanoscale materials contain several simultaneously oriented monocrystal regions called grains, and these grains are separated by high angle boundaries called the grain boundary. A fine-grained polycrystalline material is found to be stronger and harder than coarse-grained materials despite its similar chemical composition, which can be explained by the grain size and yield stress relation through the Hall–Patch relation [23]

$$\sigma = \sigma_o + \frac{K}{\sqrt{D}} \tag{2.45}$$

where σ, σ_o, K and D denote the yield stress, material constant related to the resistance of lattice to dislocation motion, Hall–Patch strengthening coefficient, and average grain size, respectively. Several articles reveal that despite the similar composition, the yield stress and microhardness of a nanoscale polycrystalline material are 2–10 times greater than those of its counterpart in a coarse-grained material.

2.7.3.3 Effect on Melting Temperature

The melting point of bulk materials has been found to be size-independent, whereas the melting point decreases with the decrease in particle size for almost all nanoscale materials. For example, the melting point of gold nanoparticles is lower than its bulk counterpart. Theoretically, the decrease of the melting point with the reduction in the size of nanoparticles could be explained by the linear relationhip between temperature and the reciprocal of particle size [24]

$$T_m = T_{bulk}\left(1 - \frac{c}{r}\right) \tag{2.46}$$

where T_m and T_{bulk} denote the melting temperature of the nanomaterial and its bulk counterpart, respectively. Furthermore, c and r represent the material constant-dependent nature of materials and the radius of the sphere, respectively. In past several decades, different theoretical models have been proposed to find the material constant c that could agree with the experimentally observed result for the size-dependent melting point of nanoscale materials.

The theoretical model utilizing the Laplace equation of the surface and Gibbs Duhem equation describes the melting temperature depression (ΔT_m) for a solid nanoparticle as given below [25]

$$\Delta T_m = T_{bulk} - T_m \approx \frac{2T_{bulk}}{\Delta H_m \rho_s} \cdot \frac{\alpha}{r} \tag{2.47}$$

where ΔH_m, ρ_s and r represent the bulk enthalpy of fusion, density of solid material, and radius of the spherical particle, respectively. The parameter α depends upon the material interfacial tension between the solid phase and its environment. This parameter may be evaluated by different models based on the melting mechanism. The three main models for describing α are (a) homogeneous melting and growth model (HGM), (b) liquid shell model (LSM), and (c) liquid nucleation and growth (LNG) model [24–26].

a. **Homogeneous Melting and Growth Model (HGM):** This model is based on the assumption of homogeneously melting the entire nanoparticle at a single melt temperature. In other words, there exists an equilibrium between the solid and liquid phases during the melting process. The parameter α_{HGM} is given by the following expression

$$\alpha_{HGM} = \sigma_{sl} - \sigma_{lv}\left(\frac{\rho_s}{\rho_l}\right)^{2/3} \tag{2.48}$$

where σ is the interfacial surface tension between the solid, liquid and vapor phase indicated with the index s, l and v respectively, whereas ρ_s and ρ_l represent the density of the material in the solid and liquid phase, respectively.

b. **Liquid Shell Model (LSM):** This model is based on the assumption of a melt process r_o thick layer on the surface of the particle prior to the core. The parameter α_{LSM} is given by the following expression

$$\alpha_{LSM} = \frac{\sigma_{sl}}{1-\frac{r_o}{r}} + \sigma_{lv}\left(1-\frac{\rho_s}{\rho_l}\right) \quad (r_o < r) \tag{2.49}$$

c. **Liquid Nucleation and Growth (LNG) Model:** This model is based on the kinetics of the melt process in spherical solid nanoparticles. According to this model, the melt process of solid into the liquid phase begins at the surface and slowly penetrates into the solid with definite activation energy. The parameter α_{LNG} in this model is given by the following expression

$$\sigma_{sl} < \alpha_{LNG} < \frac{3}{2}\left(\sigma_{sv} - \sigma_{lv}h\frac{\rho_s}{\rho_l}\right) \tag{2.50}$$

2.7.3.4 Effect of Magnetic Properties

The magnetic nanomaterial is used in drug delivery, magnetic resonance imaging (MRI), magnetic data storage, magnetic liquid (ferrofluid, i.e., stable suspension of magnetic nanomaterials), etc. Magnetic nanoparticles consisting of magnetic elements such as iron, cobalt, nickel, manganese, and their compounds are superparamagnetic because of their nanoscale particle size. Superparamagnetic nanomaterials are not magnetic at zero magnetic field but quickly get magnetized on exposure to an external magnetic field. Furthermore, on the removal of the external magnetic field, magnetization quickly reverts to a non-magnetic state. Due to the larger particle size of bulk magnetic materials, its particle generally splits into different magnetic domains separated by a domain wall. Each domain has a different direction of magnetization. However, if the approximate particle diameter is below some critical value [27]

$$d_c \approx \frac{18\sqrt{A_{ex}K}}{\mu_o M_o^2} \tag{2.51}$$

where, A_{ex}, K, μ_o and M_s are the exchange stiffness, magnetic anisotropy constant, magnetic permeability, and saturation magnetization, respectively, then the particle contains a single magnetic domain. The critical diameter is a material-dependent quantity whose value generally lies within the range of 1–1000 nm.

2.7.3.5 Effect on Optical Properties

The optical properties of nanoscale materials are most fascinating and studied with different optical spectroscopic techniques such as UV–visible spectroscopy, photoluminescence and electroluminescence spectroscopy, IR spectroscopy, etc. Based on the optical behavior of nanomaterials, it may be used as an optical detector, sensor, imaging, solar cell, and photocatalysis. The optical property of nanomaterials is dependent on several factors including the size, shape, surface characteristics, etc.; for example, the colloidal suspension of metallic gold nanoparticles is red and purple for particle dimensions of 10–20 nm and >20 nm, respectively [12,28]. However, the color changes to yellow for larger metallic gold particles. The appearance of color in metallic gold nanoparticles is because of localized surface plasma resonance (LSPRs) that occurs when the conduction band electrons collectively oscillate with the electric field of incident light. On the other hand, rod-like shaped gold nanoparticles have two LSPRs along the long (longitudinal) axis and short (transverse) axis, respectively. The optical effect in semiconducting nanoparticles is due to the electronic transition between the highest occupied molecular orbital (HOMO) called the valence band and the lowest occupied molecular orbital (LOMO) called the conduction band.

REFERENCES

[1] B. Bhushan, *Handbook of Nanotechnology*, Spinger, Berlin, Heidelberg 2004.

[2] V.V. Pokropivny, V.V. Skorokhod, Classification of nanostructures by dimensionality and concept of surface forms engineering in nanomaterial science, *Mater. Sci. Eng. C.* 27 (2007) 990–993. https://doi.org/10.1016/j.msec.2006.09.023.

[3] L.H. Madkour, *Nanoelectronic Materials*, Springer International Publishing, Cham, 2019. https://doi.org/10.1007/978-3-030-21621-4.

[4] N.C. Burton, S.A. Grinshpun, T. Reponen, Physical collection efficiency of filter materials for bacteria and viruses, *Ann. Occup. Hyg.* 51 (2007) 143–151. https://doi.org/10.1093/annhyg/mel073.

[5] C. AK, R. T, M. S, Development of DNA nanotechnology and uses in molecular medicine and biology, *Insights Biomed.* 1 (2016) 1–10. http://biomedicine.imedpub.com/development-of-dna-nanotechnology-and-uses-in-molecular-medicine-and-biology.php?aid=17814.

[6] M. Singh, M. Goyal, K. Devlal, Size and shape effects on the band gap of semiconductor compound nanomaterials, *J. Taibah Univ. Sci.* 12 (2018) 470–475. https://doi.org/10.1080/16583655.2018.1473946.

[7] R.J. Ouellette, J.D. Rawn, Structure of organic compounds, *Princ. Org. Chem.*, (2015) 1–32. https://doi.org/10.1016/B978-0-12-802444-7.00001-X.
[8] C. Kittel, *Introduction to Solid State Physics*, Eighth edition, Wiley, 2014.
[9] O. Manasreh, *Introduction to Nanomaterials and Devices*. John Wiley & Sons, 2011. https://www.wiley.com/en-in/Introduction+to+Nanomaterials+and+Devices-p-9780470927076.
[10] R.J. Singh, *Solid State Physics*, Pearson Educations, India 2012. New Delhi, India.
[11] S.O. Pillai, *Solid Stare Physics*, Eighth edition, New Age International (P) Limited, 2018. New Delhi, India.
[12] S. Mohan Bhagyaraj, O.S. Oluwafemi, Nanotechnology: The Science of the Invisible, in: *Micro and Nano Technologies, Synthesis of Inorganic Nanomaterials*, Woodhead Publishing, Sawston, 2018: pp. 1–18. https://doi.org/10.1016/B978-0-08-101975-7.00001-4.
[13] E. Pomerantseva, F. Bonaccorso, X. Feng, Y. Cui, Y. Gogotsi, Energy storage: The future enabled by nanomaterials, *Science* (80). 366 (2019) eaan8285. https://doi.org/10.1126/science.aan8285.
[14] A. Rufer, *Energy Storage- System and Components*, CRC Press, Boco Raton, FL, 2018.
[15] S.-H. Yu, S.H. Lee, D.J. Lee, Y.-E. Sung, T. Hyeon, Conversion reaction-based oxide nanomaterials for lithium ion battery anodes, *Small* 12 (2016) 2146–2172. https://doi.org/10.1002/smll.201502299.
[16] E.T. Mombeshora, V.O. Nyamori, A review on the use of carbon nanostructured materials in electrochemical capacitors, *Int. J. Energy Res.* 39 (2015) 1955–1980. https://doi.org/10.1002/er.3423.
[17] J. Lazarte, R. Dipasupil, G. Pasco, R. Eusebio, A. Orbecido, R. Doong, L. Bautista-Patacsil, Synthesis of reduced graphene oxide/titanium dioxide nanotubes (rgo/tnt) composites as an electrical double layer capacitor, *Nanomaterials* 8 (2018) 934. https://doi.org/10.3390/nano8110934.
[18] H. Choi, H. Yoon, Nanostructured electrode materials for electrochemical capacitor applications, *Nanomaterials* 5 (2015) 906–936. https://doi.org/10.3390/nano5020906.
[19] Z. Tian, X. Wang, L. Shu, T. Wang, T.H. Song, Z. Gui, L. Li, Preparation of nano BaTiO3-Based ceramics for multilayer ceramic capacitor application by chemical coating method, *J. Am. Ceram. Soc.* 92 (2009) 830–833. https://doi.org/10.1111/j.1551-2916.2009.02979.x.
[20] J.Z. Zhang, *Optical Properties and Spectroscopy of Nanomaterials*, World Scientific Publishing Co. Pte. Ltd, Singapore, 2009.
[21] J. Zhang, Y. Zhao, B. Palosz, Comparative studies of compressibility between nanocrystalline and bulk nickel, *Appl. Phys. Lett.* 90 (2007) 1–4. https://doi.org/10.1063/1.2435325.
[22] D.K. Kushvaha, B. Tiwari, S.K. Rout, Enhancement of electrical energy storage ability by controlling grain size of polycrystalline $BaNb_2O_6$ for high density capacitor application, *J. Alloys Compd.* 829 (2020) 154573. https://doi.org/10.1016/j.jallcom.2020.154573.
[23] C.S. Pande, K.P. Cooper, Nanomechanics of Hall-Petch relationship in nanocrystalline materials, *Prog. Mater. Sci.* 54 (2009) 689–706. https://doi.org/10.1016/j.pmatsci.2009.03.008.
[24] F. Gao, Z. Gu, Melting Temperature of Metallic Nanoparticles, in: *Handbook of Nanoparticles*, Springer International Publishing, Cham, 2016: pp. 661–690. https://doi.org/10.1007/978-3-319-15338-4_6.
[25] M. Zhang, M.Y. Efremov, F. Schiettekatte, E.A. Olson, A.T. Kwan, S.L. Lai, T. Wisleder, J.E. Greene, L.H. Allen, Size-dependent melting point depression of nanostructures: Nanocalorimetric measurements, *Phys. Rev. B - Condens. Matter Mater. Phys.* 62 (2000) 10548–10557. https://doi.org/10.1103/PhysRevB.62.10548.

[26] K.K. Nanda, Size-dependent melting of nanoparticles: Hundred years of thermodynamic model, *Pramana - J. Phys.* 72 (2009) 617–628. https://doi.org/10.1007/s12043-009-0055–2.

[27] S. Mørup, C. Frandsen, M.F. Hansen, Magnetic properties of nanoparticles, in: A.V. Narlikar, and Y.Y. Fu (eds), *Oxford Handbook of Nanoscience and Technology: Volume 2: Materials: Structures, Properties and Characterization Techniques*, Oxford Handbooks, online edn, Oxford Academic, 2017, https://doi.org/10.1093/oxfordhb/9780199533053.013.20 (accessed 15 August, 2022).

[28] C.L. Nehl, J.H. Hafner, Shape-dependent plasmon resonances of gold nanoparticles, *J. Mater. Chem.* 18 (2008) 2415–2419. https://doi.org/10.1039/b714950f.

3 Fundamentals of Nanomaterials

Yarazett Hernández Castillo

CONTENTS

3.1 INTRODUCTION

The first nanomaterial was formed in the 4th century by Romans; however, this application was discovered in 1990 [1], thanks to the use of transmission electron microscopy and derived from the development of nanotechnology in 1981 when the window to the nanoscale world was opened. It helps to know the nanostructure, describe it and relate it to the unexpected properties.

After years of study and analysis, nanomaterials have been defined as solid materials which are constituted by nanoparticles (NPs), minute parts of a maximum of 100 nm in at least one of its dimensions: length, wide, or thickness. However, the size of the particles depends on their shape, which in turn depends on their crystalline structure. At the same time, the crystalline structure is derived from the arrangement of the molecules to form aggregates. In this way, the interaction between metallic

DOI: 10.1201/9781003216308-3

atoms and no metallic or polymer chains are completely different; nanometers of distance provide different structures and properties to each material. This means that is important to know the composition of materials, their structure, size, shape, and dimensions of the structure since this can change from bulk to nanosize. In bulk materials, the physical, chemical, and biological properties are constant; on the contrary, in nanosized materials, they vary with the number of atoms that constitute the nanostructure; in the literature, glass transition, diffusion, viscoelastic [2], melting point, optical properties, electrical and thermal conductivities, catalytic activity, superplasticity, toughness, hardness, strength, permeability, antioxidant and antibacterial activity are some properties that are mentioned to be affected by the size reduction [3]. It is relevant how these nanomaterials had developed technology and technology to them. Advanced technology in analysis and characterization has allowed us to explore the microstructure during the formation of materials, and this helps to direct the obtention of expected properties.

The first topic of this chapter is based on the size and nanostructure dimensions; the classification of 1D, 2D, and 3D nanostructures and the main geometric forms related to each type are shown. Certain nanomaterials of the same composition present more than one type of characteristic nanostructures; for example, carbon is found as nanosheets or nanotubes. It is also important to consider the classification of nanomaterials by their composition; in the second and third topics of this chapter, two general groups of materials, inorganics and organics, are described. They are differenced by the carbon content and its arrangement in the molecular structure; the former description is about inorganic materials conformed by metals, alloys, oxides, and non-oxides inorganic nanomaterials; secondly, organic materials in the form of NPs, nanofilms, and examples of biological nanomaterials are mentioned.

Predicting the function of certain materials in a defined application is necessary to know as much as possible about the expected properties; later, by analyzing raw materials and the obtention method, assure to obtain the nanostructure related to those properties. This chapter describes the fundamental properties of different kinds of nanomaterials, from which many potential applications derive. Special artificial molecules such as dendrimers and quantum dots are mentioned throughout this chapter since they are expected to develop electrochemical methods as medical treatments; in the same way, differences between bulk and NPs are highlighted.

3.2 CLASSIFICATION AND TYPES OF NANOMATERIALS

It is important to remember that nanomaterials are those materials of almost every composition that have at least one dimension (length, width, or height) within 1–100 nm; there are exceptions of polymeric NPs which are composed of macromolecules and can have sizes between 100 and 1000 nm; above this scale, you would be talking about bulk or micromaterials. Also, the nanostructure of nanomaterials refers to the basic form of the NPs individually (cubes, spheres, nanotubes, etc.) and should not be confused with the form or shape of aggregates or cumulus of several NPs, or agglomerates. Moreover, a NP can be a nanocrystal or not, and it depends if they have a continuous atomic arrangement.

There are several classifications of nanomaterials, for example: by composition (organics and inorganics); application (superconductors, semiconductors, insulators, biomaterials, etc.); and size and dimension (0D, 1D, 2D, and 3D nanomaterials) [3]. The classification by composition is described in the next topic and several applications for these materials are described. In this topic, the classification by size and dimension is focused on for a better understanding of the kinds of nanostructures in which the nanomaterials are obtained. Figure 3.1 shows this classification. Zero-dimensional (0D) nanomaterials have three dimensions between 1 and 100 nm; the types of nanomaterials included in this classification are NPs that can be nanocapsules, core–shell NPs, and micelles, for example. One-dimensional (1D) nanomaterials have two dimensions on the nanoscale and are more than 100 nm longer. Two-dimensional (2D) nanomaterials possess just one dimension at the nanoscale (height), and the characteristic shape is of thin sheets or plates. Three-dimensional (3D) nanomaterials do not have any dimension at the nanoscale and are considered bulk materials; however, they are composed completely or partially of nanostructures before described and because of this, they are called nanostructured materials.

Among attractive 0D nanostructures are quantum dots and nanoclusters. Nanoclusters are aggregates of atoms, ions, or molecules bonded by ionic or van der Waals forces or metallic or covalent bonds; their properties change with the number of aggregated units. A kind of cluster are quantum dots, colloidal semiconductor nanocrystals conformed by some hundred atoms (diameter of 2–10 nm); they have the characteristic of emitting light at a certain wavelength depending on their size [4].

1D nanomaterials like nanowires, nanotubes, nanorods and nanobelts are different since nanowires are the longest 1D nanostructures with an aspect ratio (length/width) of 1000; nanotubes are larger than 100 nm and have a hollow morphology; nanorods have a cylindrical structure like nanowires but an aspect ratio that is ten times smaller.

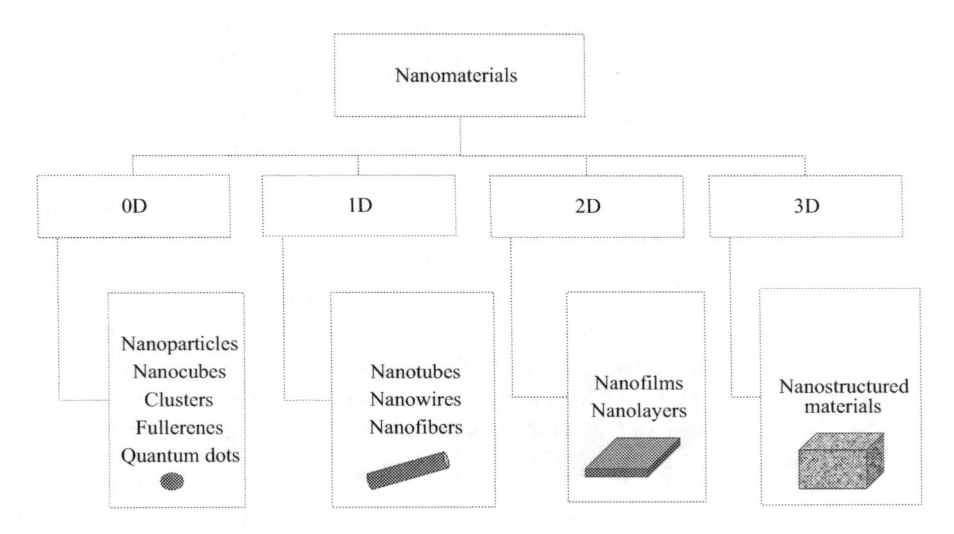

FIGURE 3.1 Classification of nanomaterials based on size and dimensions.

Nanostructured materials make up the group of 3D nanomaterials, and these can be made up of NPs, nanotubes, nanoplates of different materials or nanocrystals of the same material, or they can be nanoporous materials.

3.3 INORGANIC NANOMATERIALS

Inorganic materials refer to compounds that have a structure composed of one element, like gold (Au) or several combined elements like barium titanate ($BaTiO_3$) but in which the molecular structure is not based enterally on the arrangement of C and H atoms (in contrast to organic materials) and they can contain one C atom or more bonded to each other, like barium carbonate (BaCO3) or graphene. In this group, there are the most widely used materials that are recognized as metals, alloys, and ceramics, and there is a kind of compound called organometallic compound that contains an organic part linked to a metal. Moreover, in binary compounds, based on the representative element, inorganic materials are classified into oxides, hydrides, nitrides, phosphides, and sulfides. For that, in this topic, inorganic nanomaterials are classified as metals, alloys, oxides, and non-oxides.

These types of materials were the first to be manipulated by humans; hydroxyapatite from bones was used for tools; silica and alumina in clays to form objects; metals such as gold were used in jewelry and sacred objects. Its increasingly specialized use and application led to such great industrial and economic development that historical periods have been named after the material that drove its growth. Currently, the benefits of these materials are explored in a nanometric form, and solutions are expected to be found in the fields of environment, health, and data.

3.3.1 METAL NANOMATERIALS

In terms of chemistry, a metal is an element capable of giving up its electrons easily, which allows the flow of electrons when they form part of some compounds, and in consequence, metallic materials are characterized by having electrical conductivity. This means, they allow the passage of current or flow of electrons specifically at 0°K (absolute zero) under an applied electric field; these are known as conductors. Materials that do not conduct electricity are called insulators or non-conductors, but some of these that require an increase in temperature to pass current are called semiconductors. On the other hand, free electrons in metals also determine other well-known properties like ductility, toughness, opacity, and reflectivity. Ductility derives from the capacity of electrons to be separated from the positive ionic centers without brake electron bonds; high packaging due to unsaturated bonding results in toughness; the opacity is caused by the free electron oscillation that absorbs the energy of light, and reflectivity appears because the same electrons emit photons [5]. These are intrinsic properties that help to define a metal; however, other properties have taken relevance from the formation of metal nanomaterials.

3.3.1.1 Properties

Nanomaterials conformed by nanosized metal particles are taking relevance since 1957 when Faraday studied the interaction of light with AuNPs; since then, important

TABLE 3.1
Properties, Uses, and Applications of Most Common Metal Nanoparticles

Material	Size (nm)	Nanostructure	Property	Use	Application	References
Au	2–50	Nanoparticles	Impedance	Amplifier	Electrochemical biosensor	[6]
Ag	3.3–17.97	Semi-spheres	Binding potential	Bactericidal	Medical	[7]
Cu	4.7–17.4	Semi-spheres	Biocompatibility	Anticancer	Medical	[8]
Pt	2.2×15	Nanowires	Catalytic redox capacity	Catalyst	Dye formation reaction	[9]
Pd	50	-	Catalytic efficiency	Catalyst	Reduction of environmental contaminants	[10]

changes in their properties have been found compared with their bulk counterpart. Table 3.1 shows relevant properties, uses, and applications of some metal nanomaterials studied in recent years, for example, optical properties as surface plasmon resonance, biological activity due to biomolecule conjugation and radical oxygen species (ROS) production, high surface area, tailored shape, surface energies, redox activity, electron store capacity and quantum confinement. Next, these are briefly described.

Optical properties such as *absorption* and *dispersion* have been improved in NPs. This is because they have a high surface area, so there are more free electrons that collectively oscillate under the effect of a magnetic field, the resonance of the collective oscillations is called *surface plasmon resonance* (SPR); consequently, have been tunable by varying the size and shape and composition of the NPs. SPR is responsible for the color change of transmitted/reflected light, and also of enhance other surface signals such as photoluminescence [11].

Thermal properties are related to SPR in the following way. Metal NPs have a low quantum yield which refers to the low efficiency of photon emission compared with the photons absorbed, so a large part of the energy of the absorbed photons is converted to heat; in this way, by using a laser in SPR wavelength, the particle can be heated and applied to destroy tumors which have reduced heat tolerance. Besides, thin and planar nanostructures have shown to be more efficient in the conversion of *absorption* to heat than spherical ones due to the first having a high surface-to-volume ratio [11].

Biological activity adjudicated to metal NPs are due to *biomolecule conjugation* or capacity to binding with organic molecules. It is believed that Ag NPs release silver ions, which bind tissue proteins and trigger structural changes on the cell membrane and, as a result, affect cell permeability and the respiration process; in this step, their interaction with biomolecules can produce ROS. The shape of particles changes the surface area, and because of that, different nanostructures give different activity; also, cation size and cation charge affect the affinity of ligands [12] which have an important role in functionalization and selectivity to target biomolecules.

The *electrical properties* of NPs are better than in bulk; that is why they are used to modify electrode surfaces to enhance sensitivity. *Impedance*, for example, which is the total opposite to current that a system offers, is used to detect modifications in the electrode surface; this is because the presence of NPs bound to target biomolecules interrupts the electric field and changes the impedance [6].

Another property of metals refers to the ability to develop oxidation and reduction reactions, called *redox activity*, this quality implies the *electron store capacity* and leads the *catalytic activity* to metal NPs; although several factors can influence at the end like size, shape, composition, oxidation state and chemical/physical interparticle interactions [13], thorough analysis of kinetics to every particular system is recommended. Pt NPs, for example, enhance catalytic efficiency. which has been related to the larger surface–volume ratio and development of complex geometries in crystallographic catalyst facets planes, which should be rich in step and dangling atoms [11].

3.3.1.2 Uses and Applications

Since Faraday´s work, the study of metal NPs has not stopped. Nanomaterials are widely studied for their application in several technological fields like medicine, optics, electronics, aerospace, etc.

Possible applications in *medical diagnostics* and *therapies* have been studied due to their vast advantages as high absorption, scattering, and biocompatibility [14]. NP absorption of light can be transferred as heat and used for photothermal therapies against cancer [15]. AuNP photothermal treatments are more efficient and selective than radiofrequency, microwave, and ultrasound waves. On the other hand, SPR in very small NPs enhances thermal properties; however, the NP size increases the scattering too because of which scattering is useful in optical imaging. By using dark-field microscopy, it is possible to observe metal NPs as bright spots and fluorescent NPs, and two-photon luminescence microcopies are also benefited by SPR [11].

The biological activity of silver has been employed for centuries to treat infections [1]; actually, the effect of AgNPs against bacteria and fungi has been improved so that they can be used as *bactericides* [7], *fungicides* [16], and also as *antioxidants* [17]. One important factor to obtain success in biological applications is related to the selectivity of NPs, and this can be achieved by functionalization, which has given place to the formation of *conjugates*, for example, AuNP/drug conjugates can be used to form *drug-delivery systems*; AgNPs/antibodies conjugates present several advantages versus enzymes in immunoassays since MNPs support higher temperature than enzymes, there is no loss of activity during crosslinking, they can be detected by optical or electrochemical methods, and these could be cheaper to detect viruses antibodies [18]; AuNPs have been functionalized with folic acid to attack cancer cells because this acid is required as a ligand [11].

Another reason for the development of metallic nanomaterials is that they are more electroactive and their application on detection surfaces has been interesting in different areas of analytical sciences such as electrochemistry, where they can be used as a transducer, catalytic agent, adsorbent, analyte transporter, and markers; the increase of electrochemical activity in NPs, compared with bulk materials, has been related to the change of the NP surface reaction kinetics that is derived from the nanosize and surface excess free energy of particles [19]. To analyze the use of AuNPs as electrochemical biosensors, their impedimetric response was analyzed since an improvement in this can enhance the sensitivity of the biosensors and detect changes in the surface. Different AuNPs of sizes 2, 5, 10, 20, and 50 nm were immobilized on the carbon electrode surface by physical adsorption. Larger NPs (20 and 50 nm) provide lower impedance change and smaller particles (2–10 nm) have larger impedance change [6]. Colorimetric sensing is another application of metal NPs since their light absorption is strongly affected by the refractive index of the surrounding media [11].

Power generation processes are the main expected application for nanomaterials since they have high reactivity and can be used as catalytic materials, although longevity, high activity, sensibility, and environmental sustainability are challenges to overcome. Besides AuNPs, Pt and Pd NPs are extensively studied in the development of catalysts and electrocatalysts [13], but NPs of abundant elements such as Fe, Ti, Si, and C are also studied since they are cheaper. Cu, for example, is abundant, inexpensive, and non-toxic, and there are several ways to obtain CuNPs, which, in addition to being used for medical purposes, have applications in water treatment processes [20] and catalysis [21].

3.3.1.3 Obtention

The conventional processes involved to produce pure metals include mining, purification, and reduction. On the earth, the majority of metals are in the form of oxides and should be separated from the raw material and refined by using a reduction agent. To obtain a piece of metal for the specified application, forming operations should be elected, for example, forging, rolling, extrusion, and wire drawing operations. Further treatments are given to enhance or stabilize mechanical or electrical properties.

Among the methods used to obtain metallic NPs, we can find the liquid phase, solvated metal atom dispersion technique [22], precipitation [21], solid state [23], thermal decomposition [22], ion beam implantation [22], electrochemical deposition [24], biosynthesis [7,9], and physical adsorption [6]. To obtain NPs, bottom-up methods are preferred because physicochemical properties, size, and morphology can be controlled, for example, the chemical reduction synthetic method [8,10,22]. Another popular synthesis is the liquid phase, in these methods, a colloidal dispersion of metal is formed by hindering the diffusion of growth species; this is achieved by using low solute concentration and polymeric stabilizers; the first maintains a large diffusion distance between nuclei and the second creates a barrier that limits the diffusion, both avoid particle growth [25]. Green synthesis or biosynthesis which uses microorganisms or plants to synthesize NPs has gained popularity since it can be cheaper, environmentally friendly, and easy to scale up for larger production.

3.3.2 Alloys

The homogeneous combination of two or more metals or even metals with no metals gives place to alloys, and their relevance comes from the wide range of properties achieved. In general, there are two types of alloys: ferrous and non-ferrous alloys.

At first, iron is the main component and due to its availability on the earth's surface; these are the most used alloys; the obtention process is cheaper and is easy to obtain with different physical and mechanical properties, one example is *steel*; some drawbacks of ferrous alloys are high density, low electric conductivity, and corrosion. On the other hand, there are non-ferrous alloys, where iron is not found, and these are named based on the main metallic component. Some examples are copper alloys; copper with Zn forms brass; with Sn, Al, Si, and Ni, it forms bronze.

Returning to the topic of this chapter, nanomaterials, it should be noted that important studies have been done about nanocrystalline metallic alloys and multimetallic nanostructures that result in alloys. Chang et al. [26] summarized different nanocrystalline microstructures which have enhanced physical properties of their coarse-grained counterpart. A nanocrystalline material refers to one formed by crystallites smaller than 100 nm; these, in turn, conform to microstructures (monophasic, equiaxial, or mixed) where a high fraction of nanocrystallites is exposed on the surface. This results in states far removed from the equilibrium and high-energy systems.

On the other hand, multimetallic nanostructures are described by Ahmed et al. [27] as nanostructures constructed from two or more metals. These nanostructures present wide structure tunability and functional diversity, and four geometries are mentioned: random, cluster-in cluster, core–shell, and alloy. The differences come from the synthesis: in the first, there is not a specific orientation of atoms; in the second, one metal acts as a nanocluster and the others as binders; in the third, one metal is used to form cores and the others are clustered and form shells; in the last, alloys, all metals are assumed to nearly form at the same nucleation rate, and similarly sized particles (random alloy) or dissimilar ones (intermetallic alloy) are obtained.

Ferrando showed two kinds of nanoalloys, mixed-pattern alloys and geometrical structure alloys, in which core–shell-segregated, subcluster-segregated, mixed AB (ordered or random which also are called mixed or intermixed alloys as Hannan [27] describe), and multi-shell alloys are kinds of mixed-pattern alloys. Geometrical structure alloys are classified as non-crystalline structures formed by nanoalloy clusters like icosahedra, polytetrahedra, and polyicosahedra since these take compact shapes and have efficient packaging of atoms.

3.3.2.1 Properties and Applications

Optical, chemical, electronic, and magnetic properties have attracted the attention of scientists to study intermetallic nanostructures. Several physicochemical properties have been enhanced in comparison with monometallic nanostructures; next, some of these properties are described.

Catalytic activity is one of the most important properties of metals and is expected to be enhanced by using bimetallic or multimetallic nanostructures. Ferrando et al. [28] found in the literature that the properties that determine chemical reactivity and especially catalytic activity of nanoalloys are surface structures, compositions, and segregation properties. For example, Ni/Co alloys have been prepared to form catalysts used for the hydrogen evolution reaction (HER) in alkaline water splitting; this is a future hydrogen production option since it is an environmentally friendly, large-scale, and cost-effective process [29]. Otherwise, Ni particles alloyed with octahedral Pt–Pd nano-cages presented a highly active electro-catalyst, and the oxygen reduction reaction has been evaluated by electrochemical tests. AgCu nanoalloys have also been used in cathodes for Zn–air batteries [21].

Nanoalloys optical properties are related to SPR since they are altered by composition, segregation, or mixing of clusters. These properties can be used to tune the color of nanoalloys as the Ag/Au system which is one of the many systems studied. One possible application of this quality is to avoid counterfeiting [28].

Wettability is a property that refers to the interaction of a solid surface with a liquid; it is possible to control it by managing the energy surface and geometrical structures. This property contributes to anticorrosion, oil–water separation, and liquid transportation and is very useful in engineering and biomedicine. Superhydrophobic aluminum alloys have been prepared by immersion in HNO_3/H_2O and post-modification with stearic acid and by also using mechanical roughening, chemical etching, and decorating [30].

Enhanced magnetic measures in NPs have been related to the size reduction and coupling of 3d atoms elements as Co, which improve induced electronic spin polarization of 4d elements as Rh. Also, a corrosion effect has been detected in Au/Fe core–shell NPs, where a 1–2 nm Au shell avoids Fe oxidation and maintains magnetic properties in an inhomogeneous mixing this has been observed too. Important applications in magnetic sensors and magnetic recording are promised from giant magnetoresistance (GMR), which is the change of current flow when a magnetic field is applied; it has been observed in multi-layered materials and granular materials of magnetic clusters embedded in a solid host [28]. Hong et al. mentioned that the magnetic property of Ni/Co alloys is useful as coatings for chips and hard drives, thanks to their magnetic properties.

3.3.2.2 Obtention

The synthesis of multimetallic nanostructures is classified in successive and simultaneous methodologies [27]. The successive or seed-mediated growth method consists of the reduction of surface metal ions of pre-nucleated NPs by using reducing agents; core–shell and intermetallic alloys (an alloy with different particle sizes) are obtained by this technique. In simultaneous methodology, metal precursors are mixed in the same reaction liquor; random, cluster-in-cluster, and random nanoalloy-shaped nanostructures are obtained through this methodology. Some techniques included are co-reduction, sonochemical synthesis, thermal decomposition, and radiolytic synthesis. By using thermal decomposition, Fe/Co, Fe/Ni, Pt/Co, and Fe/Au alloyed NPs have been obtained. Also, Ag/Au, Au/Ag/Pd, and Au/Co alloys have been formed through irradiation.

Ferrando et al. [28] have explained the medium in which nanoalloys can be generated: cluster beams, colloidal solutions, and immobilized on surfaces or inside pores. Electrochemical methods are used to prepare, for example, clusters of colloidal solutions and core–shell particles. The elements used in these techniques are Fe, Ni, Pt, Pd, and Ag, among others. There are other techniques used to obtain different kinds of nanoalloys, for example, 2.8 nm particles of AgCu nanoalloy have been prepared by the laser deposition method [21]. In another case, by using ZnO nanowires as a template, Pt–Pd–Ag trimetallic-alloyed nanotubes have been obtained.

3.3.3 Metal Oxides of Transition and Non-Transition Elements

According to market research, the nanometal oxide market will grow at a compound annual growth rate between 6% and 8% until 2027; SiO_2 and TiO_2 are some of the materials more expected for growth [31]. This is due to their special chemical, electrical, mechanical, and optical properties, which are widely studied and applied in diverse science and technological fields. The study of metal oxide properties as other materials has increased since they exhibit size dependency.

Metal oxides can be formed by transition and non-transition elements; the first group refers to the elements contained in the 3d to 12d group of the periodic table, also called the 3d block, in which their valence electrons are found in the two outer shells, unlike the alkali, alkaline earth, and other metals where the valence electrons are only contained in the last shell. This difference in the location of valence electrons gives transition elements their characteristic properties: strong, lustrous, high melting and boiling points, good conductivity, and insulating properties, for example. Their properties are also related to the geometry formed around transition cations which are affected by the Coulombic interaction of O^{-2} anions [32,33].

Following this classification, some of the metal oxide nano-powders internationally commercialized are

Transition-metal oxides like TiO_2, V_2O_5, Cr_2O_3, MnO, MnO_2, Mn_2O_3, Mn_3O_4, FeO, Fe_2O_3, Fe_3O_4, CoO, Co_3O_4, NiO, CuO, ZnO, Y_2O_3, ZrO, MoO_3, PdO, Ag_2O, CdO, HfO_2, Ta_2O_5, WO_3, La_2O_3, CeO_2, Pr_6O_{11}, Nd_2O_3, Sm_2O_3, Eu_2O_3, Gd_2O_3, Tb_4O_7, Er_2O_3.

Non-transition-metal oxides like Na_2O_3, K_2O, MgO, CaO, BaO, B_2O_3, Al_2O_3, Al_2OH, Ga_2O_3, In_2O_3, SiO_2, SnO_2, PbO, Pb_2O_3, Sb_2O_3, Bi_2O_3, etc.

Others, binary oxides containing transition and non-transition elements like $BaTiO_3$, $Y_3Al_5O_{12}$, $CuGaO_2$, $PbZrO_3$, etc.

3.3.3.1 Properties and Applications

Electronic properties of transition-metal oxides have been related to the outer *d* electrons. In the form of nanocrystals, the interaction of transition-metal ions with oxygen is different than in bulk; the atomic distances are distorted on the surface predominantly and the ionic conductivity is enhanced; because of this, several studies have been carried out to develop catalytic [34], antibacterial [35], and antioxidant activity [36]. The last one is used in medicine for cancer therapy.

Furthermore, metal oxides have long been known to exhibit a wide range of electrical properties, from metallic to insulating; as well as they can also be semiconductors; in the same way, their magnetic properties vary, there are ferromagnetic and antiferromagnetic metal oxides [37]. Currently, oxides have been doped and reduced to NPs to improve electrical and magnetic properties; an example is Eu_2O_3, for which applications have been found in electronics for acoustic and data storage devices [38].

Regarding optical properties, metal oxides can act as sunscreens since they have the capability of absorbing UV light; they increase the durability of materials used in the cosmetic, agriculture, and food sector [22]. TiO_2 is a good example; Reinosa [39]

found that nanosized TiO_2 enhances their UV absorption and also increases that of micrometric ZnO when 15% TiO_2 NPs are added. Besides, rare-earth metal oxides have a high refractive index and luminescence, which are enhanced in nanometric particles; for this, they are applied in optical devices [38], labeling, and bioimaging cells [40].

Due to their UV absorption, luminescence, antioxidant, and mechanical resistance properties, some oxides are widely applied as additives or as functional fillers. CaO, Al, and MgOH are examples of functional fillers for polymers [41] applied in several industries like automotive and aerospace.

3.3.3.2 Obtention

For a long time, vapor phase processes were the most popular for obtaining metallic oxides; however, later, it was observed that the solution phase is more flexible, and therefore, methods such as sol–gel and hydrothermal methods are relevant because they do not require high temperatures [22]. In these methods, metal salts and alkoxides are precursors to metal oxides; Solvents such as water and methanol and chelating agents are required. These methods are also used to obtain $BaTiO_3$ instead of the solid-state method [42] which has been very popular but requires high temperatures. The morphologies obtained in the metal oxides are NPs, nanofibers, nanowires, nanowires, nanorings, nanoplates, nanospheres, nanotubes, nano-cages, nanohelices, spirals, and hollow spirals [22].

3.3.4 Metal Non-Oxide Inorganic Nanomaterials

Metal non-oxide inorganic compounds have ionic bonds; they are formed by a cation, which is a metallic element, and an anion, which is a non-metal atom. For example, hydrides, nitrides, phosphides, sulfides, and carbides are the group of compounds with H^{-1}, N^{-1}, P^{-1}, S^{-2}, and C^{-2} as anions, respectively. These anions are less electronegative than O^{-2} and form compounds with semiconductor properties, low cost, and catalytic properties similar to noble metals [43]. For these reasons, metal non-oxide compounds are studied as catalysts and to replace pure metal sensors. Gao et al. [43] focused on the synthesis of non-oxide transition-metal nanostructures because of this application; they summarized several methods and characteristics of materials like carbides, nitrides, sulfides, and selenides; Table 3.2 summarizes some properties, applications, and obtention methods described in that work.

To prepare metal non-oxide compounds, gas-solid reactions have traditionally been used; however, these kinds of reactions present disadvantages like the use of flammable and corrosive gases, subproducts, different end compositions, and lower catalytic activity [43]. Because of that, other methods to obtain nanostructures have been implemented by the integration of organic and inorganic single structures at the nanoscale (nanohybrid reactions); in these procedures, organic nanostructures like nanowires impregned of metal oxides are used to obtain nanotubes, for example. Another method is the urea glass route, which is very useful to obtain nitrides by the use of *N*-rich organic molecules such as urea and others.

TABLE 3.2
Properties, Application, and Synthesis of Some Metal Non-Oxide Compounds

Examples	Obtention Method	Precursors	Properties	Applications
Mo_2C	CVD	MoO_x/amine	Noble-metal-like	Electrocatalysis
WC	Sonochemistry method	Biomass	Low cost	Ammonia decomposition
W_2C	NaBet-H reduction	W and Ta/C_3N_4	Stability	
Fe_3C	Nanohybrid reactions		Element abundance	Hydrogenation
	Urea glass route		Catalytic	Cellulose degradation
Mo_2N	Nanohybrid reactions	$HMT_2(NH_4)_4$-Mo_7O_{24}	Noble-metal-like	Hydrogenation
W_2N	Urea glass route	Ni/amine	Catalytic	Oxidation of alcohols
Ni_3N	Ionothermal process	Fe/polypeptide/N_2	Lamellar morphology	
Fe_3N	Pyrolysis		Large surface area	Ammonia decomposition
Fe_7S_8	Oxygen reduction reaction	Transition metals chalcogenides	Catalytic	Electrodes
Fe_3S_4	Nanohybrid reactions		Stability	Li-ion batteries
$CoSe_2$	Pyrolysis	M_xS/amine nanowires		Water splitting

Source: Based on the Work of Gao et al. [43].

3.4 ORGANIC NANOMATERIALS

Organic materials are characterized by having a structure based on C and H atoms. Primarily, these kinds of materials are the base of life, food, and our body; secondly, the discovery of many organic compounds has revolutionized the world, beginning from petroleum used as fuel to polyolefins for commodities, from natural fibers like cotton to synthetic ones as nanocellulose, and from antibiotics to drug-delivery nano-carriers. With the study of materials at the nanoscale, their properties are improving and applications are increasing. Here, some outstanding organic nanomaterials will be described.

3.4.1 Polymeric Nanoparticles

These particular materials are built between 1 and 1000 nm (submicron colloidal particles). This is because their main components are polymeric compounds. Besides, the term NP refers to different morphologies described by Garcia et al. [44]: nanocapsules, nanospheres, nanogels, polymeric micelles, polymersomes, liquid crystals, and dendrimers [45]; these are also illustrated in Figure 3.2.

Nanocapsules, also called reservoir systems, are composed of a polymeric membrane or shell that inside has a liquid core with active substances like vaccines, drugs, genes, etc.; the shell can also contain active substances or be functionalized.

Nanospheres, also called matrix systems, are continuous macromolecular matrices, capable of adsorbing active substances on their surface or inside. Later, the

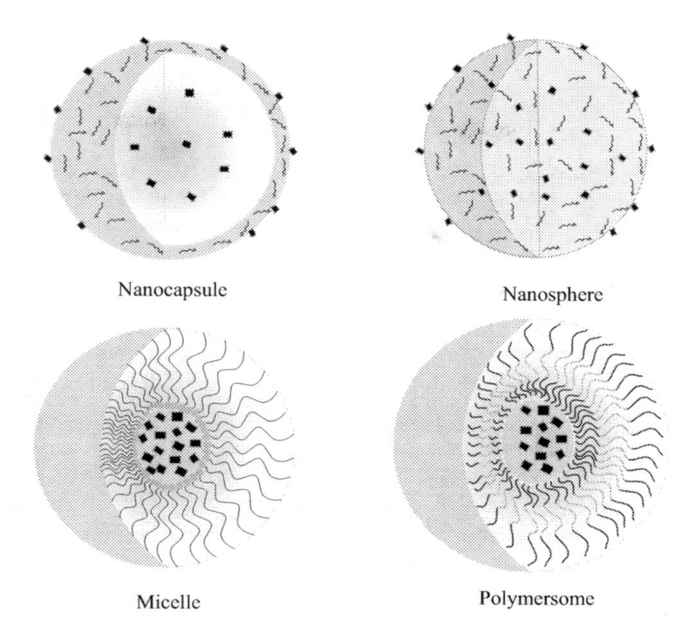

FIGURE 3.2 Nanostructures of NPs.

active substance is liberated by diffusion depending on the composition of the polymer structure which dictates the drug-release mechanism.

Nanogels are NPs of 3D networks of crosslinked polymers, which can store a high quantity of water (hydrogels) and stay in a colloidal solution; they can trap active substances, and an example is collagen, a natural polymer [46]. Amphiphilic polymer networks are required for in vivo applications as they have physical interactions and result in less stability; they also are suitable to be controlled by temperature and pH changes.

Polymeric micelles refer to spherical nanostructures (10–100 nm) composed of amphiphilic and triblock copolymers contained in an aqueous media [47]. They have high permeability, stability, biocompatibility, etc. Micelles are composed of the hydrophilic part (corona) and the hydrophobic part (core). Micellar concentration and the hydrophilic volume fraction are important factors to obtain polymeric micelles.

Polymersomes are other kinds of NPs; similar to micelles, they are a kind of core–shell nanostructure. The shell is composed of the hydrophobic region of amphiphilic triblock polymers, and the core is delimited by the hydrophilic region. High-molecular-weight polymers are needed to obtain these particles; for example, poly(2-methyl-2-oxazoline)-block-poly(dimethylsiloxane)-block-poly(2-methyl-2-oxazoline) (PMOXA-PDMS-PMOXA) is a triblock copolymer used to aquaporins polymersomes which are embedded in membranes for water treatment [48].

Dendrimers are well-defined and branched macromolecular structures. Their main characteristic is that from the core to the surface, the number of branches increases; in the core, there are hollows, where the active substances can be loaded, and the outer branches act as protection and control; the surface can also be functionalized. Some polymers employed to obtain dendrimers are poly(amide amine) (PAMAM), polyethylene glycol (PEG), polypropylene imine (PPI) dendrimers, and hyaluronic acid (HA) [44,49].

The drugs that have been used to form NPs are fenofibrate, ciprofloxacin, and curcumin, for example. Also, their application is for medical treatments like tissue regeneration, anti-inflammatory purposes, antibacterial treatment, and cancer. The identification of tumor cells has been a challenge, and through functionalized nanospheres, it has been possible to increase the selectivity and stability of the electrochemical immunosensor of tumor markers. In other applications, cinnamon oil has been loaded in chitosan nanocapsules for their use in food protection films, presenting antimicrobial activity due to the release of cinnamon oil [50].

The formulation process plays an important role to form different morphologies. Among the methods used to obtain polymeric NPs are solvent evaporation, coacervation, emulsification/solvent diffusion, nanoprecipitation, ionic, and emulsification/reverse salting-out [44,45]. Most of these involve the dispersion or emulsification of monomers and active substances, followed by a polymerization, gelation, or evaporation process, and it should not leave impurities for medical or environmental applications, so dialysis or different kinds of filtration should be implemented [44]. Moreover, as has been seen in the different kinds of NPs, the drug can be absorbed on the surface or in bulk, resulting in a complex procedure to determine the drug association mode.

Both kinds of polymers can be used, synthetic or natural, but since the main application of these materials is as nanocarriers or nano vehicles in medicine, environmental remediation and food packaging, the polymers used for these morphologies should be biocompatible, biodegradable, and resorbable to control the delivery of substances; this is useful in medicine to avoid secondary effects, optimize effects with low doses, ensure delivery in the required area.

Synthetic non-biodegradable polymers are used to form NPs, and, as has been mentioned, are poly(amide amine) (PAMAM), polyacrylamide (PAM), polypyrrole (PPy) [51], and polyacrylates [52]; biodegradable polymers mostly used are polycaprolactone (PCL), poly(lactic acid) (PLA), poly(ethylene glycol) (PEG), poly(lactic-co-glycol acid) (PLGA), chitosan, polyesters, gelatin, and alginate [44,45]; One important characteristic of biodegradable polymers is that by different mechanisms like hydrolysis, the polymers are transformed to their precursor monomers which also are biocompatible.

3.4.2 Polymeric Nanofilms

The increased interest in polymeric nanofilms begins from the study of the better polymer thin film properties than the ones of bulk polymer; these properties are glass transition, diffusion, and viscoelastic properties [2]. Polymeric nanofilms are extremely thin polymer films or sometimes called brushes [53]; Ricotti et al. [54] describe polymeric nanofilms as quasi-two-dimensional structures with a thickness between 10 and 100 nm. He fabricated PLLA microporous nanofilms to replace membranes in the development of strategies that replace in vivo studies of NP translocation and exposure; nanofilms were more permeable than commercial membranes, present cytocompatibility, and permit the cell's adhesion and proliferation. Several industries like oil, energy, chemical, clinical, and water treatment industries use polymeric membranes for gas and liquid separation processes; however, new materials are required as a solution to the problems that membranes present like free defect formation, permeability, and selectivity; for example, as a solution, polymeric nanofilms have been proposed. In the elaboration of optoelectronic devices, actuators, and sensors, intrinsically conductive polymers (ICPs) are needed [55]; thin films of ICPs can be obtained by electrochemical polymerization or electrodeposition of aromatic heterocycles like pyrrole, aniline, thiophene, carbazole, and also nanocomposites with biodegradable polymers [56].

The problem that polymeric nanofilms present are instability and dewetting at 150°C; Tumsarp et al. [57] analyzed wettability related to the distribution of ZnO NPs to reduce it; Karami et al. [58] added nanodiamond particles to decrease thin-film nanocomposites membranes wettability. However, it also depends on the obtention methods, among which grafting, layer-by-layer (LBL) assembly, spin coating, and plasma ultrasound-assisted polymerization are found [53,59]. At the same time, the formation of the thin film or nanofilm has to be elected thinking in the application, where the film will be deposited. For example, it is implemented in surface modifications that are related to areas like bioengineering, optics, sensors, and electronics [53].

3.4.3 Biological Nanomaterials

These materials are employed mainly for drug-delivery systems, and although synthetic polymers have been used, there is a great interest in the use of natural polymers. The factors that define a biological material are the size and morphology of the NPs to avoid leakage of active substances but also to assure their diffusion; surface charge and permeability that dictate stability of NPs related to preventing agglomeration through electrostatic repulsion; degree of biodegradability (decomposition into non-harmful substances), biocompatibility (do not present toxic or immunologic effects) and cytotoxicity (quality to produce cell death); drug loading; and the release profile desired [52].

Biological polymers can be synthetic or of animal or vegetal origin. The first group includes materials like polypyrrole, poly(ethylene glycol), poly(ethylene oxide), etc. [47,60]; in the second, albumin (human and bovine blood protein), gelatin (collagen derivate polypeptide), collagen (vertebrates protein), hyaluronic acid (living organisms' mucopolysaccharide), and chitosan (exoskeleton derived chitin) [61] are found. The third group includes homopolymers of 1,4-β glycoside-linked d-glucopyranose (cellulose), starch (polysaccharides from rice, potatoes, corn, and others), and soy protein (amino acids), zein (corn cell endosperm cytoplasm), etc. Both are used for the development of pharmaceutical formulations. Other biopolymers like fucoidan, carrageenan, and alginate are obtained from algae [52].

REFERENCES

[1] S. Bayda, M. Adeel, T. Tuccinardi, M. Cordani, and F. Rizzolio, "The history of nanoscience and nanotechnology: From chemical-physical applications to nanomedicine," *Molecules*, vol. 25, no. 1, pp. 1–15, 2020, doi: 10.3390/molecules25010112.

[2] R. A. L. Jones, "The dynamics of thin polymer films," *Curr. Opin. Colloid Interface Sci.*, vol. 4, no. 2, pp. 153–158, 1999, doi: 10.1016/S1359-0294(99)00025-4.

[3] A. B. Asha and R. Narain, "Nanomaterials properties". In *Polymer Science and Nanotechnology*, Elsevier, Amsterdam, 343–359, 2020.

[4] A. Rogach, *Semiconductor Nanocrystal Quantum Dots*, 2008th ed. Austria: SpringerWienNewYork, 2011.

[5] P. L. Mangonon, *Ciencia De Materiales Selección Y Diseño*, 1st ed. Mexico: Pearson Educación MExico, 2001.

[6] A. Bonanni, M. Pumera, and Y. Miyahara, "Influence of gold nanoparticle size (2–50 nm) upon its electrochemical behavior: An electrochemical impedance spectroscopic and voltammetric study," *Phys. Chem. Chem. Phys.*, vol. 13, no. 11, pp. 4980–4986, 2011, doi: 10.1039/c0cp01209b.

[7] R. A. Hamouda, M. H. Hussein, R. A. Abo-elmagd, and S. S. Bawazir, "Synthesis and biological characterization of silver nanoparticles derived from the cyanobacterium oscillatoria limnetica," *Sci. Rep.*, vol. 9, no. 1, pp. 1–17, 2019, doi: 10.1038/s41598-019-49444-y.

[8] R. Hassanien, D. Z. Husein, and M. F. Al-Hakkani, "Biosynthesis of copper nanoparticles using aqueous tilia extract: Antimicrobial and anticancer activities," *Heliyon*, vol. 4, no. 12, p. e01077, 2018, doi: 10.1016/j.heliyon.2018.e01077.

[9] R. Venu, T. S. Ramulu, S. Anandakumar, V. S. Rani, and C. G. Kim, "Bio-directed synthesis of platinum nanoparticles using aqueous honey solutions and their catalytic applications," *Colloids Surf. A: Physicochem. Eng. Asp.*, vol. 384, no. 1–3, pp. 733–738, 2011, doi: 10.1016/j.colsurfa.2011.05.045.

[10] N. Yılmaz Baran, "Generation and characterization of palladium nanocatalyst anchored on a novel polyazomethine support: Application in highly efficient and quick catalytic reduction of environmental contaminant nitroarenes," *J. Mol. Struct.*, vol. 1220, 128697, 2020, doi: 10.1016/j.molstruc.2020.128697.

[11] K. Jiang and A. O. Pinchuk, "Noble metal nanomaterials: synthetic routes, fundamental properties, and promising applications." *Solid State Phys.*, vol. 66, pp. 131–211, 2015.

[12] M. Hofmann, G. Retamal-Morales, and D. Tischler, "Metal binding ability of microbial natural metal chelators and potential applications," *Nat. Prod. Rep.*, vol. 37, no. 9, pp. 1262–1283, 2020, doi: 10.1039/c9np00058e.

[13] B. R. Cuenya, "Synthesis and catalytic properties of metal nanoparticles: Size, shape, support, composition, and oxidation state effects," *Thin Solid Films*, vol. 518, no. 12, pp. 3127–3150, 2010, doi: 10.1016/j.tsf.2010.01.018.

[14] P. K. Jain, K. S. Lee, I. H. El-sayed, and M. A. El-sayed, "Calculated absorption and scattering properties of gold nanoparticles of different size, shape, and composition: Applications in biological imaging and biomedicine," *J. Phys. Chem.*, vol. 110, pp. 7238–7248, 2006.

[15] X. Huang, W. Qian, I. H. El-Sayed, and M. A. El-Sayed, "The potential use of the enhanced nonlinear properties of gold nanospheres in photothermal cancer therapy," *Lasers Surg. Med.*, vol. 39, no. 9, pp. 747–753, 2007, doi: 10.1002/lsm.20577.

[16] Z. K. Xia et al., "The antifungal effect of silver nanoparticles on trichosporon asahii," *J. Microbiol. Immunol. Infect.*, vol. 49, no. 2, pp. 182–188, 2016, doi: 10.1016/j.jmii.2014.04.013.

[17] P. Subramanian, R. Periyannan, V. Gandhi, R. Ganesan, M. Ramar, and P. N. Marimuthu, "A green route to synthesis silver nanoparticles using sargassum polycystum and its antioxidant and cytotoxic effects: An in vitro analysis Palanisamy," *Mater. Lett.*, vol. 189, pp. 196–200, 2015, doi: 10.1016/j.matlet.2015.11.017.
[18] E. Khristunova, J. Barek, B. Kratochvil, E. Korotkova, E. Dorozhko, and V. Vyskocil, "Electrochemical immunoassay for the detection of antibodies to tick-borne encephalitis virus by using various types of bioconjugates based on silver nanoparticles," *Bioelectrochemistry*, vol. 135, p. 107576, 2020, doi: 10.1016/j.bioelechem.2020.107576.
[19] K. Brainina, N. Stozhko, M. Bukharinova, and E. Vikulova, "Nanomaterials: Electrochemical properties and application in sensors," *Phys. Sci. Rev.*, vol. 3, no. 9, 2019, doi: 10.1515/psr-2018-8050.
[20] M. F. Al-Hakkani, "Biogenic copper nanoparticles and their applications: A review," *SN Appl. Sci.*, vol. 2, no. 3, pp. 1–20, 2020, doi: 10.1007/s42452-020-2279-1.
[21] M. B. Gawande et al., "Cu and Cu-based nanoparticles: Synthesis and applications in catalysis," *Chem. Rev.*, vol. 116, no. 6, pp. 3722–3811, 2016, doi: 10.1021/acs.chemrev.5b00482.
[22] C. A. Charitidis, P. Georgiou, M. A. Koklioti, A. F. Trompeta, and V. Markakis, "Manufacturing nanomaterials: From research to industry," *Manuf. Rev.*, vol. 1, no. 11, 2014, doi: 10.1051/mfreview/2014009.
[23] V. Madhavan, P. K. Gangadharan, A. Ajayan, S. Chandran, and P. Raveendran, "Microwave-assisted solid-state synthesis of Au nanoparticles, size-selective speciation, and their self-assembly into 2D-superlattice," *Nano-Struct. Nano-Objects*, vol. 17, pp. 218–222, 2019, doi: 10.1016/j.nanoso.2019.01.010.
[24] I. Saldan et al., "Electrochemical synthesis and properties of gold nanomaterials," *J. Solid State Electrochem.*, vol. 22, no. 3, pp. 637–656, 2018, doi: 10.1007/s10008-017-3835-5.
[25] G. Cao, *Nanostructures & Nanomaterials: Synthesis, Properties & Applications*, Reprinted. Imperial College Press, 2004.
[26] I. CHANG, "Rapid solidification processing of nanocrystalline metallic alloys." In *Handbook of Nanostructured Materials and Nanotechnology*. Academic Press, 2000, 501–532.
[27] H. B. Ahmed and H. E. Emam, "Overview for multimetallic nanostructures with biomedical, environmental and industrial applications." *J. Mol. Liq.*, vol. 321, 114669, 2021.
[28] R. Ferrando, J. Jellinek, and R. L. Johnston, "Nanoalloys: From theory to applications of alloy clusters and nanoparticles," *Chem. Rev.*, vol. 108, no. 3, pp. 845–910, 2008, doi: 10.1021/cr040090g.
[29] S. H. Hong et al., "Fabrication and evaluation of nickel cobalt alloy electrocatalysts for alkaline water splitting," *Appl. Surf. Sci.*, vol. 307, pp. 146–152, 2014, doi: 10.1016/j.apsusc.2014.03.197.
[30] K. Liu and L. Jiang, "Metallic surfaces with special wettability," *Nanoscale*, vol. 3, no. 3, pp. 825–838, 2011, doi: 10.1039/c0nr00642d.
[31] I. Global Industry Analysts, "Nano-metal oxides- global market trajectory & analytics," 2021. https://www.researchandmarkets.com/reports/4805605/li-ion-battery-global-market-trajectory-and.Journal (Accessed August 16, 2022)
[32] M. Hepting, "Introduction: Transition metal oxides and their heterostructures." In *Ordering Phenomena in Rare-Earth Nickelate Heterostructures,* Cham: Springer, pp. 1–12, 2017, doi: 10.1007/978-3-319-60531-9_1.
[33] D. I. Khomskii, *Transition Metal Ions in Crystals*. Cambridge: Cambridge university press, 2014.
[34] W. Y. Hernández, O. H. Laguna, M. A. Centeno, and J. A. Odriozola, "Structural and catalytic properties of lanthanide (La, Eu, Gd) doped ceria," *J. Solid State Chem.*, vol. 184, no. 11, pp. 3014–3020, 2011, doi: 10.1016/j.jssc.2011.09.018.

[35] J. A. Mariano-Torres, A. López-Marure, M. García-Hernández, G. Basurto-Islas, and M. Á. Domínguez-Sánchez, "Synthesis and characterization of glycerol citrate polymer and yttrium oxide nanoparticles as a potential antibacterial material," *Mater. Trans.*, vol. 59, no. 12, pp. 1915–1919, 2018, doi: 10.2320/matertrans.M2018248.
[36] Y. H. Castillo et al., "Analysis of the antioxidant activity by ABTS assays of Eu doped cerium dioxide synthesized by sol-gel method," *Cer. Int.*, vol. 45, no. 2, pp. 2303–2308, doi: 10.1039/b919677n.
[37] C. N. R. Rao, "Transition metal oxides," *Annu. Revi. Phys. Chem.*, vol. 40:1, no. 595, pp. 291–326, 1989, doi: https://doi.org/10.1146/annurev.pc.40.100189.001451.
[38] S. Kumar, R. Prakash, and V. Singh, "Synthesis, characterization, and applications of europium oxide: A review," *Rev. Adv. Sci. Eng.*, vol. 4, no. 4, pp. 247–257, 2016, doi: 10.1166/rase.2015.1102.
[39] J. J. Reinosa, P. Leret, C. M. Álvarez-Docio, A. Del Campo, and J. F. Fernández, "Enhancement of UV absorption behavior in ZnO–TiO_2 composites," *Bol. la Soc. Esp. Ceram. y Vidr.*, vol. 55, no. 2, pp. 55–62, 2016, doi: 10.1016/j.bsecv.2016.01.004.
[40] G. G. Genchi et al., "comparative study Barium titanate nanoparticles : Promising multitasking vectors in nanomedicine," *Nanotechnology*, vol. 27, p. 19, 2016.
[41] R. N. Rothon, "Particulate fillers for polymers professor," *Rapra Rev. reports*, vol. 33, no. 6, pp. 2171–2183, 2000, [Online]. Available: http://www.polymerjournals.com/pdf-download/844670.pdf.
[42] E. Brzozowski and M. S. Castro, "Synthesis of barium titanate improved by modi ® cations in the kinetics of the solid state reaction," *J. Eur. Cer. Soc.*, vol. 20, no. 14–15, pp. 2347–2351, 2000.
[43] Q. Gao, N. Liu, S. Wang, and Y. Tang, "Metal non-oxide nanostructures developed from organic-inorganic hybrids and their catalytic application," *Nanoscale*, vol. 6, no. 23, pp. 14106–14120, 2014, doi: 10.1039/c4nr05035e.
[44] M. C. García, C. Aloisio, R. Onnainty, and G. Ullio-Gamboa, "Self-assembled nanomaterials." In *Nanobiomaterials.* Woodhead Publishing, 2018, 41–94..
[45] A. Zielinska et al., "Polymeric nanoparticles: Production, characterization, toxicology and ecotoxicology," *Molecules*, vol. 25, no. 16, pp. 3731–3751, 2020, doi: 10.3390/molecules25163731.
[46] I. B. Pathan, S. J. Munde, S. Shelke, W. Ambekar, and C. Mallikarjuna Setty, "Curcumin loaded fish scale collagen-HPMC nanogel for wound healing application: Ex-vivo and in-vivo evaluation," *Int. J. Polym. Mater. Polym. Biomater.*, vol. 68, no. 4, pp. 165–174, 2019, doi: 10.1080/00914037.2018.1429437.
[47] J. M. Pantshwa, P. P. D. Kondiah, Y. E. Choonara, T. Marimuthu, and V. Pillay, "Nanodrug delivery systems for the treatment of ovarian cancer," *Cancers (Basel)*, vol. 12, no. 1, 2020, doi: 10.3390/cancers12010213.
[48] R. Górecki, D. M. Reurink, M. M. Khan, V. Sanahuja-Embuena, K. Trzaskuś, and C. Hélix-Nielsen, "Improved reverse osmosis thin film composite biomimetic membranes by incorporation of polymersomes," *J. Memb. Sci.*, vol. 593, July 2019, p. 117392, 2020, doi: 10.1016/j.memsci.2019.117392.
[49] P. Kesharwani, L. Xie, G. Mao, S. Padhye, and A. K. Iyer, "Hyaluronic acid-conjugated polyamidoamine dendrimers for targeted delivery of 3,4-difluorobenzylidene curcumin to CD44 overexpressing pancreatic cancer cells," *Colloids Surf. B Biointerfaces*, vol. 136, pp. 413–423, 2015, doi: 10.1016/j.colsurfb.2015.09.043.
[50] R. R. Ferreira, A. G. Souza, and D. S. Rosa, "Essential oil-loaded nanocapsules and their application on PBAT biodegradable films," *J. Mol. Liq.*, vol. 337, p. 116488, 2021, doi: 10.1016/j.molliq.2021.116488.
[51] E. Nazarzadeh Zare, M. Mansour Lakouraj, and M. Mohseni, "Biodegradable polypyrrole/dextrin conductive nanocomposite: Synthesis, characterization, antioxidant and antibacterial activity," *Synth. Met.*, vol. 187, no. 1, pp. 9–16, 2014, doi: 10.1016/j.synthmet.2013.09.045.

[52] A. Gagliardi et al., "Biodegradable polymeric nanoparticles for drug delivery to solid tumors," *Front. Pharmacol.*, vol. 12, pp. 1–24, 2021, doi: 10.3389/fphar.2021.601626.
[53] Z. Mao, W. Tong, T. Ren, W. Zhang, S. Wu, and C. Gao, "Making polymeric nanofilms (grafting-to, grafting-from, spin coating, layer-by-layer, plasma polymerization)." In *Encyclopedia of Polymeric Nanomaterials*, Berlin Heidelberg: Springer-Verlag, 2015, doi: 10.1007/978-3-642-36199-9.
[54] L. Ricotti, G. Gori, D. Cei, J. Costa, G. Signore, and A. Ahluwalia, "Polymeric microporous nanofilms as smart platforms for in vitro assessment of nanoparticle translocation and caco-2 cell culture," *IEEE Trans. Nanobioscience*, vol. 15, no. 7, pp. 689–696, 2016, doi: 10.1109/TNB.2016.2603191.
[55] A. Palma-Cando, I. Rendón-Enríquez, M. Tausch, and U. Scherf, "Thin functional polymer films by electropolymerization," *Nanomaterials*, vol. 9, no. 8, pp. 1–13, 2019, doi: 10.3390/nano9081125.
[56] A. Karrat and A. Amine, "Recent advances in chitosan-based electrochemical sensors and biosensors," *Arab. J. Chem. Environ. Res.*, vol. 7, no. 2, pp. 66–93, 2020.
[57] P. Tumsarp et al., "Inorganic nanoparticle-blended polymer nanofilm and its wettability improvement: Film grading and dewetting cause analysis," *Appl. Surf. Sci.*, vol. 521, p. 146399, 2020, doi: 10.1016/j.apsusc.2020.146399.
[58] P. Karami, B. Khorshidi, L. Shamaei, E. Beaulieu, J. B. P. Soares, and M. Sadrzadeh, "Nanodiamond-enabled thin-film nanocomposite polyamide membranes for high-temperature water treatment," *ACS Appl. Mater. Interfaces*, vol. 12, no. 47, pp. 53274–53285, 2020, doi: 10.1021/acsami.0c15194.
[59] B. Feng, Z. Xu, J. Wang, F. Feng, L. Wang, and L. Gai, "Photoluminescent organic polymer nanofilms formed in water through a self-assembly formation mechanism," *J. Mater. Chem. C*, vol. 7, no. 11, pp. 3286–3293, 2019, doi: 10.1039/c8tc06192k.
[60] Y. Tao, E. Ju, J. Ren, and X. Qu, "Polypyrrole nanoparticles as promising enzyme mimics for sensitive hydrogen peroxide detection," *Chem. Commun.*, vol. 50, no. 23, pp. 3030–3032, 2014, doi: 10.1039/c4cc00328d.
[61] S. Sundar, J. Kundu, and S. C. Kundu, "Biopolymeric nanoparticles," *Sci. Technol. Adv. Mater.*, vol. 11, no. 1, p. 014104, 2010, doi: 10.1088/1468-6996/11/1/014104.

4 Physical Methods for Synthesis and Thin-Film Deposition

*Sachin R. Rondiya, Anurag Roy,
Ganesh K. Rahane, Ashok Jadhavar,
Mahesh M. Kamble, Puneeth Kumar P.,
Hareesh K., Mahesh P. Suryawanshi,
Nelson Y. Dzade, and Sandesh R. Jadkar*

CONTENTS

DOI: 10.1201/9781003216308-4

4.1 INTRODUCTION

The synthesis procedures of nanoparticles can be classified into two approaches – the top-down approach and the bottom-up approach (as shown in Figure 4.1) [1]. In the top-down process case, the method depends on the miniaturization of structures starting from their larger counterpart. The top-down approach by conventional lithography or inkjet printing techniques has attained the theoretical limits of about 50 nm. Various other top-down methods like ion-sputtering, vapor deposition, laser ablation, pyrolysis, etc., are well known, through their application is very limited. The top-down approach is essentially a smaller version of the nanomaterial's preparation approach, consisting of a massive division of solids into nanostructured materials or a suitable source material which then creates a structure for functionality generation from the source. The industrially accepted approach is the mechanical attrition technique. However, the process has several drawbacks like contamination, low surface area, cost-effectivity, etc. [2].

To further going down in size limit, the bottom-up approach has been considered where nanomaterial formation is associated with their assembly and construction of architectures with nanometer dimension. These are all low-cost, soft chemical processes in which sol-gel, hydrothermal, etc., are being found as some unique and facile methods for preparing nanomaterials. However, agglomeration, surface roughness, inhomogeneity of nanomaterials growth are typical problems associated with this technique [3].

However, the deposition is performed in a clean environment; the deposited film could be contamination free. This method allows atomic-level precision for the thin-film contrivance. This chapter has summarized some conventional techniques to develop thin films of nanomaterials, which will guide the rational design of a highly efficient application for their energy generation, storage, or transformation purposes.

4.2 CONVENTIONAL METHODS FOR CHEMICAL VAPOR DEPOSITION TECHNIQUES

The key distinguishing attribute of CVD is that the deposition of material onto the substrate is a multidirectional type of deposition. Considering the ways chemical reactions taking place inside the chamber, there are two popular methods available,

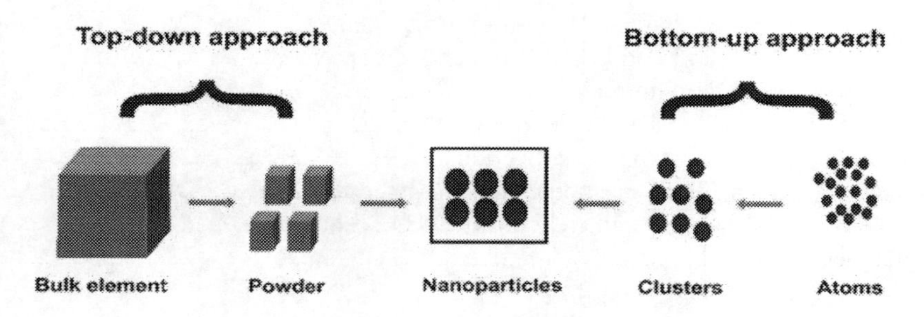

FIGURE 4.1 General synthesis approach of nanoparticles.

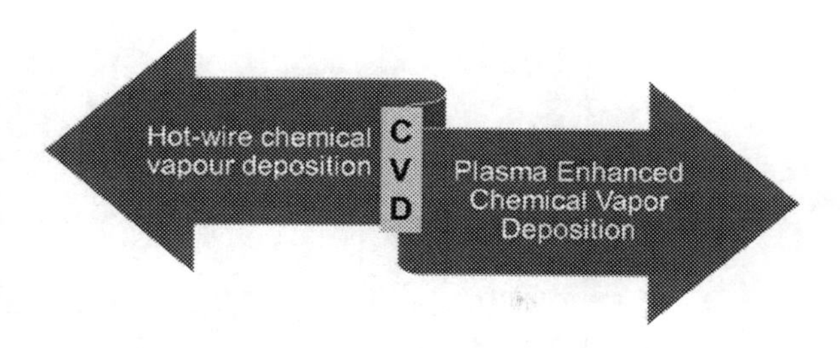

FIGURE 4.2 Two popular chemical vapor deposition (CVD) methods.

(a) hotwire and (b) plasma-enhanced chemical vapor deposition (CVD), respectively (Figure 4.2).

4.2.1 Hot-Wire Chemical Vapor Technique (HW-CVD)

HW-CVD is the steady-state decomposition of gas precursors on a heated refractory metal filament, producing radical species that react in the gaseous state and deposit onto a heated substrate. The microstructure of the subsequent films is determined by several reactor parameters such as filament temperature, growth pressure, gas flow rates, and substrate temperature. In 1979, Wiesmann et al. succeeded in depositing a-Si: H by thermal decomposition of SiH_4 from tungsten or carbon foil heated to roughly 1600°C. They were the first to introduce the HW-CVD method [4]. At present, the HW-CVD technique has been well established. It is a highly considered practical technique in industry and research to synthesize many organic and inorganic materials. This method is mainly used to fabricate first-class silicon and germanium-based thin films for photovoltaic applications. The most important advantage is the method is plasma free, highly scalable and easy to handle substrates during deposition. Besides, handing of the substrate inside the chamber is quite facile and straightforward. The schematic illustration of dual-chamber HW-CVD is shown in Figure 4.3.

4.2.1.1 Instrumentation of HW-CVD Deposition Unit

Though the schematic appears complex at first sight, it can be categorized into four major sections for easy comprehension: specifically, the load-lock chamber (LLC), the process chamber, the gas manifold and lastly, the vacuum suction. The substrates are mounted on a substrate holder and then inserted into an evacuated environment called a load lock chamber. The actual deposition takes place at the hotwire chamber, also known as the process chamber and is the most critical part of the system. The gas manifold comprises mass flow controls (MFCs) and gas-mixing chambers that connect the source-dilution and doping gas cylinders. The vacuum system consists of suction gauges, molecular turbopumps powered by rotary vane pumps, and valves. A brief discussion about each of the components is described in other sections.

4.2.1.2 Load-Lock Component

The LLC is one of the primary components in HW-CVD; the sample substrates are loaded on a substrate holder and then inserted into the LLC. This stainless steel (SS)

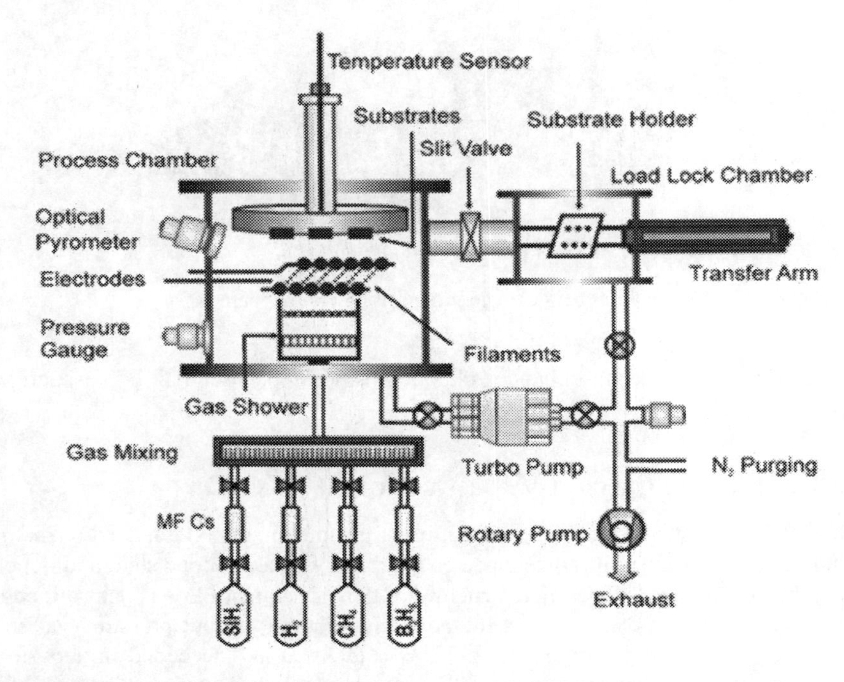

FIGURE 4.3 Graphical representation of the dual-chamber HW-CVD method [5].

chamber is mainly incorporated in the setup to avoid the hotwire chamber's exposure to the outside environment at each run during sample deposition. Some highly considered deposition substrates include Glass, c-Si, flexible foils Glass, c-Si, and flexible foils attached to an SS substrate holder. Further, the sample-substrate holder is mounted on a pneumatically powered transport arm, followed by a vacuum pump. The role of the transfer arm is to assist in placing the substrate into hotwire chamber and pull back to LLC, and vice versa. A slit valve isolates both the load lock and process chambers, and it is only opened when shifting substrates from the LLC to the hotwire chamber and vice versa, as shown in Figure 4.4.

4.2.1.3 The Process Chamber

The process chamber is considered the central part of the HW-CVD system; the substrates are moved from LLC through a slit valve into this chamber, where the complete deposition takes place. Apart from a few essential deposition parameters regulated by other sections, the process chamber and its components are responsible for the system's fundamental functionality. Figure 4.4 illustrates the HW-CVD's basic framework, which includes three primary components in a vacuum chamber: the gas manifold, hot wire assembly, and sample-substrate holder. These three components are stacked one on top of the other in a straight line. This chamber features a gas shower as an inlet and a suction port attached to a vacuum pump as an outlet. The functionality and significance of the reaction chamber's key components are discussed below. The architecture of the chamber and the arrangement of the

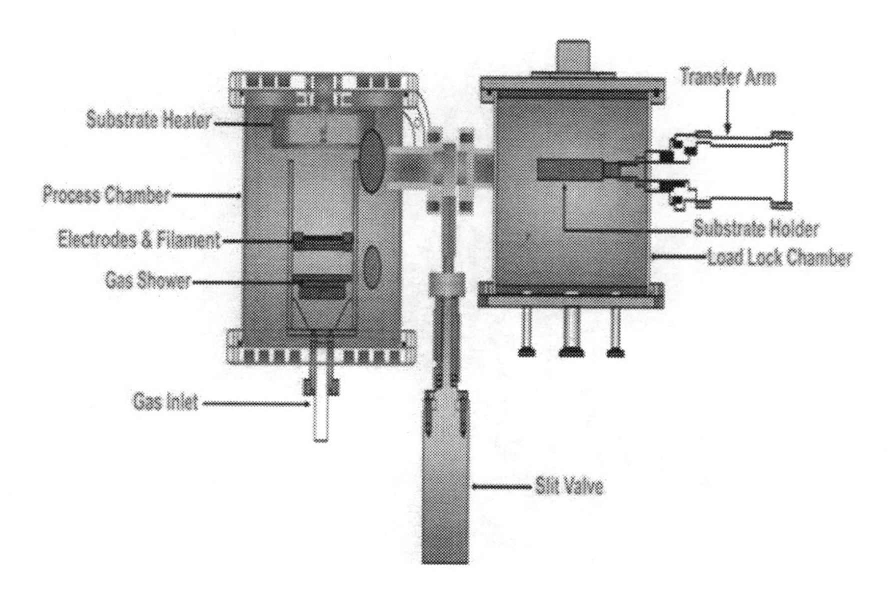

FIGURE 4.4 Schematic representation of the hotwire chamber (left) and load lock chamber (right) separated by a slit valve [6].

components within it are critical. The system offers a wide range of distance between the gas manifold, hotwire assembly, and the holder.

4.2.1.4 The Gas Shower Assembly

The design of the gas shower assembly plays a prominent role in HW-CVD since it directly affects the distribution of the incoming gas. The gas shower and hot wire assembly are shown in Figure 4.5, at which the gas mixture enters the hot-wire chamber and is spatially distributed onto the hot wires. An annular pattern offers a maximum density at the center in a point-like source and diminishes in the radial direction during the perpendicular equidistributional contours in the plane. As a result, the gas shower assembly is inadequate for large-area uniform deposition. In that circumstance, a deposition similar to that of a gas source assembly is achievable. However, there is a significant compensation for the film thickness, which is rarely accepted for industrial applications. The three-stage gas shower system ensures a consistent gas distribution over the hot wires, resulting in a homogenous large-area film deposition.

4.2.1.5 The Hot Wires

The hot wires play a pivotal role in this method, the catalytic decomposition of precursor gases at resistively heated filaments or wires. HW-CVD is also known as Cat-CVD. Various materials have been already explored for their better possible fabrication using the hot wire, anticipating that the filament material controls the dissociation of the precursor gases. Refractory metals like tantalum (Ta) [7], tungsten (W) [8], and molybdenum (Mo)s preliminary were selected, and then graphite

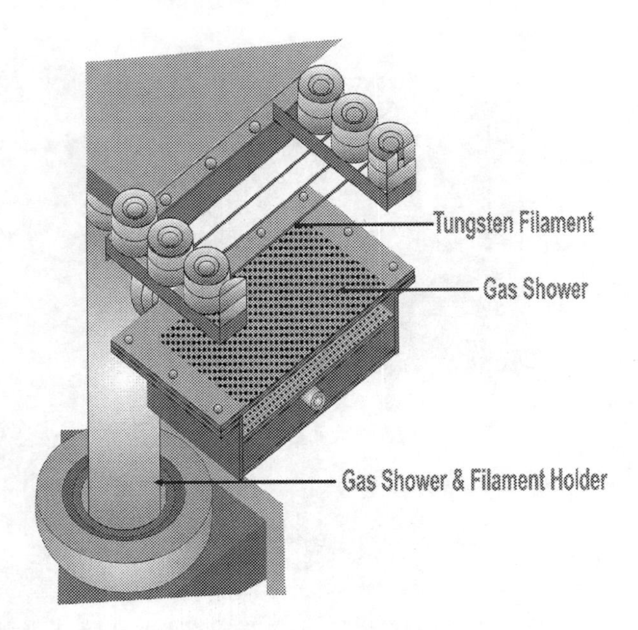

FIGURE 4.5 Schematic representation of the cross-section of the hotwire assembly [6].

(C) [9], palladium (Pa) [10], or rhenium (Re)-based materials [11], prone to catalytic activities for this purpose. However, Re and Pa exhibit an exceedingly shorter lifetime than other materials and are unsuitable for this process [8]. However, graphite or carbon-based material rolls over the materials selectivity and friendly use. Besides, *W* and Ta have been considered as alternative filament materials. The range of working temperatures used governs the choice between the two. *W* is suitable for high filament-wire temperatures (T_{fil} ~ 1800°C), *W* is a suitable alternative, as Ta tends to sag at this temperature.

4.2.1.6 The Substrate Holder

The substrates relocated from the first chamber by the transfer arm are held in this instrument section. This contains an integrated resistive heater that can increase the temperature of the substrate up to 500°C. A K-type thermocouple is used to manage the temperature inside the heater with the help of a PID controller. A cooling water facility is also available for the substrate if the substrate temperature reaches <100°C. Instead of these three main components, the process chamber also contains a pyrometer, vacuum gauges, shutter, and different viewports.

- **Pyrometer:** A noncontact thermometer records the individual wires precise temperature and monitors the deposition process.
- **High Precision Capacitance Gauge:** The pressure is measured by this device, which is attached to the chamber. Feedback is employed at the suction port, which governs the total pressure management and controls the throttle valve.

- **Shutter:** This is placed between the substrate and the hot wires and prevents undesired deposition, and they are also maintaining the hot wire temperature before and after the deposition. The remaining ports can be utilized as a viewport or have gas analyzers installed.

4.2.1.7 Gas Manifold

This section is associated with the gas source and the dilution components connected to the main chamber, streamlined through gas flow controllers (MFCs), control system valves, and the interconnected gas-mixing assembly by SS metal pipes. Also, MFCs can be used as a gas flow setting as well. There are bypass valves on every MFC line to provide safe evacuation of toxic gas from the pipelines in an MFC failure. Remember that the gases must be mixed into the main deposition chamber through the gas assembly from MFCs inside a mixing chamber. Gas mixing is also essential to make the gas homogenous and dilute throughout the deposition phenomenon.

4.2.1.8 Vacuum System

A molecular turbopump (TMP) involves an appropriate pumping capacity for high vacuum base pressure before starting the actual deposition process. It has a clean vacuum facility that avoids oil usage and eliminates any hydrocarbon deposition to the sample. A rotary backing pump is also there to support the TMP pump from the backside.

4.2.2 Thin-Film Deposition Mechanism for HW-CVD

The growth mechanism is classified into the following three sections [12] as shown in the table below.

Primary reactions	Precursor gases decomposition at the hot tungsten filament.
Secondary reactions	Followed by a successive reaction from the hot wire to the substrate, reacting with the other molecules and dissociated radicals, resulting in the formation of new species.
Tertiary reactions	It takes place at the substrate interface and highly contributes to the film growth. Tungsten filament catalyzes molecular decomposition. Interestingly, filament replacement with a heated capillary does not require deposition temperatures as high as 1900°C [13]. Thus, a catalysis reaction is highly needed in doer to decrease the decomposition temperature. For example, the Si–H bond in the silane molecule consists of the bond energy of ~4 eV; thus, breaking the Si–H bond by ~4 eV minimum is required. However, at 2000°C, the filament's thermal energy of ~0.25 eV is strong enough to break the Si–H bond, and thus catalytic thermal decomposition is highly demanding. This is because the evaporated radicals' thermal energy lies ~0.25 eV and filament catalyzes to an exothermic reaction of these species with other molecules or radicals to supply sufficient energy.

4.2.3 Parameters Involved for the HW-CVD Technique

The thin-film characteristics have a significant impact on the process parameters adopted. The significant factors that influence the overall deposition phenomenon are discussed below, and a schematic illustration has been depicted in Figure 4.6.

a. **Filament Temperature (T_{FIL}):** In terms of the structural and optical properties of the produced thin films, filament temperature (T_{FIL}) plays a significant role in the HW-CVD process. To increase the rate of sample deposition, it is required to cracking of silane (SiH_4) and methane (CH_4) molecules by raising T_{FIL}. The temperature from 1800°C to 2000°C for Si:H is suitable for forming better quality films in solar cells. While higher filament temperatures result in poor cell performance, [14] lower T_{FIL} (<1700°C) results in the formation of tungsten silicate and higher T_{FIL} (>2000°C) results in *W* evaporation and subsequent observable *W* incorporation in the film.
b. **Aperture Distance of Sample Placing:** The chamber's rate of collisions between radicals and undissociated gas molecules is affected by the filament-substrate distance. Placing substrates far away from the hot filament reduces heat radiation and lowers the temperature of the substrate. There is a trade-off between surface temperature and deposition rate parameters, and the rate of deposition, on the other hand, is dramatically lowered.
c. **Deposition Pressure (P_{dep}):** The deposition pressure is another essential factor that affects the film's growth mechanism and characteristics (P_{dep}). The rate of deposition changes proportionally with the deposition pressure. As the deposition pressure rises, the dissociation of gas molecules on the hotwire increases, increasing the rate of film-forming radicals. As a result, the rate of deposition and the pressure of deposition are proportional [15].

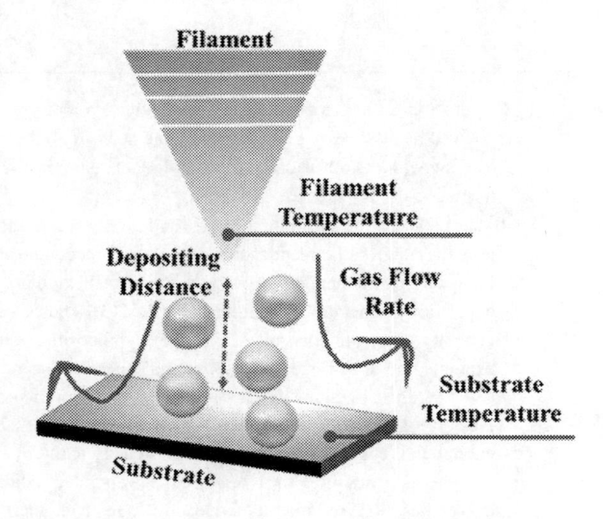

FIGURE 4.6 Schematic illustration of the significant parameters associated with the HW-CVD technique.

TABLE 4.1
Different Materials Prepared by PE-CVD

Synthesized Material	Properties	References
a-Si/SiF_2	High deposition rate, optimal bandgap and low photoconductivity	[18]
a-Si/inter-SiF_2:H	Plasma free, high material quality, and photosensitivity with low spin density	[19,20]
a-Si:H	Low photoconductivity and high photosensitivity	[21]
a-SiGe	High quality, the optimal bandgap of 1.4-1.5 eV with reasonable photoconductivity and good photosensitivity	[22]
SiN	Low temperature growth, high resistivity, and high electric breakdown	[23]
a-SiN	Smoother film surface, moderate temperature growth, increased deposition rates with low activation energy	[24]
poly-silicon	Plasma free, low temperature, micro-sized grains	[14,25]

d. **Substrate Temperature (T_{sub}):** The temperature of the substrate has a significant impact on the film structure. The substrate temperature affects the adsorbing atoms' surface mobility, the diffusion of film-forming radicals, the hydrogen concentration of formed films, and their bonding arrangement [16]. The temperature of the substrate influences the optical, structural, and optoelectronic properties of thin films. The sticking coefficient of radicals decreases as the substrate temperature rises, and vice versa. To obtain device quality thin films, an ideal substrate temperature must be determined [17].
e. **Gas Flow Rate:** The flow rates of precursor gases such as silane (SiH_4), methane (CH_4), and hydrogen (H_2) have an impact on the structural, optical, and electrical properties of thin films. To obtain device quality films, the flow rates of the gases must be optimized. A table is suitable to provide at the end of the text, related to the various materials produced by this technique. This may be effective for the readers and summarize the chapter believe so! (Table 4.1)

4.3 PLASMA-ENHANCED CHEMICAL VAPOR DEPOSITION (PE-CVD) TECHNIQUE

The PE-CVD is a conventional method used to prepare amorphous silicon (a-Si: H) and crystalline silicon (nc-Si: H) thin films due to the ability to produce highly uniform and large-area thin films. The two electrodes (plates/discs) are placed inside the process chamber, and the radio-frequency (RF) input is applied to one of the electrodes. Another one is connected to the ground as RF power is fed to one of the electrodes, plasma strikes in the process chamber. Plasma sustains in the particular low-pressure range (0.05–2 Torr). In this method, gaseous precursors are used for deposition. The radicals and ions are generated by the excitation and decomposition of the gas molecule.

The plasma excites and then generates in the chamber. The plasma provides energy to break precursor gases into corresponding radicals. When RF power is applied in between the two electrodes, high energetic electrons collide on the gases and force them to dissociate into various radicals and ions. This will begin the chemical reaction in the plasma. Plasma chemistry is enormously dependent on the process parameters of the PE-CVD.

Surface chemistry is also varied by the bombardment of the ions on the substrate. For the uniform film formation, low deposition pressure is suggested. For the fabrication of nc-Si: H films, high deposition pressure is desirable. Depending on the application, we can choose various substrates for the deposition of the a-Si:H or nc-Si: H films. The optimum substrate temperature is required to achieve desirable properties of deposited films for the solar cells. It is generally possible to deposit films from 50°C, for the device quality film substrate temperature kept between 100°C and 300°C. The PE-CVD reactor and its various parts are discussed below.

4.3.1 Different Parts of the PE-CVD Operating Chamber

The significant components of the PE-CVD system are the gas control section, deposition chamber, load lock chamber, control panel, pumping system, and throttle valve. The exhaust gases being pumped out with a neutralizing chemical scrubber. Figure 4.7 shows the schematic diagram of dual-chamber PE-CVD.

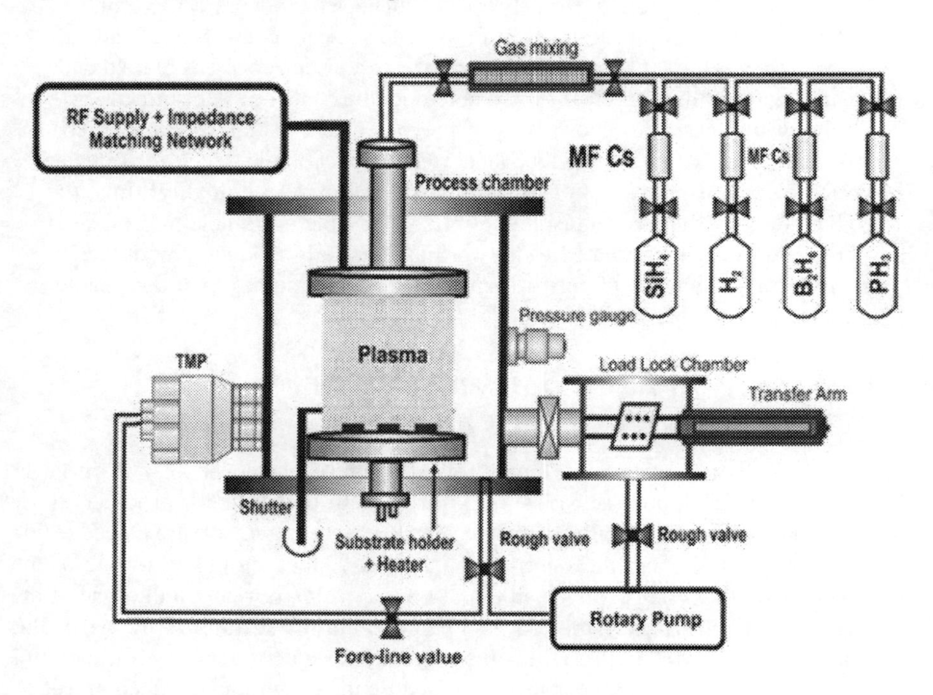

FIGURE 4.7 Schematic illustration of the PE-CVD system.

Each of the parts, as mentioned earlier, is discussed in detail in the following sections.

4.3.1.1 Gas Control Department

This is connected with various components such as gas source, pressure control system, mass flow controllers (MFCs), and the gas-mixing assembly to direct gas flow in tubes. The source gases like silane (SiH_4), hydrogen (H_2) are used as a source of silicon and hydrogen, whereas diborane (B_2H_6) and phosphene (PH_3) are used as the source of the boron and phosphorous for the doping. Careful handling of these gases is critical as they are hazardous and spontaneously flammable and comes with ultrapure grade (99.999% pure). The function of gas regulator is to control the overall gas flow. The flow rates of respective gases are controlled by MFCs with excellent accuracy. A SS-made connection tube secures rigidity and potential accidents.

4.3.1.2 Sample Deposition Unit

The vertical deposition chamber is made of SS and comprises dual electrodes in a parallel-disc arrangement in a capacitive configuration. The separation between the electrodes can be varied from 2 to 7 cm manually. In between the electrodes, shutter assembly is used for the precise control of the deposition. The upper electrode acts as the RF feed electrode, the gas shower, and the lower one acts as the substrate holder. The substrates for deposition can be fitted on the lower electrode disc. The substrate holding contains a built-in resistive heater that heats the substrate, and the gas shower evenly distributes the gas across the electrodes. Three viewports, pressure gauges, and pumping ports are also included in the deposition chamber. This process chamber also has an optional assembly of the hot wire. This assembly allowed using the system as HW-CVD or PE-CVD or a combination of both.

4.3.1.3 Load Lock Chamber Segment

The LLC is connected to the hot wire or main chamber through a slit opening. The slit valve has pneumatic control. The load lock chamber has two critical parts: chamber and transfer arm. The transfer arm has a spoon-like assembly. The transfer arm manually transfers the substrate holder from the load lock chamber to the process chamber through a slit valve. A slit valve separates the process chamber and load lock chamber. The load lock chamber is connected to the rotary pump through a roughing valve. The roughing valve has pneumatic control. Load lock has one viewport from the top side.

4.3.1.4 Pumping Unit

In PE-CVD, vacuum is the most important thing. The pumping system has two separate branches. The high vacuum branch has a molecular turbopump followed by the rotary pump connected to the process chamber through the gate or butterfly valve. Turbomolecular pump is followed by the roughing pressure created by a rotary pump inside the process chamber. After the base vacuum of the order of the 10^6 Torr, the pressure inside the chamber is controlled by the gate valve or butterfly valve during the deposition. Another branch is connected to the load lock chamber through roughing valve to obtain the roughing vacuum in the load lock chamber.

4.3.1.5 Vacuum Control Unit

The vacuum control system consists of the capacitance manometer gauge, cold cathode gauge, and butterfly valve. During the preparation of the base pressure, a cold cathode gauge is used to monitor the vacuum. Once the gases are delivered into the reaction chamber by the MFC controllers, coupled with data from the capacitance manometer, the pressure inside the chamber is monitored with the help of a butterfly valve during the process.

4.3.1.6 Exhausting Section

As we are using flammable gases as a precursor for the deposition, they need to be diluted and treated before letting them into the environment; a water scrubber device can be used to dilute the gases before they are released into the atmosphere.

4.3.2 Thin-Film Development Contrivance in the PE-CVD Technique

In PE-CVD, the thin-film deposition phenomenon consists of three main steps:

a. Production of appropriate atomic, molecular, or ionic species through dissociation reaction
b. Transport of the various species to the substrate through the gaseous medium
c. Condensation of film-forming radicals by the process of chemical/electrochemical reactions. Adsorption of the radical on the substrate through nucleation and growth mechanism.

The overall film growth process in the PE-CVD process takes place on the deposition substrate due to multiple reactions in the plasma phase, accompanied by active surface growth reactions. This includes input gas diffusion, electron-induced dissociation, gas-phase reaction, radical diffusion, and deposition [26,27]. In the following part, a basic explanation of the PE-CVD development process is provided.

Plasma Reactions: In PE-CVD, when RF input is applied between the dual electrodes, the source gases (SiH_4 and H_2) decomposition take place through the process of dissociation. Figure 4.8 illustrates how inelastic collisions with high-energetic electrons in the plasma state cause the SiH_4 and H_2 molecules to dissociate into diverse species [28].

Plasma electrons associated with various energies enable ground-state electrons of input gas species driven into their electronic excited states leading by inelastic collisions with energetic electrons. A gas molecule from a radical like SiH_4 has spontaneous excitation and forms radicals like SiH_3, SiH_2, SiH_4, H_2, and H through dissociation. Hydrogen molecules dissociate in the atomic hydrogen. In several cases, secondary dissociation is also possible. These dissociated species undergo various secondary reactions in plasma. Figure 4.9 shows the possible secondary reactions of dissociated atoms in the plasma state with SiH_4 and H_2. It also shows the concentration of each dissociated species in the steady-state plasma used to synthesize the a-Si:

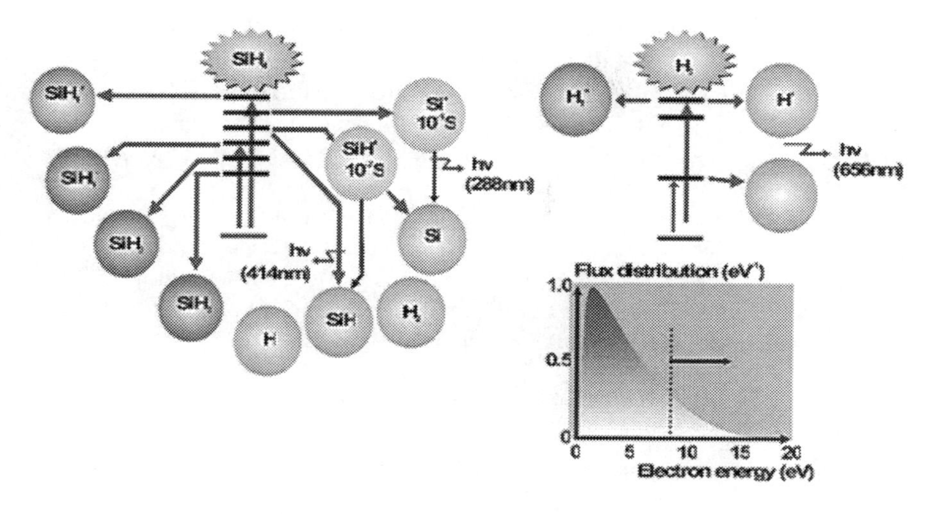

FIGURE 4.8 Dissociation pathways of SiH_4 and H_2 gases in the PE-CVD method [26].

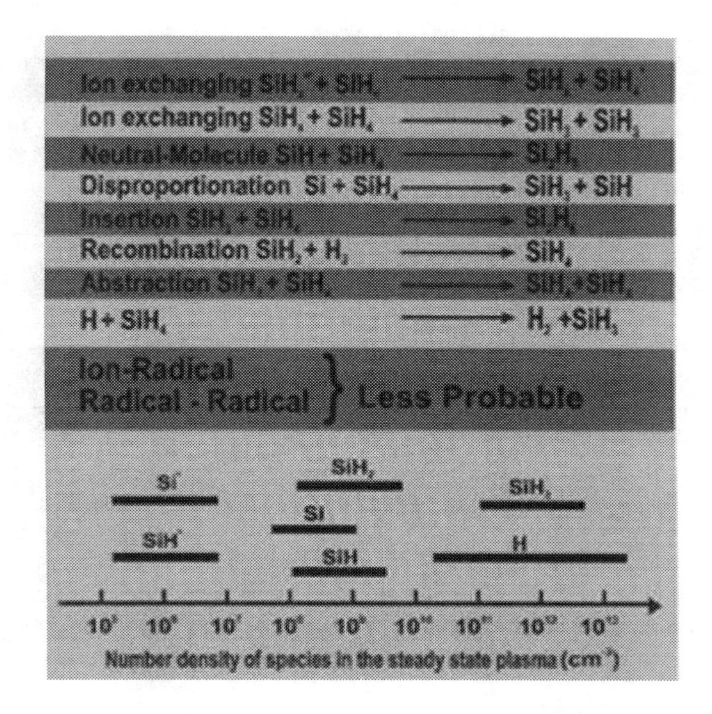

FIGURE 4.9 Possible secondary reactions of the dissociated species produced with SiH_4 and H_2 molecules with the concentration of each species in the steady-state plasma [26].

H and nc-Si: H thin films. It is clear from Figure 4.9 that SiH_3 is the only species in the formation of the a-Si:H and nc-Si:H thin films as a concentration of SiH_3 is in between the 10^{11} and $10^{13}cm^{-3}$. The density of hydrogen in the steady-state plasma varies from 10^{10} to $10^{14}cm^{-3}$. This high density is due to the hydrogen dilution in the silane. This high hydrogen dilution at constant pressure and temperature produces a highly crystalline thin film [29].

Growth Mechanism: When SiH_3 reaches the substrate's surface, it begins to diffuse across it. When this diffuses on the surface, it abstracts the hydrogen from bonded species, leading to the dangling bond formation. This dangling bond site acts as the growth site for the other SiH_3 radicals to form the Si–Si bond. This is the growth mechanism of a-Si: H thin film, controlled by the

i. SiH_3 abstract bonded H and forms SiH_4 or two SIH_3 radicals interact and form Si_2H_6.
ii. SiH_3 binds to the site comprising the dangling bond, forming the Si–Si bond.

In the growth of nc-Si: H, atomic hydrogen accessing the film-growing surface is essential. It is reported that high hydrogen is required for the formation of nc-Si: H thin films. The substrate's temperature is critical for the production of nanocrystalline grains. It is reported that crystalline volume fraction increases linearly at high hydrogen dilution up to 350(C), and however, it drastically decreases to 0 at 500°C. This result shows that optimized substrate temperature is required to form nc-Si: H thin films. The deposition rate is almost constant for the nc-Si: H thin film over the substrate temperature, but for the a-Si:H thin films, substrate temperature drastically increases after 400°C. This means that the sticking rate depends on the substrate temperature for the nc-Si: H thin film.

Based on experimental results, the formation process of μc-Si: H has the following features:

a. The film precursor is the same for a-Si:H and nc-Si: H thin films, i.e., SiH_3.
b. The formation of nc-SiH films depends on atomic H accessing the film-growing surface.
c. Above 500°C, the film becomes amorphous.
d. High energetic ions bombard the surface, which results in a reduction of crystalline volume fraction.
e. The sticking coefficient is independent of the substrate temperature, but the surface loss is probable.

Associated Growth Models: For the formation of nc-Si: H thin films, mainly three growth mechanisms have been proposed. These models are discussed below.

- **The Surface Diffusion Model:** Figure 4.10a shows the schematic of the diffusion model proposed by Matsuda [29]. The plasma results in complete bonded hydrogen surface exposure generated from the H flux resulting in local heating through H exchange reactions on the film-growing surface,

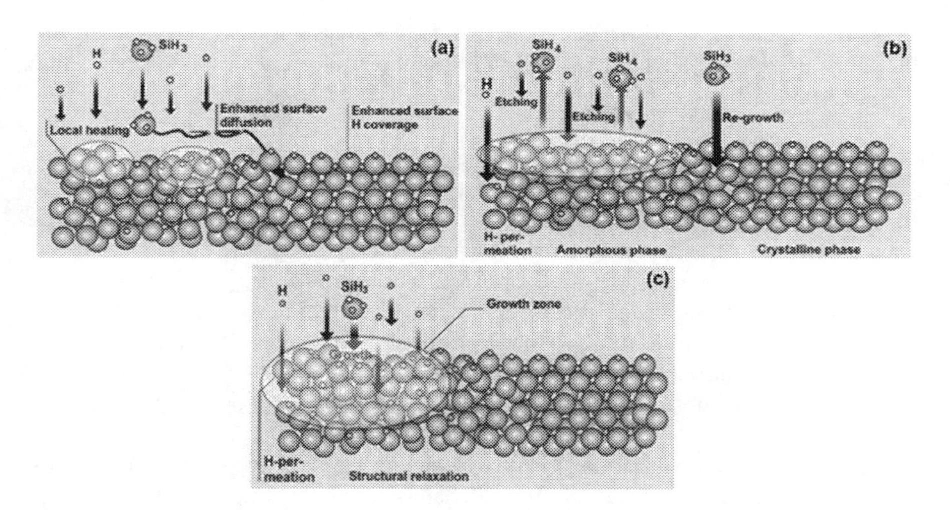

FIGURE 4.10 Schematic representation of different models for processes during the formation of nc-Si: H. (a) The diffusion model [31]. (b) The etching model [30]. (c) The chemical annealing model [26,32].

as shown in the model. These two actions enhance the surface diffusion of SiH_3 film precursors. Consequently, SiH_3 adsorbed on the surface can find energetically favorable sites, leading to anatomically ordered structures (nucleus formation). After forming the nucleus, epitaxial-like crystal growth occurs with enhanced surface diffusion of SiH_3.

- **The Etching Model:** Tsai et al. [30] observed a decrease in the growth rate of nc-Si:H films and increased hydrogen dilution. They introduce the concept of etching the deposited material. Atomic hydrogen from plasma breaks the weak Si–Si bond of the amorphous network, leading to the removal of the weakly attached Si and forms a dangling bond. This dangling bond is filled by the new-forming radical SiH_3 and forms a robust Si–Si bond, giving rise to a well-order structure and leading to nc-Si:H growth. Figure 4.10b shows the schematic of the etching model.
- **The Chemical Annealing Model:** The chemical annealing model discussed using the layer by layer of the a-Si: H thin film and hydrogen plasma treatment. Several layers are deposited on the substrate, and then these layers are exposed to the hydrogen plasma only. This procedure continues several times to fabricate desired thickness. During the plasma treatment, atomic hydrogen penetrates the film's surface and forms a good order network without removing the film's surface and developing a good order network without removing the film's surface and developing a good order network without removing the Si atom. Figure 4.10c shows the schematic of the chemical annealing model [32].

These growth models show the general character of the deposited material, and they do not include any specific parameter of the system. As a result of the differences in geometry, the process parameters may vary from system to system; this means PE-CVD machines can be designed and manufactured depending on application requirements.

4.3.3 Operation Parameters of PE-CVD Techniques

In the PE-CVD, the decomposition of the input gases is very complex. This plasma chemistry depends on the various parameters of the plasma-like power, frequency and bias voltage. At the same time, the process parameters during the deposition play an essential role in the plasma chemistry and the deposition process. Properties of the deposited substrates enormously depend on these parameters. Optimum deposition conditions are required for the desired properties of the deposited thin films. These parameters are (a) substrate temperature; (b) RF power; iii) deposition pressure; and iv) gas flow rate.

Apart from these parameters, other parameters like electrode separation, space between the electrode and wall of process chambers are also required, though not so crucial. Now let us glance at the physical effects of these variables on growth kinetics.

The surface mobility and H content of the adsorbing molecules in the thin film depend on the substrate temperature. Consequently, the film's density, morphology, and uniformity are sensitively reliant on the substrate temperature. The PE-CVD method allows choosing the substrate temperature from 50°C to 350°C temperature. The RF input decides the concentration of the plasma radicals affecting the formation of the film-forming radicals. It also has an essential role in the energy of these ions and radicals; the material properties of the deposited films strongly depend on the presence of these radicals. The number of dissociated ions or the extent of ionization is determined by the deposition pressure parameter. This has a direct impact on the film's growth rate as well as the defect density. Continuous gas feeding at constant flow rates decides the dilution of the gases in the plasma. This will determine the plasma chemistry that decides the formation of different species such as SiH_3, SiH_2. Hence for the formation of the specific species, the gas flow rate has to be optimized. This criterion is crucial in the fabrication of alloy or doped films.

As-deposited thin films' properties are strongly dependent on these various parameters due to the crucial dependence of the plasma chemistry. Each parameter has its effect, and hence for the device quality thin film, each parameter has to be optimized. Optimization of the process parameters has two different paths. In the first approach, the process parameter is determined experimentally by studying the properties of the thin film after deposition at various parameters and optimized the process parameter with desired properties of the thin film. In another path, it analyzes the plasma chemistry at a specific set of parameters using various methods like optical emission spectroscopy (OES), [33] mass spectroscopy (MS), [34] coherent antistock Raman spectroscopy [35]. OES is best suitable for these methods as it does not disturb the plasma, but it is impossible to analyze the radiative species. Thus, a combination of the OES and MS is another option to analyze the plasma chemistry. Matsuda et al. (1982) reported plasma chemistry analysis using these methods and correlated the process parameters with the film properties [36]. This study helps to understand the growth mechanism and kinetics of the plasma.

TABLE 4.2
Few Other Materials Prepared by PE-CVD

Synthesized Material	Properties	References
a-Si: H	Optimal deposition rate, low temperature, reduced surface defects	[37,38]
μc-Si: H	High crystallinity, high deposition, and low defect density	[39,40]
a-SiGe: H	Increased photoconductivity upon slow deposition rates, bandgap around <1.5 eV controlled by Ge with low spin density	[41,42]
SiN	High stability, relatively high resistivity, and breakdown strength	[43,44]
poly-Si: H	Highly stable with moderate deposition rate, device quality silicon	[14]

Regarding the glow discharge process and its associated chemistry, research and more depth understanding are required. These deposition conditions are extensive and need to be optimized carefully and maintained during the deposition. The set of the optimized parameters may not be applicable for other systems, and it just gives a guideline for the other PE-CVD system.

A table is suitable to provide at the end of the text, related to the various materials produced by this technique. This may be effective for the readers and summarize the chapter believe so! (Table 4.2)

REFERENCES

[1] G. Cao, *Nanostructures and Nanomaterials: Synthesis, Properties & Applications*, Imperial College Press, London, 2004.
[2] A. Biswas, I. S. Bayer, A. S. Biris, T. Wang, E. Dervishi, and F. Faupel, "Advances in top-down and bottom-up surface nanofabrication: Techniques, applications & future prospects," *Advances in Colloid and Interface Science*, vol. 170, no. 1–2, pp. 2–27, Jan. 2012, doi: 10.1016/j.cis.2011.11.001.
[3] A. v. Nikam, B. L. V. Prasad, and A. A. Kulkarni, "Wet chemical synthesis of metal oxide nanoparticles: A review," *CrystEngComm*, vol. 20, no. 35, pp. 5091–5107, 2018, doi: 10.1039/C8CE00487K.
[4] F. Diehl, M. Scheib, B. Schröder, and H. Oechsner, "Enhanced optical absorption in hydrogenated microcrystalline silicon: An absorption model," *Journal of Non-Crystalline Solids*, vol. 227–230, no. PART 2, pp. 973–976, May 1998, doi: 10.1016/S0022-3093(98)00203-8.
[5] V. S. Waman et al., "Nanostructured hydrogenated silicon films by hot-wire chemical vapor deposition: The influence of substrate temperature on material properties," *Journal of Nano-and Electronic Physics*, vol. 3, no. 1, p. 590, 2011.
[6] M. M. Kamble, "Synthesis and study of hydrogenated nanocrystalline silicon and its alloy by hot wire chemical vapor deposition for solar cell applications," PhD Thesis, Savitribai Phule Pune University, 2015, http://hdl.handle.net/10603/122051.

[7] H. Matsumura, "Formation of silicon-based thin films prepared by catalytic chemical vapor deposition (Cat-CVD) method," *Japanese Journal of Applied Physics, Part 1: Regular Papers and Short Notes and Review Papers*, vol. 37, no. 6 A, pp. 3175–3187, 1998, doi: 10.1143/jjap.37.3175.

[8] H. Matsumura, "Study on catalytic chemical vapor deposition method to prepare hydrogenated amorphous silicon," *Journal of Applied Physics*, vol. 65, no. 11, pp. 4396–4402, 1989, doi: 10.1063/1.343278.

[9] W. Ruihua, L. Zhiqiang, L. Li, and L. Jahe, "Study of hot wire chemical vapor deposition technique for silicon thin film," *Solar Energy Materials and Solar Cells*, vol. 62, no. 1, pp. 193–199, 2000, doi: 10.1016/S0927-0248(99)00153-1.

[10] P. A. T. T. Van Veenendaal, C. M. H. Van Der Werf, J. K. Rath, and R. E. I. Schropp, "Influence of grain environment on open circuit voltage of hot-wire chemical vapour deposited Si:H solar cells," *Journal of Non-Crystalline Solids*, vol. 299–302, no. PART 2, pp. 1184–1188, 2002, doi: 10.1016/S0022-3093(01)01086-9.

[11] K. F. Feenstra, "Hot-wire chemical vapour deposition of amorphous silicon and the application in solar cells," PhD Thesis, Utrecht University, 1998, https://www.osti.gov/etdeweb/biblio/20032621

[12] R. Zedlitz, F. Kessler, and M. Heintze, "Deposition of a-Si:H with the hot-wire technique," *Journal of Non-Crystalline Solids*, vol. 164–166, no. PART 1, pp. 83–86, 1993, doi: 10.1016/0022-3093(93)90497-L.

[13] T. Chen et al., "Microcrystalline silicon carbide thin films grown by HWCVD at different filament temperatures and their application in n-i-p microcrystalline silicon solar cells," *Thin Solid Films*, vol. 517, no. 12, pp. 3513–3515, 2009, doi: 10.1016/j.tsf.2009.01.029.

[14] R. E. I. Schropp, K. F. Feenstra, E. C. Molenbroek, H. Meiling, and J. K. Rath, "Device-quality polycrystalline and amorphous silicon films by hot-wire chemical vapour deposition," *Philosophical Magazine B: Physics of Condensed Matter; Statistical Mechanics, Electronic, Optical and Magnetic Properties*, vol. 76, no. 3, pp. 309–321, Sep. 1997, doi: 10.1080/01418639708241096.

[15] M. Komoda, K. Kamesaki, A. Masuda, and H. Matsumura, "Formation of silicon films for solar cells by the Cat-CVD method," *Thin Solid Films*, vol. 395, no. 1–2, pp. 198–201, 2001, doi: 10.1016/S0040-6090(01)01257-3.

[16] B. P. Swain and R. O. Dusane, "Effect of substrate temperature on HWCVD deposited a-SiC:H film," *Materials Letters*, vol. 61, no. 25, pp. 4731–4734, 2007, doi: 10.1016/j.matlet.2007.03.029.

[17] M. Janai et al., "Chemical vapor deposition of amorphous silicon prepared from SiF_2 gas," *Journal of Applied Physics*, vol. 52, no. 5, pp. 3622–3624, 1981, doi: 10.1063/1.329096.

[18] H. Matsumura and H. Tachibana, "Amorphous silicon produced by a new thermal chemical vapor deposition method using intermediate species SiF_2," *Applied Physics Letters*, vol. 47, no. 8, pp. 833–835, Oct. 1985, doi: 10.1063/1.96000.

[19] H. Matsumura, "Catalytic chemical vapor deposition method to obtain high quality amorphous silicon and silicon-germanium," *Journal of Non-Crystalline Solids*, vol. 97–98, no. PART 2, pp. 1379–1382, 1987, doi: 10.1016/0022-3093(87)90330-9.

[20] H. Matsumura, "Catalytic chemical vapor deposition (Ctl-cvd) method producing high quality hydrogenated amorphous silicon," *Japanese Journal of Applied Physics*, vol. 25, no. 12 A, pp. L949–L951, 1986, doi: 10.1143/JJAP.25.L949.

[21] H. Matsumura, "High-quality amorphous silicon germanium produced by catalytic chemical vapor deposition," *Applied Physics Letters*, vol. 51, no. 11, pp. 804–805, 1987, doi: 10.1063/1.98871.

[22] H. Matsumura, "Silicon nitride produced by catalytic chemical vapor deposition method," *Journal of Applied Physics*, vol. 66, no. 8, pp. 3612–3617, 1989, doi: 10.1063/1.344068.

[23] K. Yasui, H. Katoh, K. Komaki, and S. Kaneda, "Amorphous SiN films grown by hot-filament chemical vapor deposition using monomethylamine," *Applied Physics Letters*, vol. 56, no. 10, pp. 898–900, 1990, doi: 10.1063/1.102622.
[24] H. Matsumura, "Formation of polysilicon films by catalytic chemical vapor deposition (Cat-cvd) method," *Japanese Journal of Applied Physics*, vol. 30, no. 8, pp. L1522–L1524, 1991, doi: 10.1143/JJAP.30.L1522.
[25] A. Matsuda, "Thin-film silicon–Growth process and solar cell application," *Japanese Journal of Applied Physics, Part 1: Regular Papers and Short Notes and Review Papers*, vol. 43, no. 12, pp. 7909–7920, Dec. 2004, doi: 10.1143/JJAP.43.7909.
[26] A. Matsuda, "Amorphous and microcrystalline silicon," in *Springer Handbooks*, S. Kasap and P. Capper, Eds. Boston, MA: Springer, 2007, pp. 581–595. doi: 10.1007/978-0-387-29185-7_26.
[27] S. Oikawa, M. Tsuda, J. Yoshida, and Y. Jisai, "On the primary process in the plasma-chemical and photochemical vapor deposition from silane. Mechanism of the radiative species SIH*(A 2Δ) formation," *The Journal of Chemical Physics*, vol. 85, no. 5, pp. 2808–2813, 1986, doi: 10.1063/1.451038.
[28] A. Matsuda, "Formation kinetics and control of microcrystallite in μc-Si:H from glow discharge plasma," *Journal of Non-Crystalline Solids*, vol. 59–60, no. PART 2, pp. 767–774, 1983, doi: 10.1016/0022-3093(83)90284-3.
[29] A. Matsuda, "Amorphous and microcrystalline silicon," in *Springer Handbooks*, S. Kasap and P. Capper, Eds. Cham: Springer International Publishing, 2017, p. 1. doi: 10.1007/978-3-319-48933-9_25.
[30] K. Nakamura, K. Yoshino, S. Takeoka, and I. Shimizu, "Roles of atomic hydrogen in chemical annealing," *Japanese Journal of Applied Physics*, vol. 34, no. 2R, pp. 442–449, 1995, doi: 10.1143/JJAP.34.442.
[31] C. C. Tsai, G. B. Anderson, R. Thompson, and B. Wacker, "Control of silicon network structure in plasma deposition," *Journal of Non-Crystalline Solids*, vol. 114, no. PART 1, pp. 151–153, 1989, doi: 10.1016/0022-3093(89)90096-3.
[32] R. W. Griffith, "Introduction to basic aspects of plasma-deposited amorphous semiconductor alloys in photovoltaic conversion," in *Solar Materials Science*, L. E. Murr, Ed. Academic Press, 1980, pp. 665–731. doi: 10.1016/b978-0-12-511160-7.50028-4.
[33] G. Turban, Y. Catherine, and B. Grolleau, "Reaction mechanisms of the radio frequency glow discharged deposition process in silane-helium," *Thin Solid Films*, vol. 60, no. 2, pp. 147–155, 1979, doi: 10.1016/0040-6090(79)90185-8.
[34] N. Hata, A. Matsuda, and K. Tanaka, "Neutral radical detection in silane glow-discharge plasma using coherent anti-stokes raman spectroscopy," *Journal of Non-Crystalline Solids*, vol. 59–60, no. PART 2, pp. 667–670, 1983, doi: 10.1016/0022-3093(83)90259-4.
[35] A. Matsuda and K. Tanaka, "Plasma spectroscopy-Glow discharge deposition of hydrogenated amorphous silicon," *Thin Solid Films*, vol. 92, no. 1–2, pp. 171–187, 1982, doi: 10.1016/0040-6090(82)90200-0.
[36] K. Tanaka and A. Matsuda, "Deposition mechanism of hydrogenated amorphous SiGe Films," *Thin Solid Films*, vol. 163, no. C, pp. 123–130, 1988, doi: 10.1016/0040-6090(88)90417-8.
[37] Chittick RC, Alexander JH, and Sterling HF, "Preparation and properties of amorphous silicon," *Electrochem Soc-J*, vol. 116, no. 1, pp. 77–81, 1969, doi: 10.1149/1.2411779.
[38] L. Guo, M. Kondo, M. Fukawa, K. Saitoh, and A. Matsuda, "High rate deposition of microcrystalline silicon using conventional plasma-enhanced chemical vapor deposition," *Japanese Journal of Applied Physics, Part 2: Letters*, vol. 37, no. 10 PART A, 1998, doi: 10.1143/jjap.37.l1116.

[39] C. Wang, M. J. Williams, and G. Lucovsky, " The preparation of microcrystalline silicon (μ c -Si) thin films by remote plasma-enhanced chemical vapor deposition," *Journal of Vacuum Science & Technology A: Vacuum, Surfaces, and Films*, vol. 9, no. 3, pp. 444–449, 1991, doi: 10.1116/1.577430.

[40] S. Ray, A. R. Middya, S. C. De, and A. K. Barua, "Preparation of improved quality a-SiGe: H alloy films for device applications," *Solar Energy Materials*, vol. 23, no. 2–4, pp. 326–333, 1991, doi: 10.1016/0165-1633(91)90137-A.

[41] B. G. Budaguan, A. A. Sherchenkov, A. E. Berdnikov, J. W. Metselaar, and A. A. Aivazov, "The properties of a-SiC:H and a-SiGe:H films deposited by 55 kHz PECVD," *Materials Research Society Symposium - Proceedings*, vol. 557, no. 2–3, pp. 43–48, Feb. 1999, doi: 10.1557/proc-557-43.

[42] I. Kobayashi, T. Ogawa, and S. Hotta, "Plasma-enhanced chemical vapor deposition of silicon nitride," *Japanese Journal of Applied Physics*, vol. 31, no. 2R, pp. 336–339, 1992, doi: 10.1143/JJAP.31.336.

[43] Y. Kuo and H. H. Lee, "Plasma-enhanced chemical vapor deposition of silicon nitride below 250°C," *Vacuum*, vol. 66, no. 3–4, pp. 299–303, 2002, doi: 10.1016/S0042-207X(02)00134-3.

[44] J. K. Rath, F. D. Tichelaar, H. Meiling, and R. E. I. Schropp, "Hot-wire CVD poly-silicon films for thin film devices," *Materials Research Society Symposium - Proceedings*, vol. 507, no. 1, pp. 879–890, 1999, doi: 10.1557/proc-507-879.

5 Chemical Methods of Synthesis

N. Cruz-González, S.J. Montiel-Perales, N.E. García-Martínez, and Felipe Caballero-Briones

CONTENTS

5.1 SYNTHESIS OF TITANIUM OXIDE NANOSTRUCTURES BY SOLVOTHERMAL SYNTHESIS

The term hydrothermal has its origin in geology; the first to use this term was the Geologist Roderick Murchinson (1792–1871) to describe the effect that water at high temperature and pressure has on the creation of rocks and minerals [1]. The hydrothermal/solvothermal synthesis method is widely used in the synthesis of materials since this technique is easy to implement, does not require expensive and sophisticated

DOI: 10.1201/9781003216308-5

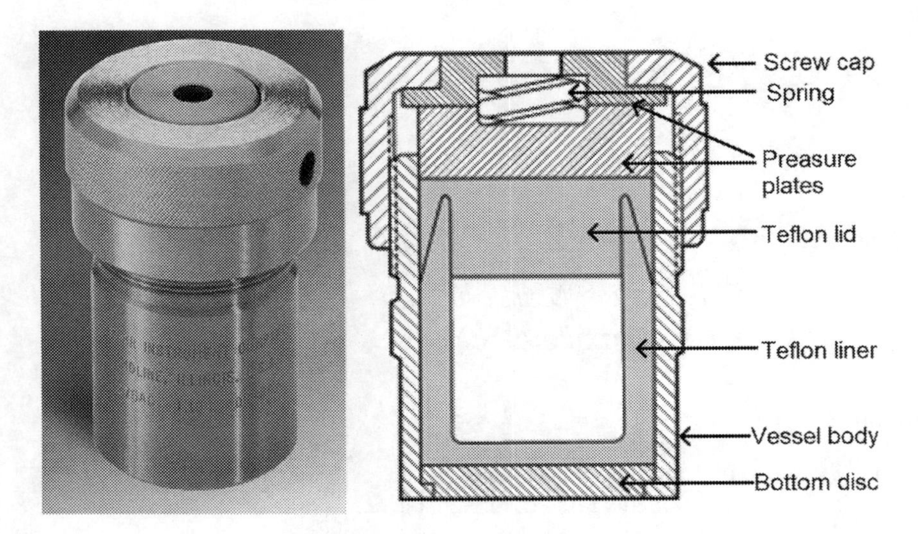

FIGURE 5.1 General-purpose autoclave used for solvothermal synthesis from Parr Instrument Company. (Reprinted with permission from [25].)

equipment, and therefore is a low-cost technique. When using water as a solvent, the synthesis is called hydrothermal, but when the solvent is an organic or inorganic substance, the synthesis is called solvothermal. Once the difference between the terms hydrothermal and solvothermal has been clarified, hereinafter will refer to it as solvothermal synthesis. The solvothermal synthesis is chemical reactions carried out in a sealed Teflon-lined steel container and heated above the boiling point of the solvent; generally, the high pressure is autogenous. By the solvothermal synthesis has been prepared a wide range of materials, such as oxide semiconductors, hybrid, graphene-like, magnetic and chalcogenide materials [2–7] in the presence of aqueous, organic or mineralized solvents under high pressure and temperature conditions to dissolve and recrystallize materials that are relatively insoluble under ordinary conditions [1]. Figure 5.1 shows a typical autoclave design such as general-purpose autoclaves.

5.1.1 Nucleation and Growth of Nanoparticles in Solution

The inorganic nanocrystal growth generally happens by bottom-up synthesis. First, the compounds are dissociated to generate the precursor atoms, followed by the precipitation step to form the nanocrystals. In the precipitation step, first occurs the particle nucleation followed by the crystal growth. In this stage, the aggregation and growth of nanoparticles (NPs) takes place.. The homogeneous nucleation takes place in a saturated solution; it happens when the solute exceeds the maximum amount that the solvent can dissolve. Upon exceeding the solubility point of the solution, the solution becomes unstable and the excess species precipitate as seeds following the LaMer model [8].

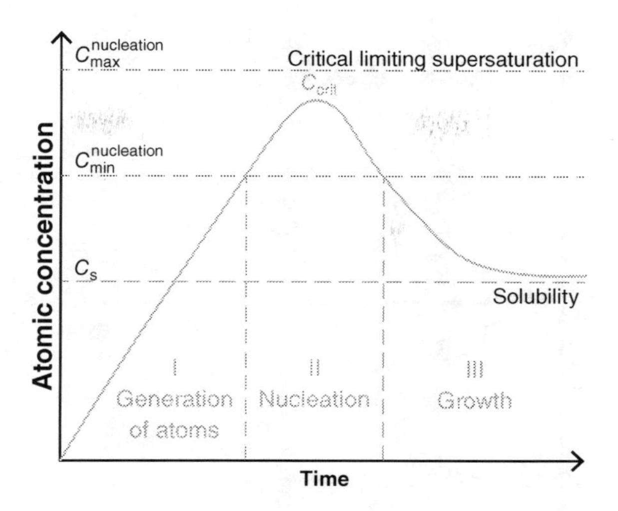

FIGURE 5.2 Scheme of the LaMer model, which describes three stages of formation nanoparticles in colloidal synthesis. (Reprinted with permission from [25].)

The LaMer model establishes that the particles' growth in solution is carried out in three stages: (I) generation of the atoms, (II) nucleation and (III) growth. Figure 5.2 shows the LaMer scheme of nucleation and growth of NPs.

I. **Generation of Atoms.** At the same time, precursors are decomposed by the effect of temperature and the concentration of atoms in the solution increases with time. In this stage, nucleation does not occur, although the solubility equilibrium (C_s) is exceeded.
II. **Nucleation.** When the solution reaches the supersaturation point, above the equilibrium point of solubility, nucleation takes place. After initial nucleation, the concentration or supersaturation of the growing species decreases until no more nuclei are formed. ($C_{min}^{nucleation}$)
III. **Growth.** When it comes, the particles' growth begins until equilibrium solubility is reached and there are no more nuclei present in the solution. ($C_{min}^{nucleation}$)

Once the minimum concentration of the nuclei has been reached, the nucleation and growth of particles occur in parallel processes but at different rates.

The formation of the nucleus is due to a change in the total free Gibbs energy (ΔG) by a spontaneous exothermic process. The energy is determined by the sum of the volume free energy (ΔG_v) with the surface free energy (ΔG_s), assuming a spherical nucleus with a radius r and γ as the surface energy per unit area, as shown in Equation (5.1).

$$\Delta G = \frac{4}{3}\pi r^3 \Delta G_v + 4\pi r^2 \gamma \tag{5.1}$$

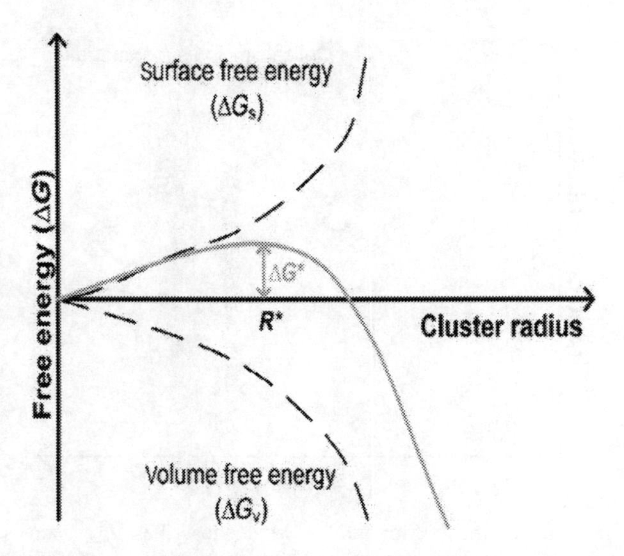

FIGURE 5.3 Plot of Gibbs free energy versus the radius of the solid cluster [9]. (Reprinted with permission from [25].)

To reach stability, the nanocrystals tend to reduce their overall free energy; the evolution of the nuclei is dependent on the competition between ΔG_v and ΔG_s. When ΔG_v decreases, the nanocrystals reach high stability, which favors the nucleation process. Conversely, when ΔG_s are increased, the solvation is promoted due to the higher surface area of the nanocrystals.

The nucleus will be stable only when its radius exceeds a critical size, R^*. If the R is smaller than R^*, the nucleus will dissolve into the solution to reduce the overall free energy; but when R is larger than R^*, the nucleus is stable and will continue to grow, as shown in Figure 5.3.

The critical radius and energy that the particles must overcome to be stable are defined by Equations (5.2) and (5.3) [8]:

$$R^* = -2\left(\gamma / \Delta G_v\right) \tag{5.2}$$

$$\Delta G^* = 16\pi\gamma / \left(3\Delta G_v^2\right) \tag{5.3}$$

Under subcritical synthesis conditions, the nucleation does not happen instantaneously; the fraction of the precursor that does not immediately precipitate will slowly crystallize around the particle nuclei and result in particle growth, giving much larger particles with wide size distributions.

5.1.2 Synthesis of TiO_2 Nanostructures

Titanium dioxide (TiO_2) is a semiconducting oxide present in nature with three crystalline structures, anatase, rutile and brookite. TiO_2 has excellent physical and chemical properties; it is used in heterogeneous catalysis as a photocatalyst [10], in solar cells for energy conversion and hydrogen production [11–13] and electrochemical devices as anode in Li-based batteries [14]; it is comparatively cheap and not toxic. TiO_2 has been fabricated in one-dimensional (1D), two-dimensional (2D) and three-dimensional (3D) nanostructures, such as nanorods, nanotubes, nanowires, nanosheets and hierarchical nanostructures [10,15–17]. The solvothermal synthesis permit obtains TiO_2 nanostructures with several morphologies and highly crystalline at mild temperatures in a one step.

In the solvothermal synthesis of TiO_2 nanostructures, there are several factors that play a role in the nucleation process and control the size, morphology and crystal structure of the material. They can be classified into two types, internal factors such as the titanium precursor, the solvent, acidic or alkaline medium and external factors such as temperature and reaction time. Below we will describe the role of each of them.

5.1.3 The Role of Solvent on Morphological and Crystalline Structure

A typical synthesis of TiO_2 nanostructures requires a titanium source, solvent and acidic or alkaline medium. The solvents used in the solvothermal synthesis can be polar or nonpolar. The solvents also offer high diffusivity, which increases the mobility of the dissolved ions and allows better mixing of the reagents [18]. Zhou et al. [16] synthesize 3D TiO_2 nanostructures from titanium *n*-butoxide ($Ti(OC_4H_9)_4$) and hydrochloric acid (HCl, 36%–38% by weight) as titanium source and acidic media, respectively. They used *n*-hexane ($CH_3(CH_2)_4CH_3$), cyclohexane (C_6H_{12}), *n*-butyl alcohol ($CH_3(CH_2)_3OH$), ethanol (CH_3CH_2OH) and water (H_2O) as solvents. The polarity of solvents increased, respectively, as mentioned. When they used n-hexane and cyclohexane as solvents (nonpolar solvents), they obtained 3D spherical structures with mean diameters of 2 and 2.5 μm, respectively; the spheres are composed of several nanorods densely closely packed forming a spherical dandelion-like structure. From polar solvents, *n*-butyl alcohol or ethanol was used; microspheres were obtained formed by a set of NPs and each particle has a diameter of about 7–6 nm, respectively. However, when water was used as a solvent, the products obtained were only bulky aggregated irregular NPs with a diameter of about 10 nm.

Generally, the hydrolysis rate of titanium ions Ti^{4+} is very fast under the aqueous solution, and the organic solvents promote slow hydrolysis of Ti-ions. In addition, alcohols are also used as oxygen donors and play a role as templates during the solvothermal synthesis of TiO_2-based nanomaterials [19]. In the reaction process, titanium ions usually undergo alcoholysis and condensation; Zhenfeng Bian et al. propose a general mechanism for solvothermal alcoholysis of Ti^{4+} ions, expressed in the following reactions [20]. The reaction rate influences the morphology of the final TiO_2 products.

$$TiX_4 + 4R - OH \rightarrow Ti(OR)_4 + 4HX; Ti(OR)_4 + TiX_4 \rightarrow 2TiO_2 + 4RX$$

$$2R - OH \rightarrow R - O - R + H_2O; TiX_4 + 2H_2The \rightarrow TiO_2 + 4HX$$

The solvents are nonpolar as *n*-hexane and cyclohexane; they are not soluble in water. In the non-aqueous solution solvothermal synthesis, the water is from an acid source, because the acid is an aqueous solution. Thus, water and solvent coexist in the system forming a liquid–liquid interface (aqueous–organic) between water and *n*-hexane or cyclohexane. The interface promotes the self-assembly of 3D nanostructures since NPs are highly mobile and can rapidly achieve an equilibrium assembly at a fluid interface [16,21].

The *n*-butoxide is highly soluble in organic solvents and slightly soluble in water, whereas HCl is the opposite. Thus, the hydrolysis and polycondensation reactions are restricted at the interface and TiO_2 crystals can nucleate and grow there. HCl supplies Cl^- ions that favor the rutile formation that usually crystallizes in a rod-like shape [22,23]. In the system analyzed, there is a small amount of water; thus, the concentrations of H^+ and Cl^- ions are relatively high at the liquid–liquid interface. The Cl^- ions are selectively adsorbed ions on rutile {110} faces and promote the anisotropic growth into nanorods along [001] orientation [23]. The nanorods aggregate into microspheres to lower their total free energy [16].

In the case of *n*-butyl alcohol and ethanol, both of them are amphiphiles, with different hydrophobic chain lengths but the same hydrophilic head group. For *n*-butyl alcohol, which is a low polar solvent, the reaction mainly occurs in the liquid–liquid interface; the amphiphiles are adsorbed on the hydrophilic surface of the particles to form an amphiphile bilayer that acts as a microreactor and soft temple for crystal growth. The microspheres are formed by the aggregation of NPs. The case of ethanol is similar to *n*-butyl alcohol, but the reaction takes place in the whole solution because ethanol is soluble in water [16].

The fast hydrolysis of titanium *n*-butoxide in water induces the precipitation of large size titania particles [16]. With a fast reaction rate, particles cannot grow with well-defined morphology and only bulky aggregated TiO_2 NPs are obtained.

It is known that high acid conditions promoted the rutile crystalline phase of TiO_2. The synthesis made with nonpolar solvents (*n*-hexane and cyclohexane) concentrates a high amount of Cl^- ions in the liquid–liquid interface. The Cl^- ions are selectively adsorbed on the (110) plane of rutile that has less superficial energy. And in the case of synthesis carried out with low polar solvents, for example a n-butyl alcohol; which is low polar and poorly soluble in water, and ethanol which is a polar solvent and soluble in water The concentration of Cl^- ions decreases due to the solubility of solvents in water. The low concentration of Cl^- ions is not enough to promote the rutile phase and therefore the anatase crystalline structure is formed. Figure 5.4 shows a scheme of three-mechanism growth of TiO_2 nanostructures in different solvents.

5.1.4 The Role of pH Solution on Crystalline Phase Transition

Zhao et al. [10] synthesize TiO_2 nanostructures by solvothermal using tetrabutyl titanate (TBT) as a titanium source, and HCl, water or NaOH aqueous solution was used as solvents to investigate the effect of an acidic or alkali environment under crystalline phase transition. Under strongly acid conditions, TiO_2 nanostructures crystallize in the rutile phase, due to high concentration of Cl^- ions that were selectively adsorbed on the (110) plane of rutile. When Cl^- ions concentration decreased, a mix

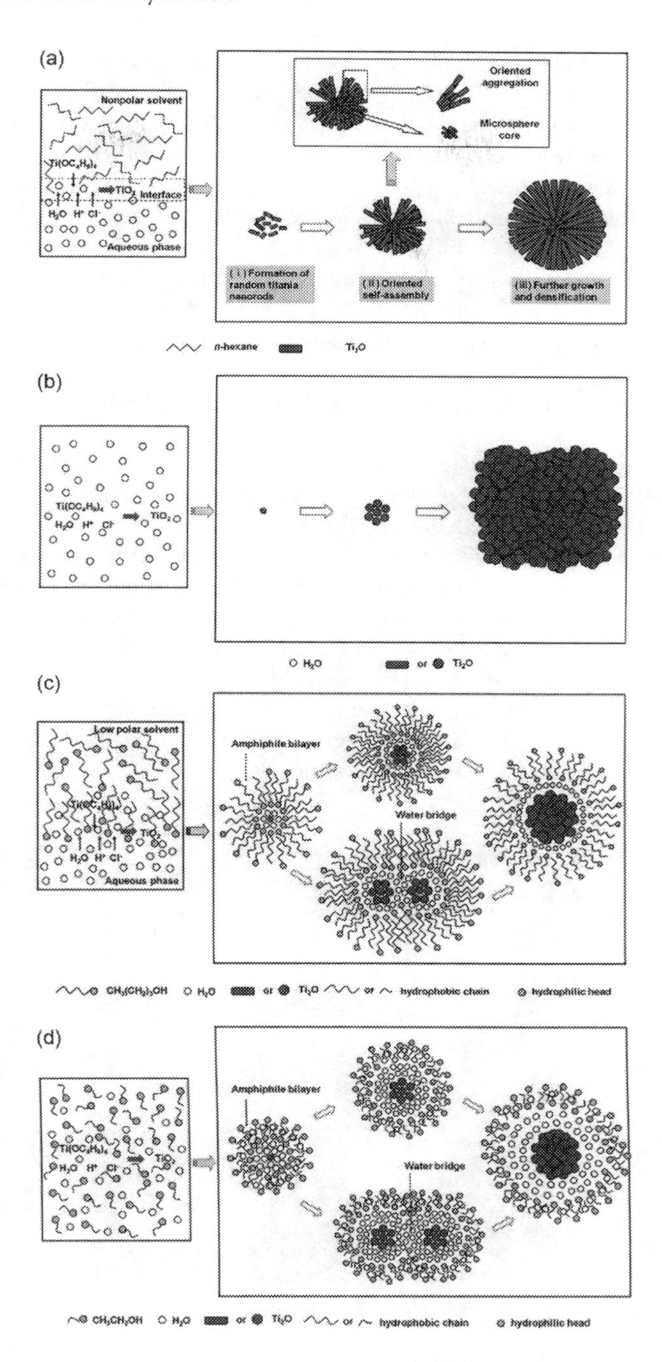

FIGURE 5.4 Scheme of the growth mechanism of the self-assembly of nanorods into 3D dandelion-like microspheres in a nonpolar solvent (a), self-assembly of nanoparticles into microsphere in an amphiphilic solvent in pure water (b), ethanol and (c) *n*-butyl alcohol (d). (Reprinted with permission from [16].)

of rutile, anatase and brookite phases was observed. The intensity of the (100) anatase plane was increased when Cl^- ions concentration decreased because the low acid conditions go against the formation of a rutile TiO_2 polymorph. Acid or alkali low concentrations solution around the neutral pH values could be advantageous for the formation of an anatase TiO_2 polymorph. Crystalline phase transformation continued to alkaline pH values; when Na^+ ions concentration increased, pure brookite was obtained. The evolution of phase transformation continues with the increased concentration of Na^+ ions; for stronger alkaline solutions, the formation of sodium dititanate ($Na_2Ti_2O_5$) or sodium trititanate ($Na_2Ti_3O_7$) was observed.

The formation of metal oxides from metal alkoxides can occur through hydrolysis and the polycondensation process; Zhao et al. [10] describe the mechanism formation of TiO_2 nanostructures with different crystalline phases as follows:

i. In the beginning of TBT hydrolysis, some octahedral monomers of $[Ti(OH)_2(OH_2)_4]^{2+}$ or edge-sharing dimers are generated under acid or near neutral hydrothermal conditions (Figure 5.5a1 or b1).
ii. The monomers and dimers combine with each other through oxolation or olation to form several polymeric nuclei joined equatorially (Figure 5.5a2) or by the edges (Figure 5.5b2). The monomers or dimers that form linear chains produce rutile-type nuclei (Figure 5.5a1–a3), while skewed chains produce anatase-type nuclei (Figure 5.5b1–b3).
iii. Both rutile-type and anatase-type nuclei grow when they exceed a critical size and become stable and follow growing to form rutile (Figure 5.5a3–a4) or anatase (Figure 5.5b3–b4) crystallites.

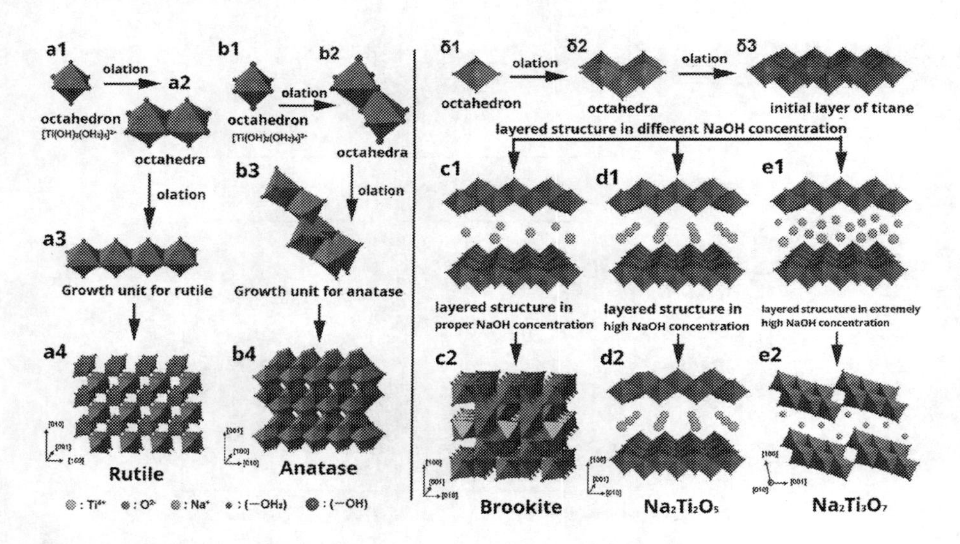

FIGURE 5.5 Mechanism of the evolution of crystal growth of rutile, anatase, brookite, dititanate $Na_2Ti_2O_5$ and trititanate $Na_2Ti_3O_7$ under acid or alkali solutions. (Reprinted with permission from [10].)

iv. When TBT was added to a strong alkali solution, it promotes the formation of layered titanate structure was probably formed by the condensation of TiO_6 octahedral monomers (Figure 5.5δ1–δ3).

v. The crystal growth of brookite TiO_2, titanate $Na_2Ti_2O_5$ and trititanate $Na_2Ti_3O_7$ should be followed by Ostwald's step rule [24].

vi. When NaOH concentration is increased into the solution, titanate interlayers are intercalated by Na^+ ions in the initial reaction process. However, the amount of Na^+ ions is too deficient to stabilize the interlayered structure. The small amount of Na^+ ions in the layered structure promotes that the structural transformation was delayed and deflected, which results in a special lattice shear in the combination process of layers and the final formation of pure brookite TiO_2 (Figure 5.5c1 and c2).

vii. For high NaOH concentration (0.5 M–6 M) (Figure 5.9d), Na^+ ions maintain the layered structures and produce the formation of sodium titanate during the reaction process (Figure 5.5d1 and d2).

viii. When the concentration of NaOH is higher than 8 M, dititanate interlayers cannot endure more Na^+ ion intercalations so that the orthorhombic lamellar structures of titanate $Na_2Ti_2O_5$ should condense themselves to result in the formation of monoclinic layered structures of trititanate $Na_2Ti_3O_7$, (Figure 5.5e1 and e2).

5.1.5 Effect of Reaction Time on Growth of TiO_2 Nanowires

In solvothermal synthesis, the hydrolysis of reactants can be fast or slow. If the nucleation occurs at a fast rate, all the nuclei have similar size, since they are formed under the same conditions and growth at the same time. If the nucleation occurs at a slower rate, NPs will usually become larger with variable sizes [25]. Thus, the reaction time plays an important role in the growth of structures because their shape and crystallinity evolution with them. Figure 5.6 shows an illustration of two nucleation kinetic rates.

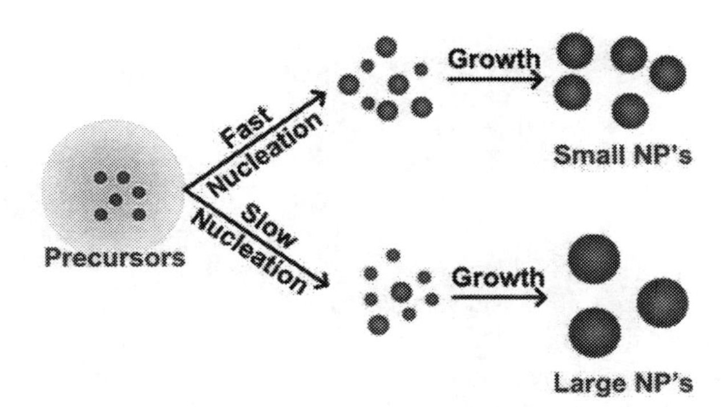

FIGURE 5.6 Schematic representation of two types of nucleation kinetics and the influence on the final nanoparticle size. (Reprinted with permission from [25].)

The crystal growth generally takes place on the lowest surface's energy, which has the fastest-growing velocity along the plane, but the slowest growing velocity perpendicular to the plane. It is known that the (110) plane is the lowest surface energy in rutile [26]. Wei et al. [11] reported that the growth of TiO_2 nanowire array (TNA) synthesized by solvothermal conditions (180°C) grew along the (002) direction. Figure 5.6 shows the evolution of the surface shape of TNA from needle like to rectangle when the reaction time was increased from 0.5 h to 6 hours.

The formation of nanowire structure happens at different growth velocities along axial and lateral directions. At the beginning of the reaction, the axial growth is faster than lateral growth; then, the stone-like bulks are formed growing with one-dimensional structure tendency (Figure 5.7a1–b3). By increasing the reaction time, the stone-like bulks change their shape to the thin nanowire; the edges were not well defined (Figure 5.7c1–d3). For longest times, the TNA increased their size and the surface was transformed into a rectangular shape (Figure 5.7e1–e3).

5.1.6 Effect of Temperature on the Growth of TiO_2 Nanowires

In the conventional solvothermal reactor as shown in Figure 5.1, the heat transference is carried on convection to the inner of solution; this process can be slow and delay homogenizing the temperature in the whole solution. The temperature and pressure conditions facilitate the dissolution of the chemical reagents and the production of the products by crystallization. The amount of thermal energy provided to the system determines that the mobility with the chemical species is moved inner the solution.

Figure 5.8 shows the evolution of the shape of TNA with temperature [11]. At low temperatures around 140°C–160°C, the nuclei formed during the hydrolysis and polycondensation processes do not have enough kinetic energy to move and interact with each other; thus, only bulks of nanorods not uniform in shape are found (Figure 5.8a–d). When the temperature increases (180°C to 220°C), the mobility of nuclei increases; therefore, there is a great interaction between the particles, and the axial or lateral growth is promoted; the average width of a single wire increases and the nanowires grow more uniformly (Figure 5.8e–j).

5.2 COLLOIDAL SYNTHESIS OF CdSSe NANOPARTICLES

In recent years, nanocrystalline semiconductors, also called quantum dots (QDs), have been seen as promising for the next generation of solar cells [27].

The great interest in solar cells from QDs arises on the one hand, from the possibility of modulating the width of the band gap of the semiconductor by varying the size of the QD [28], which allows choosing the wavelength of light absorption and even preparing the so-called panchromatic cells that use the entire solar spectrum and, on the other hand, the phenomenon called multiple exciton generation (MEG) can be used to achieve higher efficiency in solar cells beyond the limits of 32% theoretical efficiency of silicon solar cells [29]. In addition, there are simple preparation methods to achieve their synthesis and achieve size control.

Among the most studied QDs are those of CdSe (cadmium selenide) since it has been reported that it has optical properties that make it an attractive material in

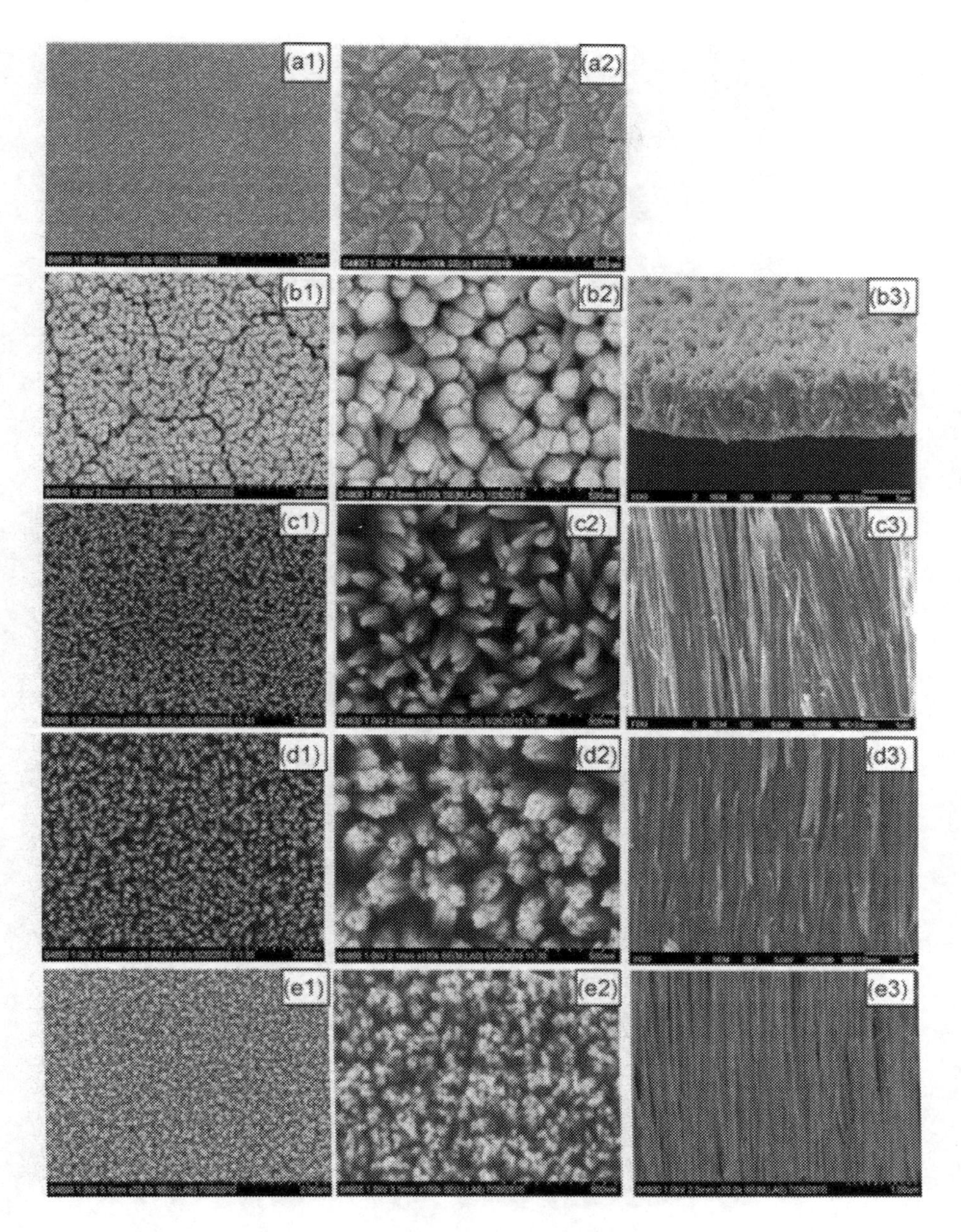

FIGURE 5.7 FE-SEM images of TiO_2 nanowire arrays fabricated at 180°C for different durations: (a1 and a2) 0.5 hours, (b1–b3) 1 hours, (c1–c3) 2 hours, (d1–d3) 4 hours and (e1–e3) 6 hours. (Reprinted with permission from [11].)

applications of conversion of solar energy to electric because the value of its band gap is around the 1.70 eV, close to the optimum for solar cells and because it is a good photoconductive material.

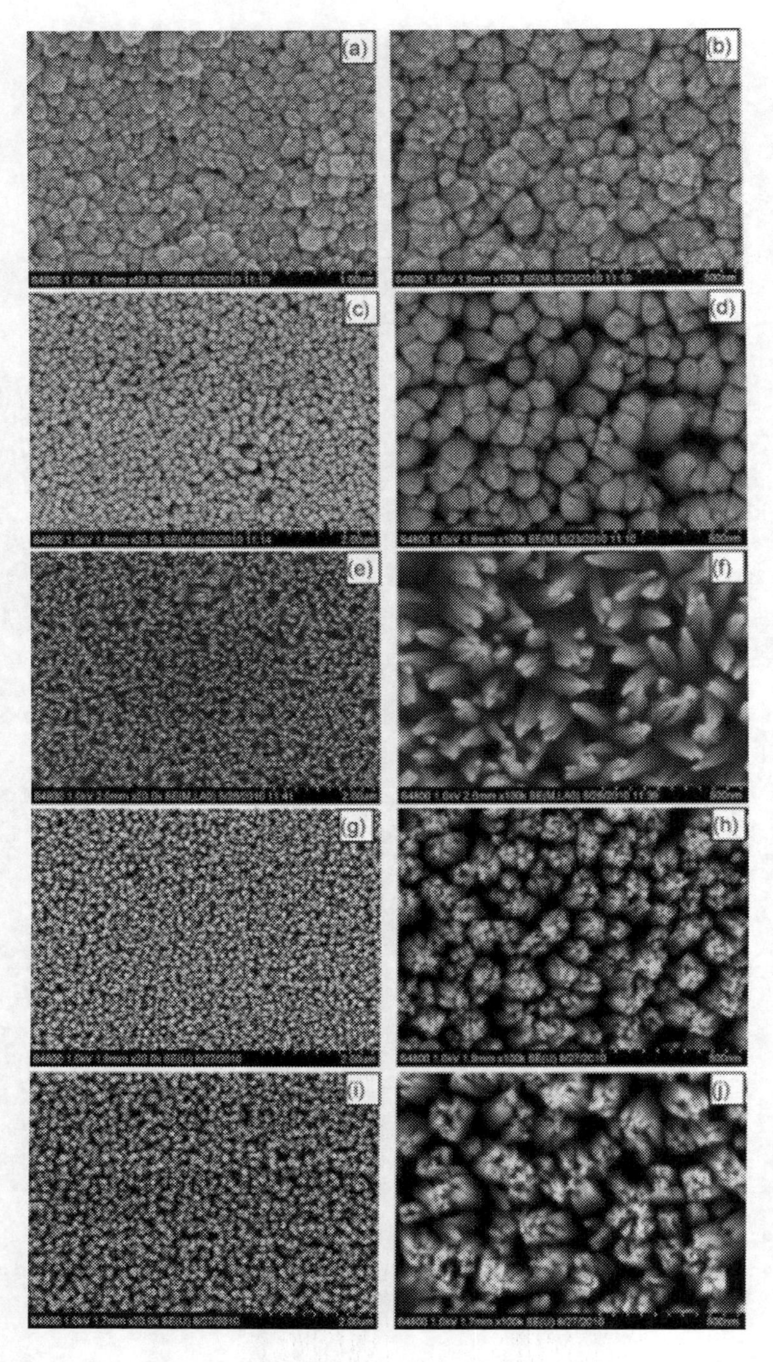

FIGURE 5.8 FE-SEM images of TiO_2 nanowire arrays fabricated for 2 hours at different temperatures: (a and b) 140°C, (c and d) 160°C, (e and f) 180°C, (g and h) 200°C, (i and j) 220°C. (Reprinted with permission from [11].)

The third generation volumetric heterojunction (BH) solar cells are composed of an active layer formed of a polymeric matrix that is capable of conducting holes, in which the nanostructures or QDs are embedded; this layer is the one that is responsible for absorbing light to transform it into charge carriers [30]. So far, in the prototypes of volumetric heterojunction solar cells, maximum efficiencies of the order of 6% energy conversion have been obtained [31]. To increase these efficiencies, there are fundamental research challenges such as particle size control, particle size distribution and controlled agglomeration, as well as the electrical interaction between NPs. Low conversion efficiencies have been related to the lack of optimization of the nanostructure surface. In some investigations, it has been observed that the formation of segregated phases of nanostructures within the polymeric matrix, mainly due to its poor dispersion, substantially affects the performance of this type of photovoltaic devices [32].

To prepare active layers in bulk heterojunction solar cells, NPs are usually synthesized by simple chemical methods, and then their surface is functionalized with a ligand that stabilizes them and allows charge exchange and finally they are dispersed in polymer solutions. The types of ligands used in functionalization are either short chain molecules or branched ones that surround the NP and allow the flow of electrons. Some of the molecules used as molecular spacers are thiols and dithiols that strongly bind with the CdSe surface and that can also present the length that is adequate to electrical conduction for charge transport by tunnel effect [33].

Once the functionalized NPs have been dispersed in the polymeric solution, the hybrid material film can be prepared by deposition methods such as spin coating or doctor blade, among others [34].

In this section, colloidal synthesis, which is based on the formation and growth of nuclei in a liquid media which could be aqueous or not, will be studied. The heterogeneous formation of nuclei is ruled by the ionic concentration of the species and the diffusion coefficient of the ions in the media. The rate of nuclei formation can be controlled by complexing agents or by changing the colligative properties of the media. In the present section, the principles of colloidal synthesis of NPs will be presented. As an example, the synthesis of $CdS_{1-x}Se_x$ NPs by direct reaction of the precursors in a media with different viscosities will be discussed.

In colloidal synthesis, there are three components of great importance, the precursors that are the reagents from which the NPs will be formed, the surfactants that direct the shape and size of NPs and the solvents that act as the reaction medium. Obviously, the choice of precursors, surfactants and solvents is made according to the material and type of morphology desired. The synthesis reaction can be carried out through different mechanisms including sol–gel, thermal decomposition, chemical reduction or oxidation, galvanic replacement, or controlled precipitation. In the described example, controlled precipitation will be discussed.

Colloidal Synthesis

Colloidal synthesis is a method that has attracted attention due to the possibility of obtaining high-quality materials at a relatively low price, thanks to the use of relatively simple to obtain reagents and equipment. On the other hand, this technique allows an excellent control over the morphology and dimensions of the

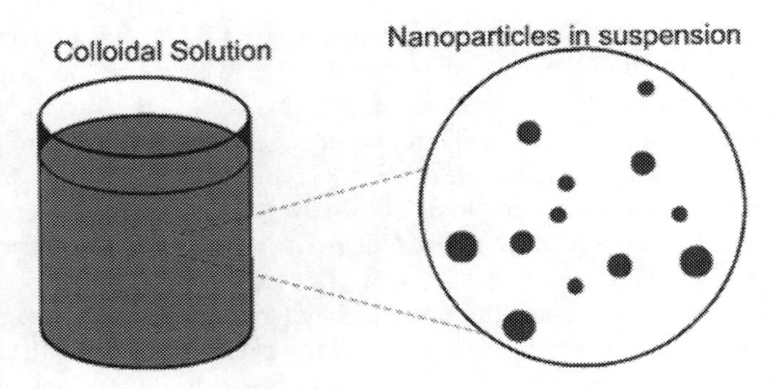

FIGURE 5.9 Scheme of suspended nanoparticles in a colloidal solution. (Adapted from [37] with permission from Elsevier.)

nanostructures to be obtained [35]. In colloidal synthesis, NPs are dispersed within a dispersing medium as shown in Figure 5.9; if this medium is a liquid, the colloidal system is called “sol”. They are usually perfectly dispersed systems with high surface energy, which is equivalent to the free energy required to increase the surface area of the dispersion medium. This method is widely used in the synthesis of group II–VI nanostructures [36] .

As mentioned above, there are various mechanisms in which the reaction takes place in colloidal syntheses; the purpose of the text is to understand the principles of colloidal synthesis by controlled precipitation.

5.2.1 Controlled Precipitation Method

Controlled precipitation is one of the most commonly used colloidal synthesis techniques; this formation of colloidal semiconductor nanocrystals follows a bottom-up approach and can be described by the model of LaMer and Dinegar [38]. The synthesis of nanocrystals occurs in three phases: the nucleation phase, the growth phase and the ripening phase of Ostwald. Figure 5.10 depicts the three steps of the process.

Forming colloids is an old process that may be traced back to Michael Faraday in 1857, when he synthesized gold sols [39]. The quick injection of a solution containing reactive compounds from groups II and VI into a heated and constantly stirring solvent containing molecules that can coordinate with the precipitated QD particles' surface is a fairly frequent approach for precipitating colloidal QDs II–VI [36,40]. This results in an initially large number of nucleation centers, and the coordinating ligands of the heated solvent prevent particle growth through Ostwald ripening, which can be defined as the formation of larger particles at the expense of smaller particles in order to reduce the higher surface free energy associated with smaller particles. Selective precipitation, in which a non-solvent is slowly added to a colloidal solution of particles, causing the larger particles to precipitate, can improve the size distribution of QD particles, as the solubility of molecules with the same type of chemical structure decreases as size increases. The size distribution of colloidal II–VI QDs can be reduced to a small percentage of the mean diameter by repeating

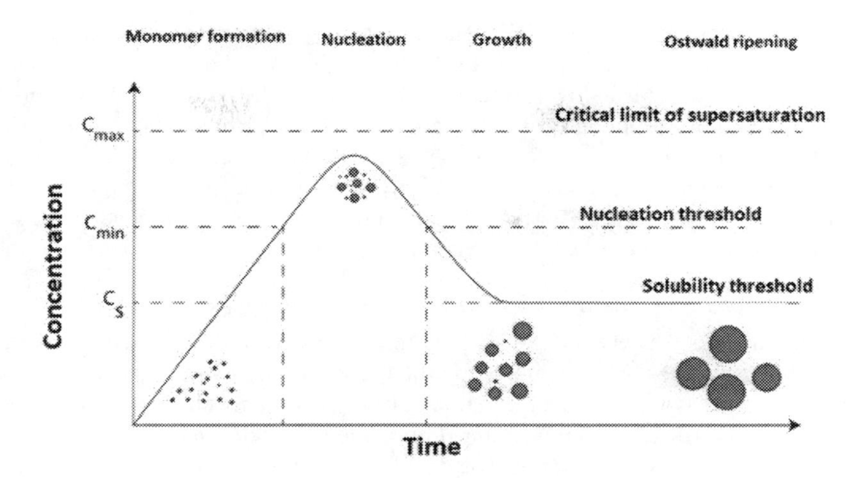

FIGURE 5.10 Scheme process nucleation and growth of nanocrystals. (Adapted with permission from [38]. © 1950 American Chemical Society.)

this method several times. To successfully use the colloidal chemical method for QD synthesis, it is critical that the stabilizer and solvent do not decompose during the reaction time. This will ensure that QDs are soluble after synthesis and will avoid the formation of different trapping states on the surface. Colloidal III–V QDs are more difficult to synthesize than II–VI QDs. This is because III–V semiconductors are more covalent compounds that must be synthesized at elevated temperatures. Colloidal synthesis of QDs becomes more difficult as the semiconductor series progresses and becomes more covalent [36,37].

5.2.1.1 Principles of the Controlled Precipitation Method

Precipitation is the formation of a new phase from an apparently homogeneous phase; chemical supersaturation is the driving force behind the appearance of a precipitate. Precipitation can be thought of as a rapid crystallization, the rate of which is determined by the supersaturation level at which it occurs. The mechanisms and relative rates of nucleation, growth, maturation and phase transformations are expected to determine the properties of the precipitates. On the other hand, precipitation occurs as a result of a series of reactions occurring within the solution, which must take into account the structural and compositional changes that it may undergo in the solid phase.

Some important characteristics of the precipitation process are as follows: Precipitation occurs in systems containing compounds that are relatively insoluble. This low solubility enables high supersaturation values; high supersaturation values result in a rapid nucleation rate; the above characteristic of the nucleation speed enables the formation of a large number of crystals but limits their growth; a small number of precipitated crystals can result in secondary processes such as aging, agglomeration, coagulation and so on, all of which have an effect on the particle size distribution of the precipitates; the system's oversaturation, which is required for precipitation to occur, is caused by chemical reactions occurring within the system (reactive crystallization), and precipitation is typically carried out at a constant temperature, not necessarily at low-temperature values [41]. The nucleation or formation of crystals within a solution enables precise control of the precipitated number of particles, size and morphology. Regrettably, the mechanisms underlying particle formation remain poorly understood. This is partly due to the fact that solid phase nuclei range in diameter from 5 to 20 Å, making them too large to use individual atomistic concepts but too small to use thermodynamic concepts.

5.2.1.2 Balance of Chemical Species

When the interaction between the ions and molecules within the system is considered, it becomes possible to comprehend the existence of chemical species such as complex ions, polynuclear species and small polymers, as well as their evolution toward crystal formation. At room temperature, the ions and/or molecules of the solute are in constant motion and thus within the coordination sphere of the other ions or molecules. This facilitates the grouping of species that are always present in the solution regardless of concentration [42]. Precipitation is accomplished by reacting cationic and anionic precursor solutions. Water is the most frequently used solvent. It is critical to know the solubility values of the compounds that react at the working temperature in order to achieve the initial supersaturation required for precipitation to occur. The literature contains several solubility values for inorganic compounds, particularly in binary mixtures [43]. Additionally, because the substances commonly used as precursors are partially soluble, their solubility values are expressed as solubility products, with water's activity set to 1.

5.2.1.3 Stages of Precipitation

Precipitation is a two-stage process, the first of which is the nucleation of the solid phase and the second of which is the growth of these nuclei. However, in reality, a large number of factors act in the process that allows other threads to be distinguished [44], including (a) an optically homogeneous solution in a saturated or supersaturated state that is extremely sensitive to changes in concentration and temperature. These changes produce complex ions and polynuclear species and (b) the formation of aggregates of the aforementioned chemical species, which can be stable or unstable and are referred to as embryos or germs. They lack a stable internal crystal structure; (c) consolidation of nuclei as growth units, units with an internal crystalline structure; (d) conformation of primary particles with colloidal individuality, with a diffuse layer surrounding them; and (e) formation of secondary structures due to weak agglomeration of primary particles, resulting in structures with a "sponge"-type texture. The composition scheme of the five transition states is depicted in Figure 5.11.

Nucleus Formation

Chemical species form in a supersaturated system as a result of changes in composition and/or temperature; the species interact, forming groups that can be dissolved and later reassembled into phase solids; these groups are the embryos of the phase solid. When complex ion concentrations are high enough, clusters grow large

Constituents and fundamental components

(a) Complex

Complex: Mononuclear
+X

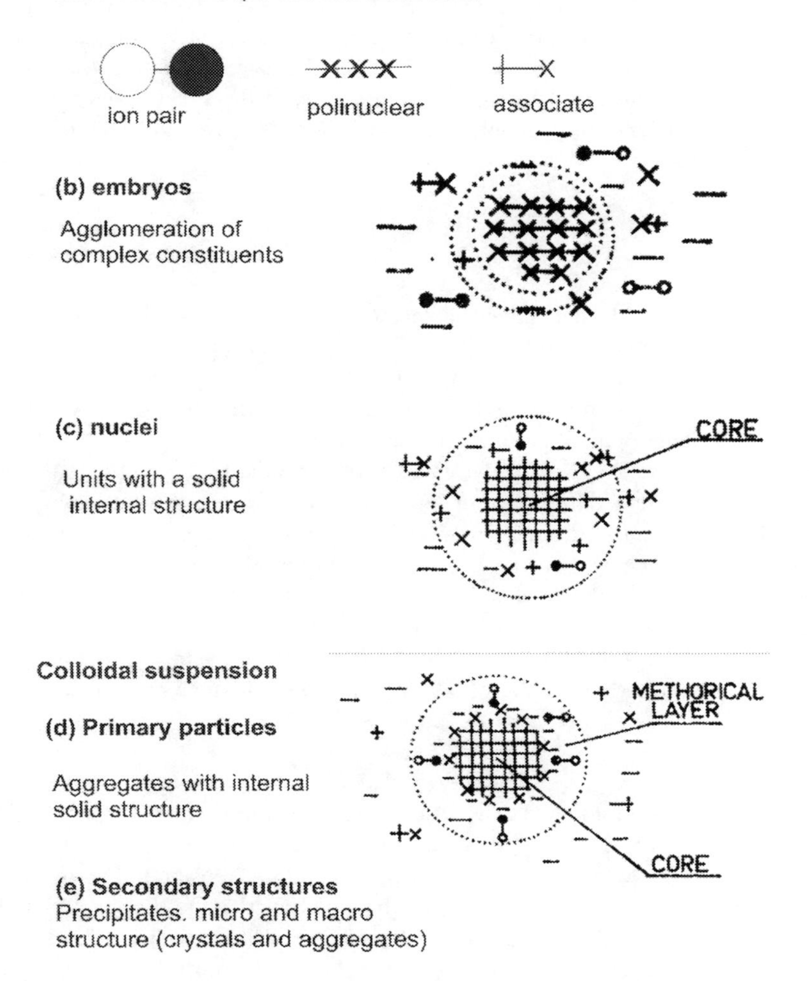

FIGURE 5.11 Scheme of the composition of the five sub-systems or stages that can occur during precipitation. (Adapted with permission from [45]. CC BY 4.0 ©1970. Croatica Chemica Acta.)

enough to reach a critical size and consolidate the nuclei, which are irreversible units that grow spontaneously. The first step in the formation of nuclei is the grouping of chemical species via Van der Waals type bonds and hydrogen bridges. In ionic species, the mechanism can be one of two types: (a) the formation of a small crystal assembly with strong bonds, which is modeled as "an embryo in a cavity" or (b) the formation of a diffuse grouping of solvated ions, which is nearly equal to its initial state and has weak bonds [41].

The supersaturation of nuclei in the medium leads to precipitation or nucleation [46]. These groups of atoms are thermodynamically unstable due to surface tension, which is the energy associated with the asymmetry of interatomic forces exerted at the interface.

As nanocrystals have a very high surface/volume ratio, this surface energy is not negligible in the thermodynamic equilibrium of the entire system, so that nucleation takes place, the free energy variation of the system must be favorable; this variation results from an opposition between the free energy of the surface and the crystallization energy, which, respectively, unfavors or favors nucleation according to the equation:$(\gamma)(\Delta G_V)$

$$\Delta G = \frac{4}{3}\pi r^3 \Delta G_V + 4\pi r^2 \gamma$$

This equation shows the dependence determined by the radius of the formed crystal, called the critical radius, to the radius of a nucleus that has a free energy of zero ($\Delta G = 0$). A nucleus with this radius will have the same chances to redissolve in the monomer as it does to grow as a NP. The larger nuclei will have more possibilities to grow from the energy point of view and vice versa [47].

Homogeneous nucleation (in the absence of pre-existing nucleation sites) has a high activation energy barrier, requiring extremely high supersaturation levels. The ideal nucleation step is instantaneous and unique, in order to reduce the number of solvated atoms below the nucleation threshold and thereby disadvantage nucleation thermodynamically. In this case, the system switches to a growth mode, in which remaining or newly generated nuclei are deposited preferentially in previously formed agglomerates rather than forming new ones. The crystals formed as a result of so-said nucleation, which shares a common growth history and ensures greater crystal homogeneity.

Growth

Once the agglomerate reaches a certain size, structural rearrangement becomes too energy-intensive, which stabilizes the crystal structure. The agglomerate forms a nucleus, which grows to form a NP either through the deposition of atoms in solution or monomers on the surface, or through coalescence with other nuclei. As long as zero-valent atoms are formed, NPs will grow until they reach equilibrium with the atoms on the surface and the atoms in the solution. If appropriate surface ligands are not used to stabilize these nuclei, they will tend to aggregate, reducing the total surface area and thus the total energy [47,48].

The nuclei will grow in proportion to the concentration of free monomers in the system. If this concentration is significantly greater than the solubility of the NP, any monomer that approaches

the surface will be automatically integrated; thus, growth occurs in diffusion equilibrium. On the other hand, if the solution concentration is extremely low, solution-precipitation equilibrium prevails and the reaction is in reaction equilibrium.

In diffusion mode, the rate of monomer arrival to the surface depends on the monomer concentration and the size of the solid. This results in the size distribution. As a result, a size distribution occurs. Additionally, slow growth ensures growth homogeneity and prevents local disturbances from interfering with the growth of specific crystals. Additionally, rapid initial growth enables the concentration to rapidly fall below the saturation point, avoiding prolonged nucleation [47].

Ostwald Ripening

In contrast, the reaction equilibrium results in Ostwald ripening, which elongates the crystal size dispersion in general. This elongation is caused by the imbalance between the atoms on the particle's surface and the atoms in solution. Due to the fact that the smaller NPs are more soluble than the larger ones, the NPs are eventually dissolved and reprecipitated on the latter; the large NPs then grow to the benefit of the small ones [49]. The smaller crystals act as fuel for the growth of bigger crystals. A colloidal particle grows through a process of monomer diffusion toward the surface followed by a monomer reaction at the nanocrystal's surface. The coarsening effects of mass transport or diffusion are frequently referred as Ostwald ripening process. In other words, if the reaction is not stopped after the growth stage, a final stage takes place: Ostwald ripening. The small crystals, having a high surface energy, will dissolve and the monomers formed will merge on the larger NPs, which will thus grow. This step promotes the polydispersity of the NPs in solution. It is therefore necessary to stop the reaction before ripening begins.

5.2.2 Case Study: Synthesis of CdS_{1-x} Se_x Nanoparticles by Direct Reaction of the Precursors in a Media with Different Viscosities

Martinez-Garcia in 2017 [50] carried out the synthesis of CdSe NPs by the controlled chemical precipitation method from their precursors in glycerin and gelatin as dispersing media, in order to control the size and shape of the NP. For the synthesis, the concentrations of the precursors and the reaction media were varied.

5.2.2.1 Synthesis of CdS_{1-x} Se_x Nanoparticles

The synthesis was carried out as follows:

For the selenium source solution, aqueous solutions containing sodium sulfide (N_2S) and selenium (Se) were prepared, while for the cadmium precursor solution, different aqueous solutions of $CdCl_2$ were prepared.

Two different reaction media were carried out, powdered porcine gelatin and purified glycerin. The reaction medium was dissolved in a certain amount of water to obtain different concentrations (gelatin 1%–5% mm^2 and glycerin 50%–100% vol.) to study its influence on the size of NPs.

TABLE 5.1
Conditions of the Reaction Baths Prepared for the Synthesis of the Nanoparticles

Bath	It^{2-} (mL)	Cd^{2+} (mL)	$[Se^{2-}]$ (M)	$[Se^{2+}]$ (M)
1	0.6	0.4	0.018	0.013
2	1	0.6	0.024	0.016
3	1.4	0.85	0.030	0.019

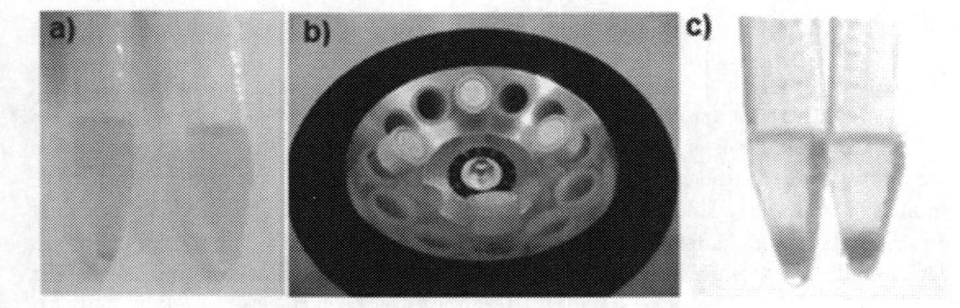

FIGURE 5.12 Photographs of the sequence of the separation of the nanoparticle solution in two tubes (a) to centrifuge (b) and decant (c).

These reaction media were heated to 50°C and reserved. For the synthesis of NPs, different volumes of the selenium and cadmium precursor solutions were added to 10 mL of the reaction medium to obtain the concentrations shown in Table 5.1.

After the reaction media were prepared, they were withdrawn from the thermostatic bath and immediately the Se^{2-} precursor solution was added, followed by the Cd^{2+} precursor solution. After this, the reactor was put in a water bath at 18°C for 3 minutes to stop the reaction and keep the particle size.

Then, two 15 mL Falcon tubes were prepared, one of which contains CdSe NPs that have just been produced. Two volumes of a solvent were added to each tube (in this example, 5 mL of acetone or 5 mL of ethanol) to separate NPs from the reaction media. To remove the residual solvent, the tube was thoroughly shaken and put in a centrifuge for 10 minutes at 4000 rpm. The precipitate was detected at the bottom of the Falcon tube after the time had passed. Five milliliters of the solvent were added to each tube, and the centrifuge was used for 10 minutes. Afterward, the solvent was removed and 5 mL of the supernatant was decanted. The as-synthesized particles, the centrifuge and the decanted NPs are shown in Figure 5.12.

5.2.2.2 Functionalization of Nanoparticles with Molecular Spacers

The prepared NPs were functionalized with different molecular spacers to study the effect of chain length and the terminal groups on the stability of NPs and on the film-forming capacity. The molecular spacers shown in Table 5.2 were used.

TABLE 5.2
Molecular Separators Used in the Development of the Work

Molecule	Formula
Hexanedithiol	$SHCH_2(CH_2)_4CH_2SH$
Octanedithiol	$HSCH_2(CH_2)_6CH_2SH$
Propanethiol	$CH_3CH_2CH_2SH$

Dithiols have polar groups (–SH) at the ends, capable of joining chalcogenide surfaces, while thiol has only one. It is expected that dithiols would form a network of NPs joint through NP–S–NP bridges.

Preparation of Molecular Separator Solutions

All the solutions used to functionalize NPs were prepared with a concentration of 0.2 mM. For the preparation of solutions, the necessary amount of thiol was measured with a precision pipette, placed in a volumetric flask and then diluted in 50 mL of isopropyl alcohol. This solution was kept covered until its use.

Nanoparticle Functionalization

The NPs produced under the varied reaction conditions were resuspended in 1 mL of solvent, after which the precipitate is entirely resuspended in the solvent. Subsequently, 1 mL of the solution of molecular separators was added to the tubes and left to sit until a colloid formed at the bottom of the tubes. The supernatant was decanted, and the material was recovered from the bottom of the tube. Figure 5.13 shows the functionalized NP colloid.

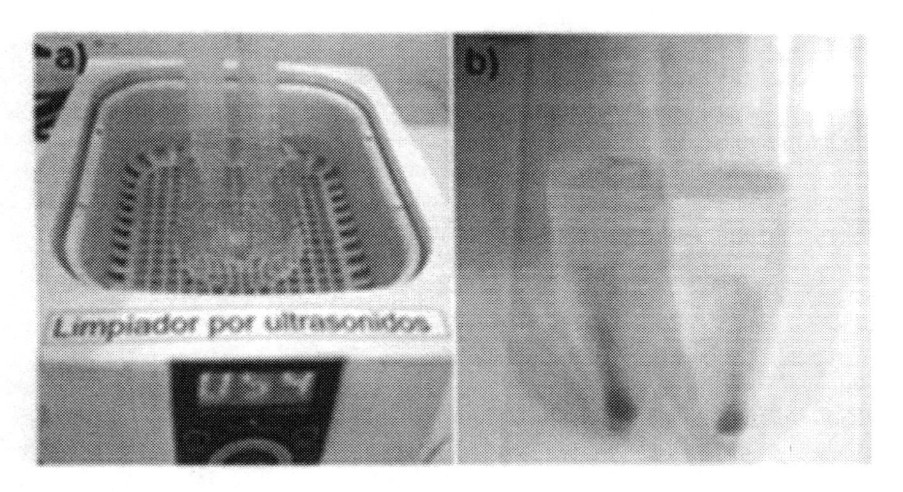

FIGURE 5.13 Functionalization of CdSe nanoparticles subjected to sonication (a) and the formation of the colloid at the bottom of the tubes (b).

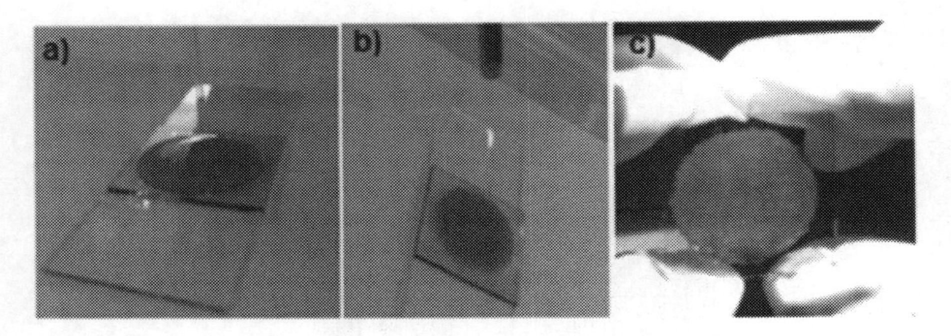

FIGURE 5.14 Photographs of the deposit of the functionalized nanoparticles (a), dried with nitrogen (b) and the ready film (c).

Nanoparticle Film Deposition

After the NPs' functionalization, they were deposited on a glass substrate by the drop-casting technique (Figure 5.14a), for which an aliquot of the functionalized NP solution was taken with a Pasteur pipette and dropped onto the substrate, then allowed to dry for 3 minutes and finally dried under a nitrogen flux (Figure 5.14b,c).

5.2.2.3 Characterization

To corroborate the presence of CdSe and the size of the NPs, a sample prepared in glycerin and functionalized with octanedithiol (ODT) was studied by Transmission Electron Microscope (TEM). Figure 5.15 depicts and describes bright-field and dark-field images obtained in various parts of the TEM grid. The obtained NPs are crystalline, as evidenced by the presence of crystalline planes in images (f), (g) and (e) have a regular size distribution of about 5 nm. The NPs are embedded in a diffuse mass that corresponds to ODT in images (a)–(d), resulting in the expected three-dimensional structure.

Energy-dispersive X-ray (EDS) spectra were collected at various points in the measured images to determine the composition of the observed material. C, S, Se, Cd, O, Cl and Si were detected, but the latter three were not quantified because they were due to the preparation. Given that the S/C ratio in the ODT is one-tenth, the remaining S was attributed to the Na_2S incorporated during the synthesis. Figure 5.16 shows the TE images with the boxes where the EDS spectra were acquired. Images A1 and A2 show the NP and the thiol; therefore, the percentage of C is low and the ratio of S:C is 1:10. Image A3 shows the NP and the membrane of the grid and the C2 spectrum corresponds only to the membrane that is made of carbon. With the results obtained, it was estimated that the average composition of the NPs is $CdS_{(0.7-0.8)}Se_{(0.3-0.2)}$.

Figure 5.17 shows the TEM images of CdSSe NPs that have been functionalized with hexanedithiol. The NPs show a larger size dispersion than when functionalized with ODT. There is no evidence for the development of a three-dimensional network, and it is also clear that NPs are more polyhedral than spherical. Additionally, this colloidal solution is not as stable over time as ODT, which remains stable without agglomeration for several weeks, but the hexanethiol solution remains stable for just 1 or 2 days.

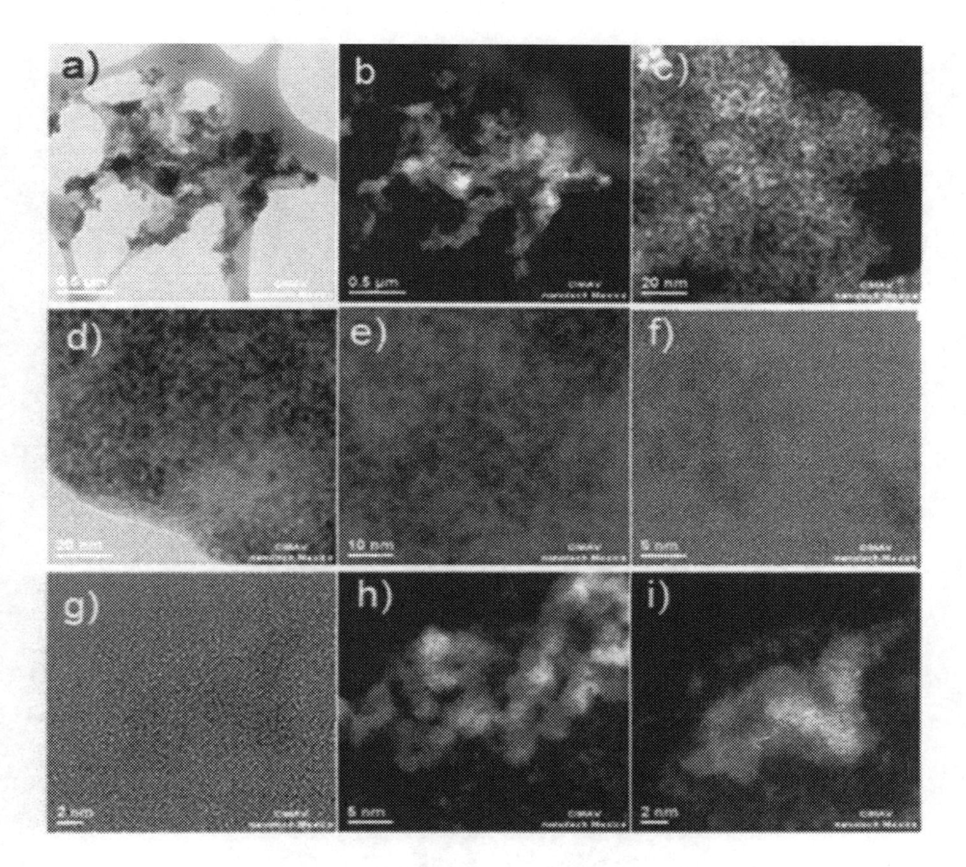

FIGURE 5.15 TEM images at different magnifications of a suspension of ODT-functionalized nanoparticles. (a) Bright-field image and (b) dark field where you can see the ODT network that binds the nanoparticles, (c) dark-field image of a cluster of nanoparticles, (d) bright-field image where you can see the nanoparticles embedded in the ODT colloid, (e) bright-field imaging with nanoparticles of approximately uniform size, (f) ~5 nm nanoparticles, (g) 5 nm cluster of nanoparticles where crystalline planes can be observed, (h) bright-field imaging and (i) dark field of the same area shown by the crystalline planes of nanoparticles of less than 5 nm.

Following, the crystal structure of the CdSSe casted film was analyzed by X-ray diffraction as shown in Figure 5.18. The diffractogram shows peaks at two thetas 26°, 42°, 46° and 50° associated with the hexagonal structure of $CdS_{0.75}Se_{0.25}$, which agrees with the composition measured by EDS.

Diffuse transmittance measurements in an integrating sphere were made on three series of NP films functionalized with ODT that were produced in glycerin medium and three series that were generated in gelatin media and coated on glass. Figure 5.19 illustrates the transmittances of the two groups. The low transmittance is a result of the films' roughness. The spectra of samples prepared in glycerin have an absorption edge at 550 nm, whereas samples prepared in gelatin exhibit an absorption edge

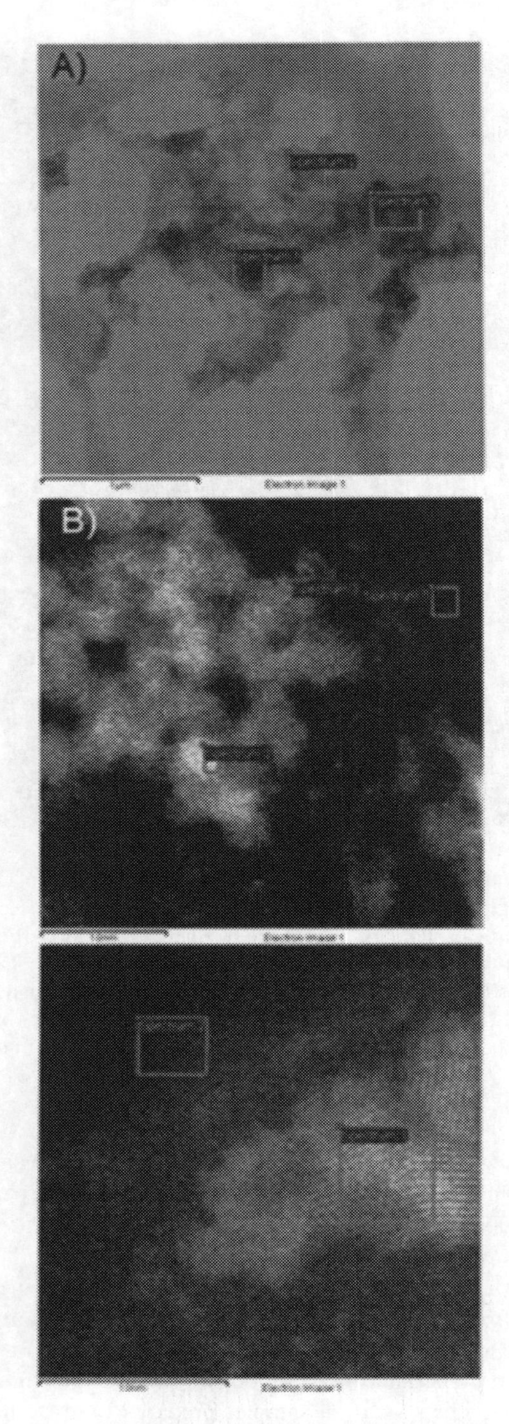

FIGURE 5.16 TEM images in which the EDS spectra were acquired.

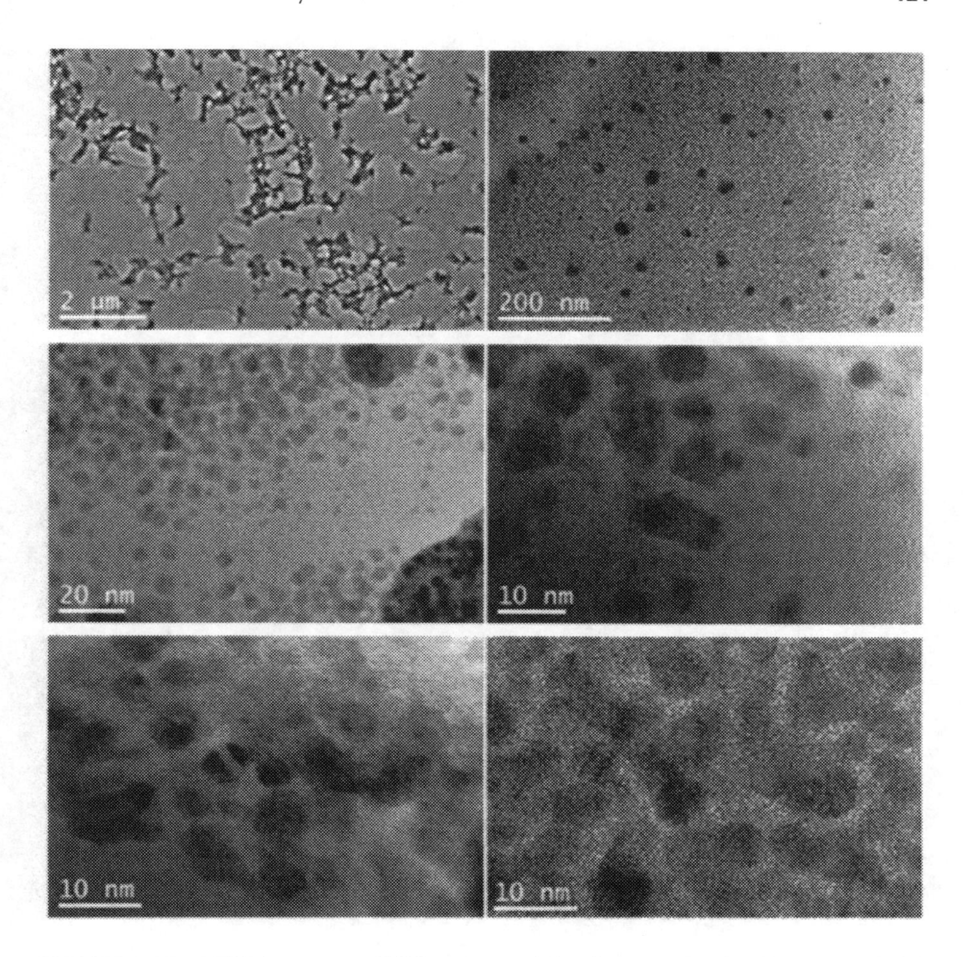

FIGURE 5.17 TEM images of CdSSe nanoparticles functionalized with hexanedithiol. (a) Image that presents a region of the sample in which particles can be observed in a non-continuous thiol network, (b) image where different sizes of particles can be observed, (c) in this image nanoparticles of different sizes can be observed, (d) image where 10 nm nanoparticles and their crystalline planes are observed, (e) image with nanoparticles of approximately 5 nm, (f) image that presents crystalline planes of nanoparticles of approximately 10 nm and (g) image of 5 nm nanoparticles and agglomeration and coalescence of some of them.

near 600 nm, indicating that the former has a smaller mean NP size or a higher sulfur content, as the CdS gap is 2.4 eV and the CdSe gap is 1.7 eV.

In Figure 5.20, the Raman spectra of the films denoted 2A, 2B and 2C that were synthesized in glycerin (A-100%, B-80% and C-50%) and functionalized with ODT are presented. The spectra were normalized to the peak at around 200 cm^{-1}. The optical longitudinal phonon (LO) of CdSe is present in the spectrum at 200 cm^{-1}, and a peak at 285 cm^{-1} is detected as well, representing the LO of CdS. As can be observed, the relative intensity of CdS and CdSe LO peaks fluctuate with the viscosity of the

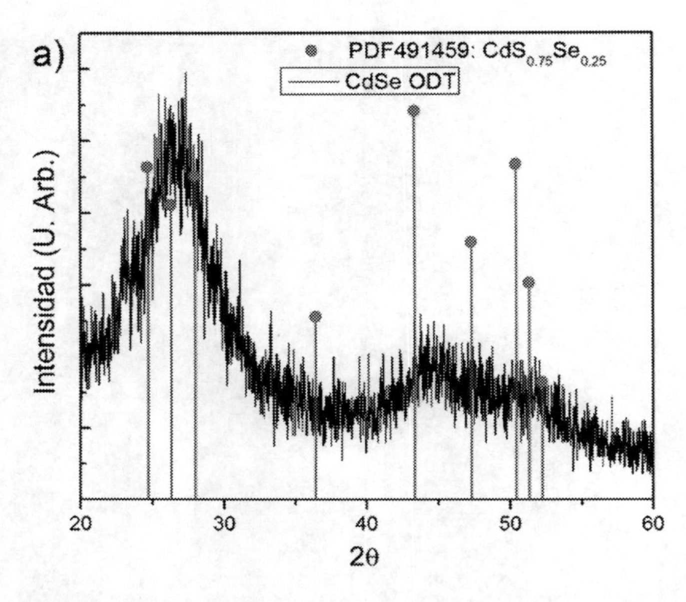

FIGURE 5.18 Diffractograms of a CdSSe film functionalized with ODT and deposited on a glass substrate.

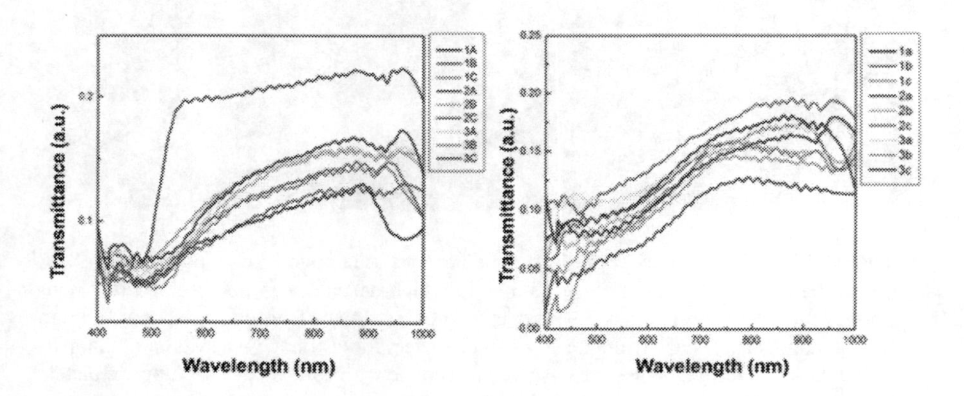

FIGURE 5.19 Diffuse transmittance of ODT-functionalized CdSSe films with (a) glycerin and (b) gelatin as the medium.

bath (from A to C, the glycerin content decreases), indicating that a more extensive analysis of the film composition is required.

The transmittance spectrum shows that samples prepared in gelatin have absorption peaks at shorter wavelengths, indicating smaller NP sizes, whereas films prepared in glycerin exhibit a stronger effect of precursor content. This demonstrates that gelatin is a more effective medium for the production of NPs than glycerin. The viscosity reduction appears to have the expected effect of increasing the size distribution since peaks show at longer wavelengths.

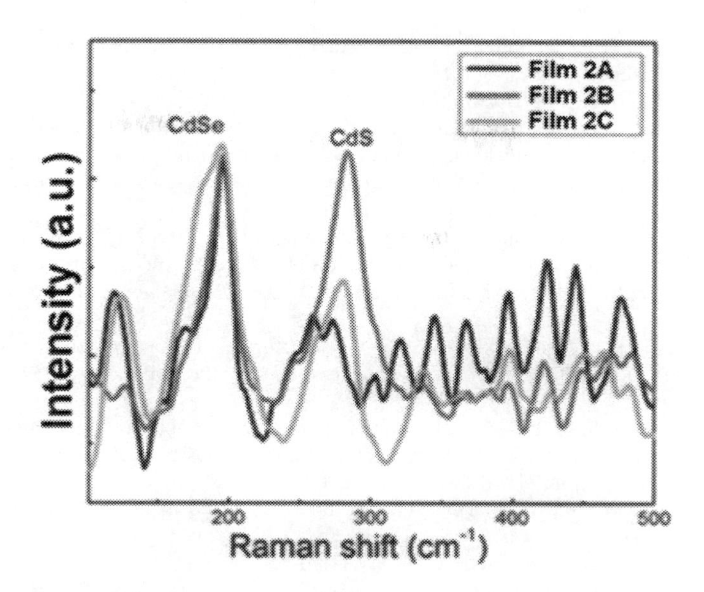

FIGURE 5.20 Raman spectrum of films 2A, 2B and 2C deposited on glass.

All of these results demonstrate that it is possible to synthesize and functionalize CdSSe films using the controlled precipitation method; it is also clear that the medium in which the reaction occurs has a significant effect on the particle size obtained; viscosity reduction appears to have the expected effect of increasing the size distribution.

5.3 CONCLUDING REMARKS

The controlled precipitation method allows producing NPs with a relatively inexpensive and simple procedure that allows the precise control of the particle's morphology and size by varying the concentration, reaction time and the use of molecular spacers.

ACKNOWLEDGMENTS

Supported by CONACYT Science of Frontier Grant 40798 and by SIP-IPN 20211513 project.

REFERENCES

[1] K. Byrappa and T. Adschiri, "Hydrothermal technology for nanotechnology," *Prog. Cryst. Growth Charact. Mater.*, vol. 53, no. 2, pp. 117–166, 2007.

[2] V. M. Ramakrishnan, M. Natarajan, A. Santhanam, V. Asokan, and D. Velauthapillai, "Size controlled synthesis of TiO_2 nanoparticles by modified solvothermal method towards effective photo catalytic and photovoltaic applications," *Mater. Res. Bullet.*, vol. 97. pp. 351–360, 2018.

[3] F. He, F. Ma, T. Li, and G. Li, "Solvothermal synthesis of *N*-doped TiO_2 nanoparticles using different nitrogen sources, and their photocatalytic activity for degradation of benzene," *Cuihua Xuebao/Chinese J. Catal.*, vol. 34, no. 12, pp. 2263–2270, 2013.
[4] M. Wang, Z. Wang, L. Wei, J. Li, and X. Zhao, "Catalytic performance and synthesis of a Pt/graphene-TiO_2 catalyst using an environmentally friendly microwave-assisted solvothermal method," *Cuihua Xuebao/Chinese J. Catal.*, vol. 38, no. 10, pp. 1680–1687, 2017.
[5] M. Z. N. Cruz, O. Calzadilla, J. Roque, F. Chalé, J.K. Olarte, M. Meléndez, "Study of the effect of TiO_2 layer on the adsorption and photocatalytic activity of TiO_2–MoS_2 heterostructures under visible-infrared light," *Int. J. Photoenergy*, vol. 2020, pp. 1–9, 2020.
[6] R. S. Dubey and S. Singh, "Investigation of structural and optical properties of pure and chromium doped TiO_2 nanoparticles prepared by solvothermal method," *Results Phys.*, vol. 7, pp. 1283–1288, 2017.
[7] S. S. Kalanur, S. H. Lee, Y. J. Hwang, and O. S. Joo, "Enhanced photoanode properties of CdS nanoparticle sensitized TiO_2 nanotube arrays by solvothermal synthesis," *J. Photochem. Photobiol. A Chem.*, vol. 259, pp. 1–9, 2013.
[8] R. H. D. Victor and K. Lamer, "Theory, production and mechanism of formation of monodispersed hydrosols," *J. Am. Chem. Soc.*, vol. 76, no. 11, pp. 4847–4854, 1950.
[9] D. Erdemir, A. Y. Lee, and A. S. Myerson, "Nucleation of crystals from solution: Classical and two-step models," *Acc. Chem. Res.*, vol. 42, no. 5, pp. 621–629, 2009.
[10] B. Zhao, L. Lin, and D. He, "Phase and morphological transitions of titania/titanate nanostructures from an acid to an alkali hydrothermal environment," *J. Mater. Chem. A*, vol. 1, no. 5, pp. 1659–1668, 2013.
[11] Z. Wei, R. Li, T. Huang, and A. Yu, "Fabrication of morphology controllable rutile TiO_2 nanowire arrays by solvothermal route for dye-sensitized solar cells," *Electrochim. Acta*, vol. 56, no. 22, pp. 7696–7702, 2011.
[12] S. Kathirvel, C. Su, Y. J. Shiao, Y. F. Lin, B. R. Chen, and W. R. Li, "Solvothermal synthesis of TiO_2 nanorods to enhance photovoltaic performance of dye-sensitized solar cells," *Sol. Energy*, vol. 132, pp. 310–320, 2016.
[13] A. Trenczek-Zajac, M. Radecka, and M. Rekas, "Photoelectrochemical properties of Nb-doped titanium dioxide," *Phys. B Condens. Matter*, vol. 399, no. 1, pp. 55–59, 2007.
[14] Y. Liu and Y. Yang, "Recent progress of TiO based anodes for Li ion batteries," *J. Nanomater.*, vol. 2016, no. 2, pp. 310–320, 2016.
[15] J. Du, J. Zhang, Z. Liu, B. Han, T. Jiang, and Y. Huang, "Controlled synthesis of Ag/TiO_2 core-shell nano wires with smooth and bristled surfaces via a one-step solution route," *Langmuir*, vol. 22, no. 3, pp. 1307–1312, 2006.
[16] J. Zhou, G. Zhao, B. Song, and G. Han, "Solvent-controlled synthesis of three-dimensional TiO_2 nanostructures via a one-step solvothermal route," *Cryst. Eng. Comm.*, vol. 13, no. 7, pp. 2294–2302, 2011.
[17] H. G. Yang et al., "Anatase TiO_2 single crystals with a large percentage of reactive facets," *Nature*, vol. 453, no. 7195, pp. 638–641, 2008.
[18] M. Yoshimura and K. Byrappa, "Hydrothermal processing of materials: Past, present and future," *J. Mater. Sci.*, vol. 43, no. 7, pp. 2085–2103, 2008.
[19] N. Pinna and M. Niederberger, "Surfactant-free nonaqueous synthesis of metal oxide nanostructures," *Angew. Chemie–Int. Ed.*, vol. 47, no. 29, pp. 5292–5304, 2008.
[20] Z. Bian, J. Zhu, and H. Li, "Solvothermal alcoholysis synthesis of hierarchical TiO_2 with enhanced activity in environmental and energy photocatalysis," *J. Photochem. Photobiol. C Photochem. Rev.*, vol. 28, pp. 72–86, 2016.
[21] O. D. Velev, K. Furusawa, and K. Nagayama, "Assembly of latex particles by using emulsion droplets as templates. 2. Ball-like and composite aggregates," *Langmuir*, vol. 12, no. 10, pp. 2385–2391, 1996.

[22] H. Li, X. Duan, G. Liu, X. Jia, and X. Liu, "Morphology controllable synthesis of TiO_2 by a facile hydrothermal process," *Mater. Lett.*, vol. 62, no. 24, pp. 4035–4037, 2008.
[23] J. G. Li, T. Ishigaki, and X. Sun, "Anatase, brookite, and rutile nanocrystals via redox reactions under mild hydrothermal conditions: Phase-selective synthesis and physicochemical properties," *J. Phys. Chem. C*, vol. 111, no. 13, pp. 4969–4976, 2007.
[24] R. A. Van Santen, "The Ostwald step rule," *J. Phys. Chem.*, vol. 88, no. 24, pp. 5768–5769, 1984.
[25] D. Nunes, A. Pimentel, L. Santos, P. Barquinha, L. Pereira, E. Fortunato, and R. Martins, Synthesis, design, and morphology of metal oxide nanostructures. In: *Metal Oxide Nanostructures.*. Elsevier, pp. 21–57 (2019).
[26] U. Diebold, "The surface science of titanium dioxide," *Surf. Sci. Rep.*, vol. 48, pp. 53–229, 2003.
[27] I. Robel, V. Subramanian, M. Kuno, and P. V. Kamat, "Harvesting light energy with CdSe nanocrystals molecularly linked to mesoscopic TiO_2 films," *J. Am. Chem. Soc.*, vol. 13, pp. 295–303, 2006.
[28] T. Hanrath, D. Veldman, J. Choi, C. G. Christova, M. M. Wienk, and R. A. Janssen, "PbSe nanocrystal network formation during pyridine ligand displacement," *J. ACS Appl. Mat. Int.*, vol. 1, pp. 244–260, 2009.
[29] M. T. Trinh, A. J. Houtepen, J. M. Schins, T. Hanrath, J. Piris, and W. Knulst, "Multiple exciton generation in PbSe nanorods," *Nano Lett.*, vol. 8, pp. 1713–1722, 2008.
[30] S. Emin., "Quantum dot solar cells," *J. Am. Chem. Soc.*, vol. 6, pp. 513–518, 2011.
[31] S.H. Park, A. Roy, S. Beaupre, S. Cho, N. Coates, J. S. Moon, D. Moses, M. Leclerc, M., Lee, and K. Heeger, Bulk heterojunction solar cells with internal quantum efficiency approaching 100%," *Nature*, vol. 3, pp. 297–303, 2009.
[32] I. Domingo, A. Garza, F. Cienfuegos, L. Chávez, "Applications of nanotechnology in alternative energy sources," *Engineering*, vol. 13, pp. 53–62, 2010.
[33] J.M. Luther, M. Law, Q. Song, C.L. Perkins, M.C. Beard, A.J. Nozik, "Structural, optical, and electrical properties of self-assembled films of PbSe nanocrystals treated with 1,2-ethanedithiol," *ACS Nano*, vol. 2, pp. 271–280, 2008.
[34] K.W. Johnston, A.G. Pattantyus-Abraham, J.P. Clifford, S.H. Myrskog, S. Hoogland, H. Shukla, E.J, Klem, L. Levina, E.H. Sargent, "Efficient Schottky-quantum-dot photovoltaics: the roles of depletion, drift and diffusion," *Appl. Phys. Lett.*, vol. 92, pp. 420–431, 2008.
[35] An, Kwangjin, et al. "Colloid chemistry of nanocatalysts: A molecular view," *J. Coll. Inter Sci.*, vol. 373, pp. 1–13, 2012.
[36] A. J., Nozik and O. I. Mićić. "Colloidal quantum dots of III–V semiconductors," *MRs Bull.*, vol. 23, pp. 24–30, 1998.
[37] R. Gillibert, et al. "Explosive detection by surface enhanced Raman scattering," *TrAC Trends Anal. Chem.*, vol. 105, pp. 166–172, 2018.
[38] V. K. LaMer, Victor K., and R. H. Dinegar. "Theory, production and mechanism of formation of monodispersed hydrosols," *J. Am. Chem. Soc.*, vol. 72, no. 11, pp. 4847–4854, 1950.
[39] Michael Faraday, X. "The Bakerian lecture. Experimental relations of gold (and other metals) to light," *Phil. Trans. Royal Soc. London*, vol. 47, pp. 145–181, 1857.
[40] C. B. Murray, *Synthesis and Characterization of II–VI Quantum Dots and Their Assembly into 3D Quantum Dot Superlattices*, (Ph.D. Thesis, Massachusetts Institute of Technology) (1995).
[41] J. E. Rodriguez Paez, "Synthesis of ceramic powders by the precipitation method," *Bull. Soc. Esp. Ceram. Glass*, vol. 3, pp. 173–184, 2001.
[42] A. G. Walton, et al. *The Formation and Properties of Precipitates.* Ed. 23. Interscience Publishers, New York (1967).

[43] O. Sohnel, J. Garside. *Precipitation: Basic Principles and Industrial Applications.* Butterworth-Heinemann, Oxford (1992).

[44] B. Tezak, "Coulombic and stereochemical factors of colloid stability of precipitating systems," *Disc. Faraday Soc.* vol. 42 pp. 175–186, 1966.

[45] B. Tezak, "Methorics of the precipitation from electrolytic solutions," *Croat. Chem. Acta*, vol. 40, no. 2, pp. 63–78, 1968.

[46] S. G. Kwon and T. Hyeon. "Formation mechanisms of uniform nanocrystals via hot-injection and heat-up methods," *Small*, vol. 7, no. 19, pp. 2685–2702, 2011.

[47] F. Magnan, *Synthesis and Characterization of Colloidal Systems with Core/ Shell Geometry of Indium Silica Towards a Biodetection Application*, (Master's thesis, Laval University, Quebec, Canada) (2018).

[48] Y. Xia, Y. Xiong, B. Lim, and S. E. Skrabalak, Shape-controlled synthesis of metal nanocrystals: Simple chemistry meets complex physics?, *Angewand Chemie International. Ed.*, vol. 48, pp. 60–103, 2009.

[49] T. Sugimoto, "Preparation of monodispersed colloidal particles," *Adv. Colloid Interface Sci.*, vol. 28 pp. 65–108, 1987.

[50] N. E Martínez-García, *Synthesis and Functionalization of Semiconductor Nanoparticles with Potential Application to Third Generation Solar Cells*, (Master's thesis, IPN, Altamira, Mexico) (2014).

6 Electronic and Mechanical Properties of Nanoparticles

R.M. Mehra

CONTENTS

6.1 INTRODUCTION

Science and technology of nanomaterials is an interdisciplinary area of research and development and has the potential to revolutionize the ways in which materials and products can be created and nature of functionalities that can be accessed, as shown in Figure 6.1. It is already achieved a significant commercial impact, which will surely increase in the near future.

This chapter focuses on the size-dependent electrical, optical and mechanical properties of nanomaterials. Uncovering the nature of the hardness and elastic modulus of nanoparticles enables the successful design of the particles for specific

DOI: 10.1201/9781003216308-6

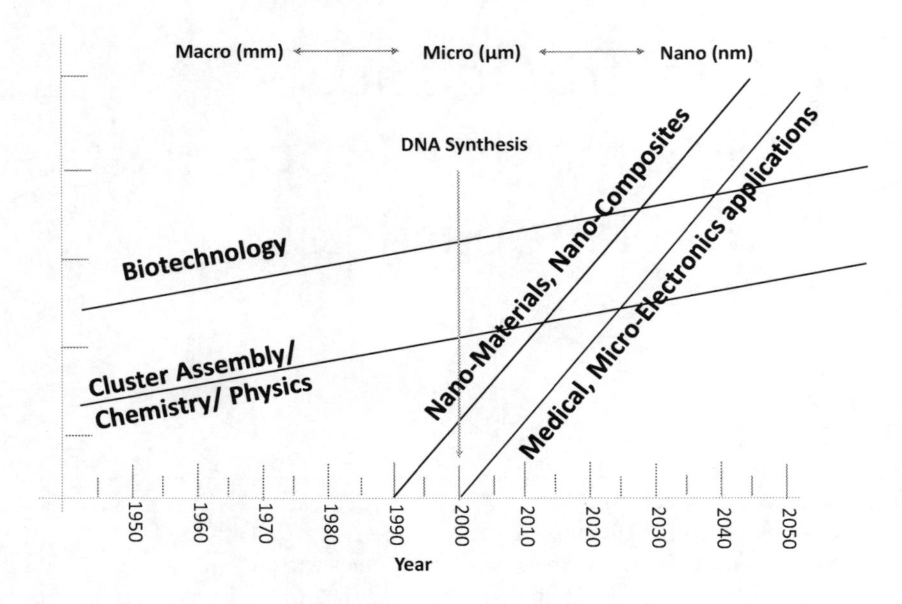

FIGURE 6.1 Science and technology from macro to nano.

applications. A brief note on the definition, classification, synthesis and methods of preparing nanostructured has also been included in the chapter for completeness.

6.2 WHAT ARE NANOMATERIALS?

A material that has at least one dimension in nanoscale below 100 nm, in a three-dimensional space, is termed as nanomaterial [1]. However, it is well known that different materials diverge from their equivalent bulk materials at different particle sizes. At nanometer scale, materials exhibit unique electrical, optical and magnetic properties which have great potential for impacts in electronics, medicine and other fields. As materials get smaller, transforming from bulk material to nanoparticle, their properties alter. It is well known now that as materials reach the nanoscale, they no longer hold the same properties as their bulk material counterparts. The uniqueness in the electrical, optical and mechanical properties of nanomaterials arises due to the volume, surface and quantum effects of nanoparticles. It is these properties that have made nanoparticles an increasingly popular field of study. To take advantage of the nanoparticles' potential, their properties need to be fully understood.

6.3 CLASSIFICATION OF NANOMATERIALS

When a given material has, at least, one dimension in nanoscale, the electronic wave vectors become quantized and the system exhibits discrete energy levels. When the dimension of the system in one dimension is lower than de Broglie wavelength (λ_{dB}), we call it a two-dimensional (2D) system. Graphene, one atom thick layer exfoliated from graphite, is an example of a 2D system as well as semiconductor superlattices.

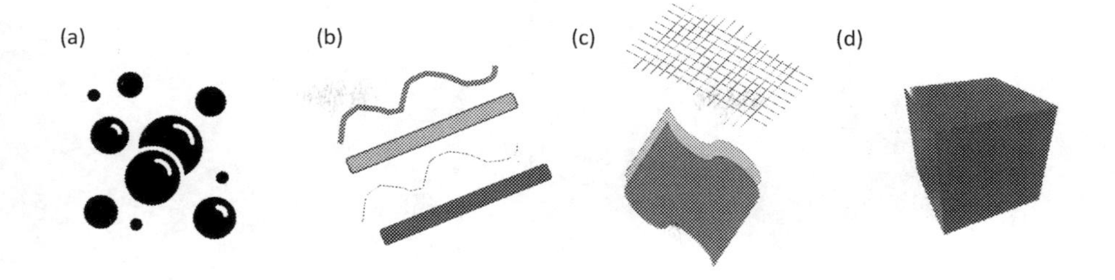

FIGURE 6.2 (a) 0D spheres and clusters. (b) 1D nanofibers, wires, and rods. (c) 2D films, plates, and networks. (d) 3D nanomaterials.

One-dimensional (1D) system is obtained when two dimensions are lower than λ_{dB} values (nanowires and nanotubes). When all three dimensions are lower than $\lambda_{dB,}$ we have 0D systems known as quantum dots (Figure 6.2). They exist in single, fused, aggregated or agglomerated forms with spherical, tubular and irregular shapes. Common types of nanomaterials include nanotubes, dendrimers, quantum dots and fullerenes.

Nanomaterials are of two types: nanostructured materials and nanostructured elements. In nanostructured materials, its structural dimensions are nanoscale [2,3]. Their electrical and mechanical properties are dependent on their size, but for many nanoparticles, these are still unknown. Researchers are committed to understanding how the nanoscale versions of materials behave, and what properties they have, to exploit them and establish new and useful future applications with new capabilities. To create successful applications, research must focus on the mechanical properties of nanomaterial, such as elasticity, hardness, interfacial adhesion and friction, and how these properties evolve with size.

6.4 METHODS FOR CREATING NANOMATERIALS AND NANOSTRUCTURES

There are several physical, chemical methods to prepare/synthesize nanomaterials. Their basic aim is to have a better control over the particle size, distribution and improved properties [4]. Top-down (size reduction from bulk materials) and bottom-up (material synthesis from atomic level) are the two main approaches for nanomaterial synthesis, as shown in Figure 6.3.

In the TOP-down process, the bulk material is made smaller by breaking up larger particles in to smaller particles using physical processes like crushing, milling or grinding. This process suffers with (a) results in non-uniformly shaped materials (b). Very difficult to obtain very small particles even with high energy consumption (c). Imperfection of the surface structure. The physical properties and surface chemistry of nanostructures and nanomaterials are significantly altered due to such imperfections.

In the bottom-up process, nanomaterial is obtained by chemical synthesis, i.e., atom-by-atom, molecule-by-molecule or cluster-by-cluster. This process is very useful for preparing most of the nanoscale materials having uniform size, shape and distribution [5].

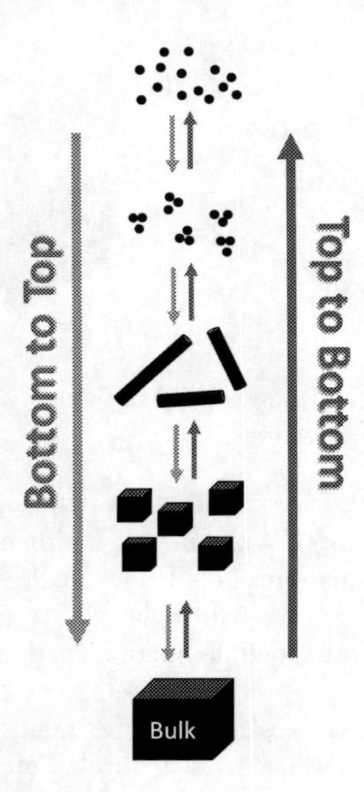

FIGURE 6.3 Top-down and bottom-up preparative methods of nanoparticles.

6.4.1 Mechanical Grinding

The top-down method is quite a simple and inexpensive method to make nanocrystalline materials in very large quantity. However, it suffers with serious problems such as:

1. Milling media and atmosphere may introduce contamination.
2. Nanocrystalline structure may not be perfect.

High energy shaker, planetary ball, or tumbler mills are the main constituents of mechanical milling apparatus. Nanoparticles are produced by the shear action during grinding. This process is not very suitable for the production of non-oxide materials. For such materials, the milling is to be performed in an inert atmosphere. Nanocrystalline elemental or compound and homogeneous alloy powders can easily be obtained by this method, as shown in Figure 6.4.

6.4.2 Wet Chemical Synthesis of Nanomaterials

Wet chemical synthesis of nanomaterials can be achieved by

1. By electrochemical etching of single crystals in an aqueous solution for producing nanomaterials (top-down approach). Synthesis of porous silicon by electrochemical etching of crystalline silicon is one of the important examples.

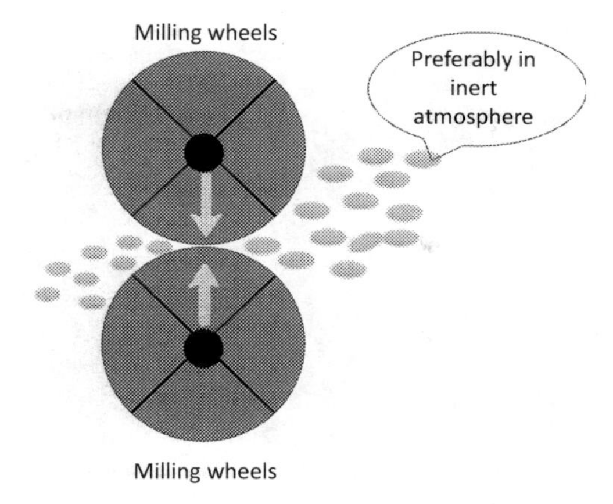

FIGURE 6.4 Mechanical grinding/milling.

2. By the sol–gel method (bottom-up). In this method, precursors containing desired materials are mixed in a controlled manner to form a colloidal solution. The sol–gel method of synthesizing nanomaterials is widely employed to prepare oxide materials. The two distinct steps, namely, formation of a colloidal suspension (**sol**) and gelation of the sol to form a network in a continuous liquid phase (**gel**), of the sol–gel process, are shown in Figure 6.5.

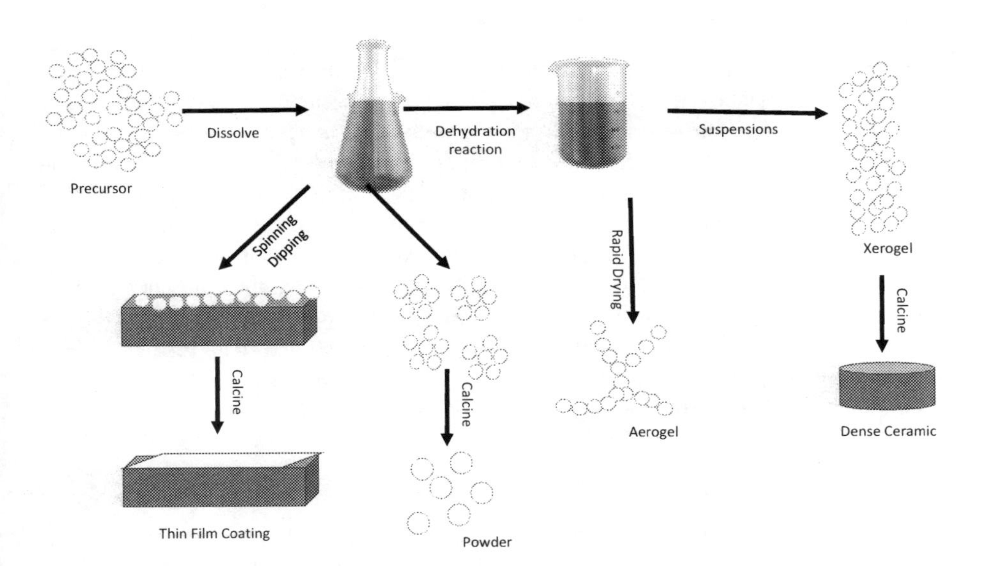

FIGURE 6.5 Distinct steps of the sol–gel process for the synthesis of nanomaterials.

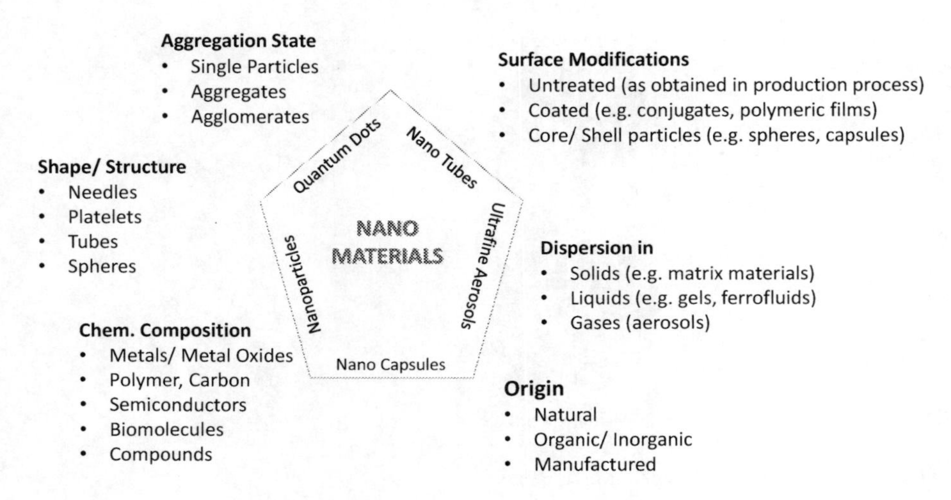

FIGURE 6.6 Characterization parameters of nanomaterials.

6.5 CHARACTERIZATION PARAMETERS OF NANOMATERIALS

The relevant parameters for the characterization of nanomaterials are shown in Figure 6.6. Nanoparticles tend to stick together and form aggregates and agglomerates under ambient conditions. These aggregates and agglomerates can have various structures such as dendritic structure, chain, or spherical with sizes normally in the micrometer range [6,7].

6.6 PROPERTIES OF NANOMATERIALS

The properties of nanomaterials are significantly different from those of bulks materials. Nanomaterials, due to their nanometer size, possess an extremely large surface area to volume ratio, large fraction of surface atoms, high surface energy, spatial confinement and reduced imperfections as compared to the corresponding bulk materials. Large surface area to volume ratio results in more "surface"-dependent material properties. These properties of nanomaterials give rise to various important applications such as metallic nanoparticles as very active catalysts and chemical sensors. Further, there is bandgap enhancement due to quantum/spatial confinement. The electronic and optical properties of the nanomaterials can be tailored by controlling the size of the nanomaterial, thereby modifying the energy band structure and charge carrier density in the materials. Quantum dots and quantum wires are very promising optoelectronics applications such as lasers and light-emitting diodes.

6.6.1 Electronic Properties of Nanomaterials

The surface-to-volume ratio increases rapidly when particle size decreases. Considering the nanoparticle as sphere of diameter D, the surface-to-volume ratio is proportional to D^{-1}. The surface atoms have different properties because they present an asymmetric interaction with atmosphere and the bulk material. Thus, the size-induced properties in nanomaterials depend basically on (a) the surface phenomena and (b) quantum confinement effects. The asymmetry causes lattice distortion due to which the properties exhibit a gradient near the surface up to a critical length l_c. The value of l_c depends on the potential interaction between the constituents of [8,9]. Carbon nanotubes and carbon nanomaterials are model systems for illustrating size-induced effects. Several authors [10–13] have discussed the role played by the surface atoms and the quantum confinement in nanomaterials. The structural and morphological properties change due to surface atoms as compared to their counterpart bulk structures leading to changes in lattice parameters, which can result in new phases and morphologies of the nanomaterial. Quantum confinement is responsible for unique electronic properties of 0D and 1D materials. Carbon nanotubes (1D) show metallic or semiconductor behavior depending on the geometry. It is expected that carbon nanomaterials will be very useful in future for the development of tissue engineering to support neuronal regeneration and for the production of multifunctional human–brain interfaces [8].

6.6.2 Electrical Properties

Atoms have discreet energy states. In having several atoms together, their atomic orbitals split and produce a number of molecular orbitals proportional to the number of atoms. In solids (bulk materials), there are so many atoms; the difference in energy between them becomes very small, so that forming bands of energy rather than the discrete energy levels, so nearly continuous bands of states are formed. Valence and conduction bands are formed with a bandgap between them. Quantum confinement (the spatial confinement of electron–hole pairs (excitons) in one or more dimensions within the material) leads to a collapse of the continuous energy bands of a bulk material into discrete, atomic-like energy levels, so a discrete absorption spectrum, compared to the continuous absorption spectrum of a bulk semiconductor.

Quantum confinement is more prominent in semiconductors because they have a bandgap. Metals do not have a bandgap. The regime of quantum confinement length scale ranges from 2 to 25 nm for typical semiconductors of groups IV, III–V and II–VI.

Carbon nanotubes are a good example to study the dependence of conductivity upon dimensions like diameter or area of cross section and twist in the rod. It is found that conductivity changes with change in the area of cross section of the carbon nanotube. It is also observed that conductivity also changes when some shear force (in simple terms twist) is given to nanotube. Further, the conductivity of a multiwalled carbon nanotube is different than that of single nanotube of the same dimensions. The carbon nanotubes can act as a conductor or semiconductor.

6.6.3 Optical Properties

Nanomaterials have very useful and fascinating optical properties which depend on parameters such as feature size, shape, surface characteristics. Based on the optical properties, nanomaterials have several applications such as optical detector, laser,

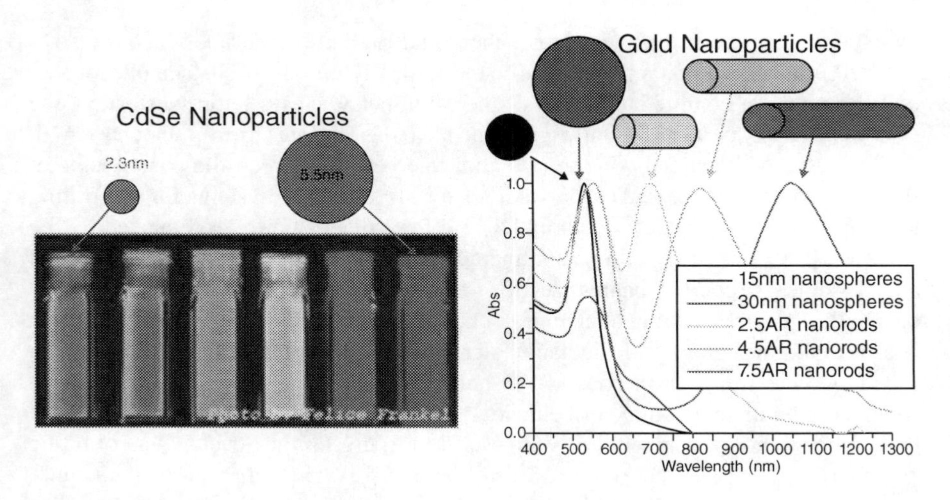

FIGURE 6.7 (a) Fluorescence emission of (CdSe) quantum dots of various sizes and (b) absorption spectra of various sizes and shapes of gold nanoparticles. (Reproduced from S. Eustis and M. A. El-Sayed, *Chem. Soc. Rev.*, 2006, **35**, 209-217 with permission from the Royal Society of Chemistry.)

sensor, imaging, display, solar cell, photocatalysis, photo-electrochemistry and biomedicine.

Figure 6.7a shows the change in the optical properties of CdSe semiconductor nanoparticles as the size alters from 2.3 to 5.5 nm. Figure 6.7b illustrates the absorption spectra of various sizes and shapes of gold nanoparticles [14].

6.6.4 Size Effect on Optical Properties

The role of surface plasmons and quantum confinement effects on the electrical properties of nanomaterials has already been discussed in previous sections. Metals in their bulk form have overlapping valance and conduction bands. In the nanometer range, metal nanoparticles can develop a bandgap and can exhibit semiconducting or insulating properties. On decreasing the size, the quantum confinement effect leads to: (a) increase in bandgap energy and (b) band levels get quantized (discrete). In semiconductor quantum dots, a blue shift to higher energy is observed in optical absorption and emission as the size of the dots decreases.

The effective bandgap of nanoparticle of radius r is given by Burstein–Moss effect [15,16].

$$E_g^{nano} = E_g^{bulk} + \frac{h^2}{8m_0r^2}\left(\frac{1}{m_e^*} + \frac{1}{m_h^*}\right) - \frac{1.8e^2}{4\pi\varepsilon\varepsilon_0 r}$$

E_g^{nano} is the effective band gap energy of particle of radius r
ε is the dielectric constant (bulk)
m_e is the mass of free electron

me^* is the effective mass of electron
m_h^* is the effective mass of hole

Semiconductor compound nanomaterials (SCNs) such as CdSe, CdTe, ZnS, ZnSe and ZnTe have extensive applications due to their size-dependent band gap. Madan Singha et al. [17] have developed a model, based on the cohesive energy of the nanocrystals compared to the bulk crystals, and have formulated the size- and shape-dependent band gap energy of SCNs in different sizes and shapes, viz., spherical nanosphere, nanowires and nanofilms. This theoretical estimation of the energy band gap is reasonably consistent with the available experimental data SCNs.

The structures, densities of states, absorption spectra and optical gaps of spherical ZnSe and CdTe semiconductor quantum dots in the size range 0.3–2.2 nm have been estimated by Sufian Alnemrat et al. [18] using the first-principles density functional and time-dependent density functional computational methods. The computed optical gaps in surface-passivated ZnSe and CdTe quantum dots of various sizes are found to be consistent with the quantum confinement effect.

Several researchers have been experimentally studying the variation of optical band gap of various SCNs having different size and structures. A detailed description of method of synthesis, characterization and measurements has been cited in the literature for different materials such as CdSe nanospheres/films [19], CdS nanoclusters [20], CdSe nanowires [21], CdTe nanocrystalline films [22], CdTe quantum dots [23], CdTe nanowires [24], ZnS nanospheres [25], [26], ZnSe nanospheres [27,28] and ZnTe nanospheres, nanowires and nanofilms [29].

Sasha Corer and Gary Hodes [19] have synthesized studied CdSe films by chemical solution deposition and have shown that these films possess a nanocrystalline structure and exhibit quantum size effects due to the small crystal size, which results in a blue shift of the optical spectra. Li et al. [21] have reported the variation of band gaps of CdSe quantum dots with diameters varying from 3.0 to 6.5 nm. An empirical pseudopotential calculation performed on CdSe quantum dot nanorods has been presented by them to provide a qualitative explanation for the dependence of band gap on width and length. Yitzhak Mastai and Gary Hodes [22] have described the deposition of CdTe nanocrystalline films by electrodeposition from a dimethylsulfoxide solution of tri-(n-butyl) phosphine telluride and cadmium perchlorate at 100°C and showed that these films exhibit size quantization. They obtained stoichiometric films using pulse-reverse plating with an average crystal size which could be controlled, by varying the pulse duty cycle, from 4 to 7 nm. All the films exhibited blue shifts which correlated with the crystal size and could be explained by quantum confinement effect. Masumoto and Sonobe [23] have shown that size-dependent quantized electronic levels of CdTe quantum dots shift monotonously without any crossing or anticrossing, reflecting the rather simple valence-band structure of CdTe. Li and Wang [24] systematically calculated the electronic states of group III–V (GaAs, InAs, InP, GaN, AlN and InN) and group II–VI (CdSe, CdS, CdTe, ZnSe, ZnS, ZnTe and ZnO) quantum dots and wires. They reported exciton energies as functions of the quantum dot sizes and quantum wire diameters for all the above materials and found a good agreement between calculated and experimental results. Rossetti et al. [25] have observed quantum size effects on the fundamental band gap of cubic CdS crystallites in the absence of adsorbed polymeric molecules. They explained their results in terms of the crystallite molecular orbitals and an elementary confined electron and hole model. Inoue et al. [26] have synthesized ZnS microcrystals ~4 nm diameters with a narrow size distribution modified with ethanethiol, butanethiol and octanethiol. The bandgap was found to be 3.96 eV for the octanethiol capping and 3.94 and 3.96 eV for the capping with ethanethiol and butanethiol, respectively. Mazher et al. [27] have synthesized self-organized Q-ZnSe films with narrow particle size distribution using a wet chemical route. The dependence of strain, morphology and luminescence properties on the particle size have been investigated using X-ray diffraction, atomic force microscopy, transmission electron microscopy and optical absorption and luminescence spectroscopy. Blue shift in the optical absorption edge and PL maxima has been observed with the variation in particle size. Margaret A. Hines and Philippe Guyot-Sionnest [28] have synthesized relatively monodisperse, highly luminescent ZnSe nanocrystals in a hexadecylamine/trioctylphosphine coordination solvent with tunable sample sizes. Pure band-edge fluorescence size tunable between 2.8 and 3.4 eV is obtained at room temperature. Jun et al. [29] have obtained controlled ZnTe nanocrystals, in shape and size, using a simple one-pot synthesis process with [Zn(TePh)2] *Tetramethylethylenediamine* (TMEDA) as a monomeric molecular precursor. The shape and size have been controlled by varying the growth temperature or the templating surfactants and observed the quantum size effects.

6.7 MECHANICAL PROPERTIES OF MATERIALS

The mechanical properties are a measure of strength and lasting characteristics of the material under the action of external forces. Most of the bulk materials are characterized for their mechanical properties such as strength, elasticity, plasticity, hardness, toughness, brittleness, stiffness, ductility, malleability, cohesion, impact strength, fatigue and creep. In case of nanomaterials, the study of following properties is of interest.

- Elastic properties
- Hardness and strength
- Ductility and toughness
- Creep of nanocrystalline materials and ductility

6.7.1 Mechanical Properties of Nanomaterials

Mechanical properties of nanoparticles deals with the influence of porosity and grain size, superplasticity, polymer-based nanocomposites filled with platelets and carbon nanotube-based composites. Filling polymers with nanoparticles or nanorods and nanotubes, respectively, leads to significant improvements in their mechanical properties. The larger the particles of the filler or agglomerates, the poorer are the properties obtained. Although the best composites are those filled with nanofibers or nanotubes, experience teaches but such composites generally have the least ductility. Composites consisting of a polymer matrix and defoliated phyllosilicates exhibit excellent mechanical and thermal properties.

The nanosize of the nanomaterials tends to modify many of the mechanical properties of nanostructured materials from the bulk materials. An enhancement of mechanical properties of nanomaterials generally results from structural perfection of the materials [3]. Improvements in mechanical property have resulted in major interest in nanocomposite in various automotive and general industrial applications [30,31].

6.7.2 Elastic Properties

Early measurements performed on nanostructured materials prepared by gas condensation method depicted lower values of the elastic constants like Young's modulus as compared to the values for conventional grain size materials. While many reasons were attributed for the low values of *E*, Kristc and coworkers suggested that the presence of extrinsic defects such as pores and cracks were responsible for the low values of *E* in nanostructured materials compacted from powders. However, Wong et al. observed that nanocrystalline nickel powder produced by electroplating with negligible porosity levels had an *E* value comparable to fully dense conventional grain size nickel. Subsequent work on porosity-free materials has supported these conclusions, and it is now believed that the intrinsic elastic moduli of nanostructured materials are essentially the same as those for conventional grain size materials until the grain size becomes very small, less than 5 nm. This is illustrated for nanocrystalline Fe prepared by mechanical attrition and measured by a nano-indentation technique [31].

6.7.3 Hardness and Strength

Among many of the novel mechanical properties of nanostructured materials, high hardness has been discovered from many nanostructured material systems. A variety of super hard nanocomposites can be made of nitrides, borides and carbides by plasma-induced chemical and physical vapor deposition. In the appropriately synthesized systems, the hardness of the nanocomposite exceeds significantly than that of mixtures in bulk. Superhardness also comes from pure nanoparticles. For example, Gerbericha reports the superhardness from the nearly spherical, defect-free silicon nanospheres with diameters from 20 to 50 nm of up to 50 GPa, fully four times greater than the bulk silicon [32,33].

Hardness and strength of conventional grain size materials (grain diameter, $d > 1$ μm) is a function of grain size. The dependence of yield stress on grain size in metals is well established in the conventional polycrystalline range (micrometer and larger-sized grains). Yield stress, for materials with grain size *d*, is found to follow the Hall–Petch relation [34]. This gives an inverse relationship between delta yield strength and grain size to some power, *x*.

6.7.4 Ductility and Toughness

In the conventional grain size (>1 μm) regime, usually a reduction in grain size leads to an increase in ductility. Thus, one should expect a ductility increase as the grain size is reduced to nanoscale. On a very basic level, mechanical failure, which limits ductility, is an interplay or competition between dislocations and cracks (Thomson [35]). Nucleation and propagation of cracks can be used as the explanation for the fracture stress dependence on grain size (Nagpal and Baker [36]). Grain size refinement can make crack propagation more difficult and, therefore, in conventional grain size material, increase the apparent fracture toughness. However, the large increases in yield stress (hardness) observed in nanocrystalline materials suggest that fracture stress can be lower than yield stress and therefore result in reduced ductility. Koch [37] identified three major sources of limited ductility in nanocrystalline materials, namely:

i. Artifacts from processing (e.g., pores);
ii. Tensile instability;
iii. Crack nucleation or shear instability.

The results of ductility measurements on nanocrystalline metals are mixed and are sensitive to flaws and porosity, surface finish and method of testing (e.g., tension or compression testing). In tension, for grain sizes <30 nm, essentially brittle behavior has been observed for pure nanocrystalline metals that exhibit significant ductility when the grain size is conventional.

6.8 CREEP OF NANOCRYSTALLINE MATERIALS

Creep in coarse-grained materials has been widely studied for approximately one century and accurate models exist to capture deformation features and to explain mechanisms involved therein. Creep in nanocrystalline materials has been studied only in recent years owing to several complications involved.

The limitation of synthesizing bulk nanomaterial free of defects (porosity and impurities) with uniform grain size distribution could provide reliable data to explain the deformation process.

The significant increase in the volume fraction of grain boundaries and intercrystalline defects such as triple lines and quadruple junctions renders the creep mechanism complicated and leads to associated challenges in developing a model that could explain the deformation process.

Grain growth occurs at much lower temperature as compared to coarse-grained materials limiting the testing temperatures to a low fraction of the melting point.

Diffusion creep is considered to be significant because the volume fraction of grain boundaries is high [38–41]. There is no evidence that the grain boundaries in nanostructured materials are significantly different from the ones in conventional polycrystals. One can assume safely a boundary a few atomic distances thick. However, it is safe to assume that the number of grain-boundary ledges is reduced because of the size. The reorientation/rotation of the nanosized boundaries is also significantly enhanced in comparison with conventional boundaries.

Yield stress: The yield strength of nanocrystalline materials has been measured and there is a consensus that the H–P relationship breaks down with a decrease in slope in the 1 μm–100 nm range.

6.9 DUCTILITY

Nanocrystalline metals are characterized by a low work-hardening rate, which is a direct consequence of the low density of dislocations encountered after plastic deformation. This low work-hardening rate leads to tensile instability and a low tensile ductility. There are reports of increased ductility in nanocrystalline metals; the increased ductility that is exhibited in some cases comes, basically, from the inhibition of shear localization.

Creep: There are several studies in the literature [31], reporting a decrease in creep resistance by virtue of Coble creep in which the creep rate is proportional to d-3. Nevertheless, conflicting results report a creep resistance much higher than the Coble prediction. This could be due to the contamination of the grain boundaries with impurities, which act as "brakes" to grain-boundary sliding.

REFERENCES

[1] C. Wu, W. X. Ma, Y. P. Chen, Y. Li, Y. Chen, and J. Li, "Nano materials and its application in space," *Appl. Mech. Mater.*, 2014, vol. 482, pp. 34–37, doi: 10.4028/www.scientific.net/AMM.482.34.

[2] R. Dingreville, J. Qu, and M. Cherkaoui, "Surface free energy and its effect on the elastic behavior of nano-sized particles, wires and films," *J. Mech. Phys. Solids*, Aug. 2005, vol. 53, no. 8, pp. 1827–1854, doi: 10.1016/j.jmps.2005.02.012.

[3] Y. Koutsawa, S. Tiem, W. Yu, F. Addiego, and G. Giunta, "A micromechanics approach for effective elastic properties of nano-composites with energetic surfaces/interfaces," *Compos. Struct.*, Jan. 2017, vol. 159, pp. 278–287, doi: 10.1016/j.compstruct.2016.09.066.

[4] S. Shibata, K. Aoki, T. Yano, and M. Yamane, "Preparation of silica microspheres containing Ag nanoparticles," *J. Sol-Gel Sci. Technol.*, 1998, vol. 11, no. 3, pp. 279–287, doi: 10.1023/A:1008610413657.

[5] H. Hahn, "Gas phase synthesis of nanocrystalline materials," *Nanostructured Mater.*, Jan. 1997, vol. 9, no. 1–8, pp. 3–12, doi: 10.1016/S0965-9773(97)00013-5.

[6] P. Sharma and M. Bhargava, "Applications and characteristics of nanomaterials in industrial environment," *Int. J. Civil, Struct. Environ. Infrastruct. Eng. Res. Dev.*, 2013, vol. 3, no. 4, pp. 63–72, [Online]. Available: http://www.tjprc.org/view_paper.php?id=2701.

[7] H. Hofmann, "Advanced nanomaterials ours support," *Powder Technol.*, vol. 1, pp. 1–270, 2009.

[8] *E. Roduner,* Nanoscopic Materials. Royal Society of Chemistry, *Cambridge, 2007.*

[9] G. Cao, Nanostructures and Nanomaterials. Published By Imperial College Press And Distributed By World Scientific Publishing Co., 2004.

[10] A. G. S. Filho and S. B. Fagan, "Nanomaterials Properties," in Nanostructured Materials for Engineering Applications, Springer, Berlin Heidelberg, 2011, pp. 5–22.

[11] A. J. Siitonen, D. A. Tsyboulski, S. M. Bachilo, and R. B. Weisman, "Surfactant-dependent exciton mobility in single-walled carbon nanotubes studied by single-molecule reactions," *Nano Lett.*, May 2010, vol. 10, no. 5, pp. 1595–1599, doi: 10.1021/nl9039845.

[12] R. Sanjinés, M. D. Abad, C. Vâju, R. Smajda, M. Mionić, and A. Magrez, "Electrical properties and applications of carbon based nanocomposite materials—An overview," *Surf. Coatings Technol.*, Nov. 2011, vol. 206, no. 4, pp. 727–733, doi: 10.1016/j.surfcoat.2011.01.025.

[13] R. Rauti, M. Musto, S. Bosi, M. Prato, and L. Ballerini, "Properties and behavior of carbon nanomaterials when interfacing neuronal cells—How far have we come?," *Carbon*, Mar. 2019, vol. 143. Pergamon, pp. 430–446, doi: 10.1016/j.carbon.2018.11.026.

[14] S. Eustis and M. A. El-Sayed, "Why gold nanoparticles are more precious than pretty gold—Noble metal surface plasmon resonance and its enhancement of the radiative and nonradiative properties of nanocrystals of different shapes," *Chem. Soc. Rev.*, 2006, vol. 35, no. 3, pp. 209–217, doi: 10.1039/b514191e.

[15] E. Burstein, "Anomalous optical absorption limit in InSb [4]," *Physical Review*, Feb. 1954, vol. 93, no. 3. American Physical Society, pp. 632–633, doi: 10.1103/PhysRev.93.632.

[16] T. S. Moss, "The interpretation of the properties of indium antimonide," *Proc. Phys. Soc. Sect. B*, Oct. 1954, vol. 67, no. 10, pp. 775–782, doi: 10.1088/0370-1301/67/10/306.

[17] M. Singh, M. Goyal, and K. Devlal, "Size and shape effects on the band gap of semiconductor compound nanomaterials," *J. Taibah Univ. Sci.*, Jul. 2018, vol. 12, no. 4, pp. 470–475, doi: 10.1080/16583655.2018.1473946.

[18] S. Alnemrat, Y. Ho Park, and I. Vasiliev, "Ab initio study of ZnSe and CdTe semiconductor quantum dots," *Phys. E Low-Dimensional Syst. Nanostructures*, Mar. 2014, vol. 57, pp. 96–102, doi: 10.1016/j.physe.2013.10.037.

[19] S. Gorer and G. Hodes, "Quantum size effects in the study of chemical solution deposition mechanisms of semiconductor films," *J. Phys. Chem.*, 1994, vol. 98, no. 20, pp. 5338–5346, doi: 10.1021/j100071a026.

[20] T. Vossmeyer et al., "CdS nanoclusters—Synthesis, characterization, size dependent oscillator strength, temperature shift of the excitonic transition energy, and reversible absorbance shift," *J. Phys. Chem.*, 1994, vol. 98, no. 31, pp. 7665–7673, doi: 10.1021/j100082a044.

[21] L. S. Li, J. Hu, W. Yang, and A. P. Alivisatos, "Band gap variation of size- and shape-controlled colloidal cdse quantum rods," *Nano Lett.*, Jul. 2001, vol. 1, no. 7, pp. 349–351, doi: 10.1021/nl015559r.

[22] Y. Mastai and G. Hodes, "Size quantization in electrodeposited CdTe nanocrystalline films," *J. Phys. Chem. B*, Apr. 1997, vol. 101, no. 14, pp. 2685–2690, doi: 10.1021/jp963069v.

[23] Y. Masumoto and K. Sonobe, "Size-dependent energy levels of CdTe quantum dots," *Phys. Rev. B - Condens. Matter Mater. Phys.*, Oct. 1997, vol. 56, no. 15, pp. 9734–9737, doi: 10.1103/PhysRevB.56.9734.

[24] J. Li and L. W. Wang, "Band-structure-corrected local density approximation study of semiconductor quantum dots and wires," *Phys. Rev. B - Condens. Matter Mater. Phys.*, Sep. 2005, vol. 72, no. 12, p. 125325, doi: 10.1103/PhysRevB.72.125325.

[25] R. Rossetti, R. Hull, J. M. Gibson, and L. E. Brus, "Excited electronic states and optical spectra of ZnS and CdS crystallites in the ≈15 to 50 Å size range—Evolution from molecular to bulk semiconducting properties," *J. Chem. Phys.*, Aug. 1985, vol. 82, no. 1, pp. 552–559, doi: 10.1063/1.448727.

[26] H. Inoue, N. Ichiroku, T. Tsukasa, T. Sakata, H. Mori, and H. Yoneyama, "Photoinduced electron transfer from zinc sulfide microcrystals modified with various alkanethiols to methyl viologen," *Langmuir*, Dec. 1994, vol. 10, no. 12, pp. 4517–4522, doi: 10.1021/la00024a022.

[27] J. Mazher, A. K. Shrivastav, R. V. Nandedkar, and R. K. Pandey, "Strained ZnSe nanostructures investigated by X-ray diffraction, atomic force microscopy, transmission electron microscopy and optical absorption and luminescence spectroscopy," *Nanotechnology*, Mar. 2004, vol. 15, no. 5, pp. 572–580, doi: 10.1088/0957-4484/15/5/030.

[28] M. A. Hines and P. Guyot-Sionnest, "Bright UV-blue luminescent colloidal ZnSe nanocrystals," *J. Phys. Chem. B*, May 1998, vol. 102, no. 19, doi: 10.1021/jp9810217.

[29] Y. W. Jun, C. S. Choi, and J. Cheon, "Size and shape controlled ZnTe nanocrystals with quantum confinement effect," *Chem. Commun.*, Jan. 2001, no. 1, pp. 101–102, doi: 10.1039/b008376n.

[30] O. I. Sekunowo, S. I. Durowaye, and G. I. Lawal, "An overview of nano-particles effect on mechanical properties of composites," *World Acad. Sci. Eng. Technol. Int. J. Anim. Vet. Sci.*, Nov. 2015, vol. 9, no. 1, pp. 1–7, doi: 10.5281/ZENODO.1097327.

[31] T. Hasan, "Mechanical properties of nanomaterials—A review," *Int. J. Adv. Res. Innov. Ideas Educ.*, 2016, vol. 2, no. 4, pp. 1131–1138, doi: 16.0415/IJARIIE–3018.

[32] S. Veprek and A. S. Argon, "Mechanical properties of superhard nanocomposites," *Surf. Coatings Technol.*, Sep. 2001, vol. 146–147, pp. 175–182, doi: 10.1016/S0257-8972(01)01467-0.

[33] W. W. Gerberich et al., "Superhard silicon nanospheres," *J. Mech. Phys. Solids*, Jun. 2003, vol. 51, no. 6, pp. 979–992, doi: 10.1016/S0022-5096(03)00018-8.

[34] S. N. Naik and S. M. Walley, "The Hall–Petch and inverse Hall–Petch relations and the hardness of nanocrystalline metals," *J. Mater. Sci.*, Nov. 2020, vol. 55, no. 7. Springer, pp. 2661–2681, doi: 10.1007/s10853-019-04160–w.

[35] R. Abbaschian, L. Abbaschian, and R. E. Reed-Hill, *Physical Metallurgy Principles*. 4th Edn. pp. 1–769, 2009, Accessed: Sep. 19, 2021. [Online]. Available: https://www.elsevier.com/books/physical-metallurgy/cahn/978-0-444-89875-3.

[36] P. Nagpal and I. Baker, "The effect of grain size on the room-temperature ductility of NiAl," *Scr. Metall. Mater.*, Dec. 1990, vol. 24, no. 12, pp. 2381–2384, doi: 10.1016/0956-716X(90)90097–Z.

[37] C. C. Koch, "Ductility in nanostructured and ultra fine-grained materials—Recent evidence for optimism," *J. Metastable Nanocryst. Mater.*, 2003, vol. 18. Trans Tech Publications Ltd, pp. 9–20, doi: 10.4028/www.scientific.net/JMNM.18.9.

[38] J. Deng, D. L. Wang, Q. P. Kong, and J. P. Shui, "Stress dependence of creep in nanocrystalline Ni–P alloy," *Scr. Metall. Mater.*, Feb. 1995, vol. 32, no. 3, pp. 349–352, doi: 10.1016/S0956-716X(99)80063-6.

[39] H. Hahn and R. S. Averback, "Low-temperature creep of nanocrystalline titanium(IV) oxide," *J. Am. Ceram. Soc.*, Nov. 1991, vol. 74, no. 11, pp. 2918–2921, doi: 10.1111/j.1151-2916.1991.tb06863.x.

[40] G. W. Nieman, J. R. Weertman, and R. W. Siegel, "Mechanical behavior of nanocrystalline Cu and Pd," *J. Mater. Res.*, 1991, vol. 6, no. 5, pp. 1012–1027, doi: 10.1557/JMR.1991.1012.

[41] P. G. Sanders, M. Rittner, E. Kiedaisch, J. A. Weertman, H. Kung, and Y. C. Lu, "Creep of nanocrystalline Cu, Pd, and Al–Zr," *Nanostructured Mater.*, Jan. 1997, vol. 9, no. 1–8, pp. 433–440, doi: 10.1016/S0965-9773(97)00096-2.

7 Various Characterization Methods

F. Ruiz-Perez, S.J. Montiel-Perales, R.V. Tolentino-Hernandez, M.S. Ovando-Rocha, Felipe Caballero-Briones, and F.J. Espinosa-Faller

CONTENTS

DOI: 10.1201/9781003216308-7

7.1 INTRODUCTION

In the center of Materials Science lies the characterization. Materials are designed following a purpose, i.e., an application, an entire device or the quest for fundamental knowledge; the synthesis variations and further modifications of the as-prepared materials, look for the modulation of the structure and stoichiometry, which in turn modifies the electronic structure and then the electronic, magnetic, optic and morphology properties. Some of the properties can be studied while the preparation or postpreparation treatments are in course; other can be assessed while the materials or devices are operating or being influenced by external conditions such as light, electric or magnetic fields, force, heat and so on. The *ex-situ*, *in-situ* or *in-operando* properties are compared with the desired functionality and then the cycle preparation-structure-properties-functionality continues. Thus, the understanding of the fundamentals and capabilities of the characterization techniques is of capital importance. This chapter describes some instrumental techniques for materials characterization. The principles and applications of Scanning Probe Microscopies, Raman spectroscopy (RS), X-ray photoelectron spectroscopy (XPS) and X-ray absorption spectroscopy (XAS) will be revised, with emphasis in experimental details and selected experiments, to give the reader both the general and specific panorama of these important techniques.

7.2 SCANNING TUNNELING MICROSCOPY

Nowadays, the scanning probe microscopy (SPM) family of techniques is an important tool for the characterization of materials in nanoscience. In general, in SPM, the principle is to move a tip in the close vicinity of the surface of a sample to measure various properties with molecular or even atomic scale. The first technique of SPM family, the scanning tunneling microscope (STM), was developed in 1981 by Gerd Binning and Heinrich Rohrer; this invention allowed both to win the Nobel Prize in Physics of 1986. Figure 7.1a shows the essential components. Basically, a probe tip usually made of W or Pt-Ir alloy, mounted on a piezoelectric system to provide three-axis (X, Y, Z) movement, a conductive sample, the feedback control system and the processing and display system equipment [1]. This technique provides atomic resolution images of metal and semiconductor surfaces, like double-layer graphene on SiO_2 ($G/G/SiO_2$) in Figure 7.1b.

7.2.1 Theory: Principle of Tunneling

When a voltage between the tip and the sample is applied, approaching the tip toward the sample without direct contact (approximately 0.5–1 nm) induces a current flow, this is called *tunneling current.* At this separation distance, in base to quantum

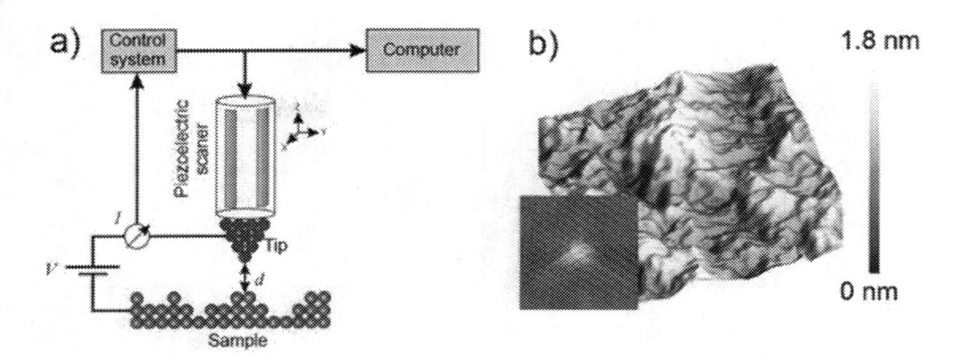

FIGURE 7.1 (a) Schematic of scanning tunneling microscope. (b) STM topography of a double-layer graphene on SiO_2(G/G/SiO_2). (Image taken as original, Y. Jiang et al., "Inducing Kondo screening of vacancy magnetic moments in graphene with gating and local curvature," Nat. Commun., vol. 9:2349, no. 1, pp. 1–7, 2018) [2].

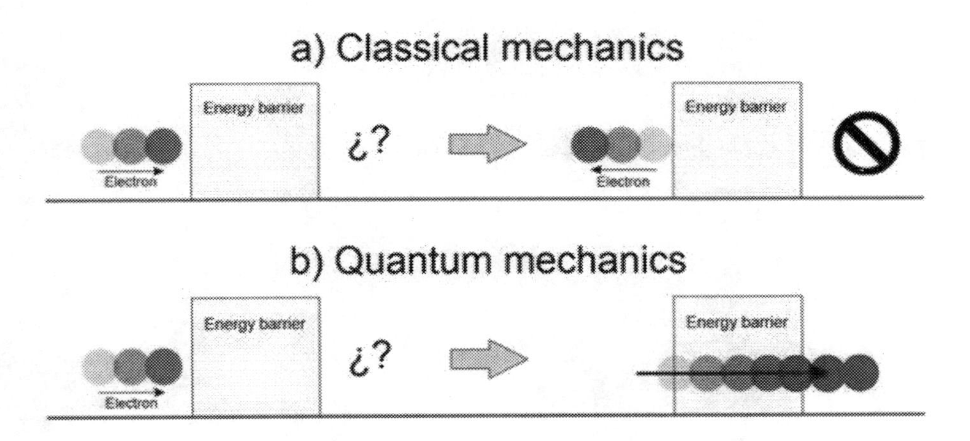

FIGURE 7.2 Difference between classical and quantum electron theory.

mechanics, small particles exhibit wave properties, thus allowing a finite probability that the electron can tunnel the potential barrier between the tip and the sample surface. In classical mechanics, this is impossible as depicted in Figure 7.2.

STM needs a conductive sharp tip close of a conductive sample which creates a metal-insulator-metal junction. In a metal, the electron energy level fills the energy levels up to the Fermi energy. To eject an electron from the metal, an additional energy Φ (called work function) is necessary. With a small tip to sample separation and applying a voltage, the electrons can "tunnel" the barrier and travel from sample to tip or vice versa in function of the polarity voltage.

In section A in Figure 7.3, the electron can be reflected by the barrier, or tunnel through section B, or complete the tunneling process in section C. Section A is described by Schrödinger's equation:

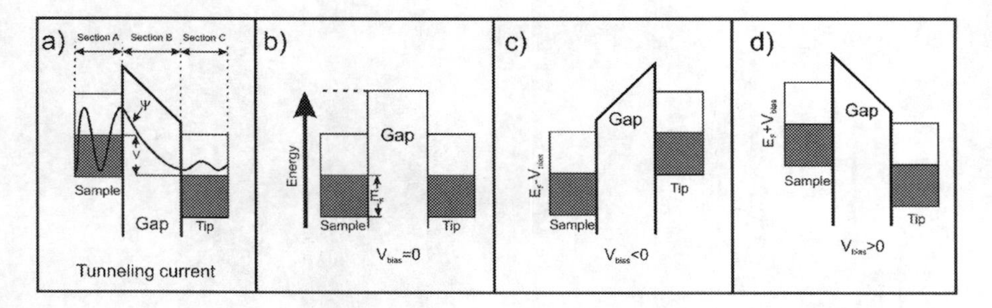

FIGURE 7.3 Metal-insulator-metal junction schematic. (a) Tunneling current generation applying a bias voltage (eV) between sample-tip. (b) With a small sample-tip distance, the wave (electrons) can tunnel to the tip and produce a "tunneling current." (c) At $V_{bias} = 0$, the electrons cannot move in any direction from the Fermi level (gray area). (c) When voltage < 0, the tunneling electron direction is from the occupied state of the tip to unoccupied (blank space) state of the sample. (d) On the other hand, at voltage > 0, the electrons of the occupied state of the sample can tunnel to the unoccupied state of the tip.

$$-\frac{h^2}{2m}\frac{d^2\psi_A}{d^2x} = E\psi_A \tag{7.1}$$

The equation that describes the wave function in section B is

$$-\frac{h^2}{2m}\frac{d^2\psi_{SA}}{d^2x} + V_0 = E\psi_{SB} \tag{7.2}$$

Finally, for the section C:

$$-\frac{h^2}{2m}\frac{d^2\psi_{SC}}{d^2x} = E\psi_{SC} \tag{7.3}$$

where m and E are the mass and energy of an electron, respectively. The expressions for the traveling wave in the three sections are:

$$\psi_{SA} = e^{ikx} \tag{7.4}$$

$$\psi_{SB} = B'e^{ikx} + C'e^{ikx} = Be^{-\xi x} + Ce^{-i\xi x} \tag{7.5}$$

$$\psi_{SC} = De^{ikx} \tag{7.6}$$

where the wave vector k is $\sqrt{\frac{2mE}{\hbar^2}}$ and ξ is equal to $\sqrt{-k'^2} = \sqrt{\frac{[2m(V_0 - E)]}{\hbar^2}}$.

$$\hbar = \frac{Plank's\ constant}{2\pi}$$

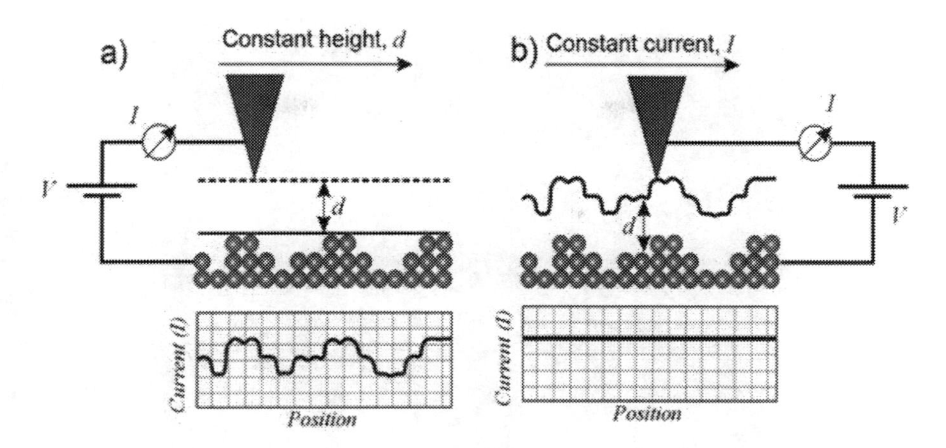

FIGURE 7.4 STM operation modes. (a) Constant height. (b) Constant current.

The tunneling current value is proportional to a tip-sample separation (the distance varies 0.1 Å the tunneling current change by 20%). The STM images are a representation of its magnitude in the *X–Y* plane. Due to the current-distance dependence, the information it provides is made up of topographic changes and surface electronic anisotropy [3].

7.2.2 Operation Modes

There are two configurations in which STM can be used, as shown in Figure 7.4. The first, and most used operation mode is the constant current mode. In this mode, the value of the tunneling current is kept constant by modifying the height of the tip over the surface of the sample by using the feedback control system and the piezoelectric system. The second mode of operation is constant height. In this mode, as the name indicates, the height of the tip is kept constant while the surface of the sample is scanned and the variation in the value of the tunneling current is measured at the chosen bias voltage; the feedback control system is not in operation in this mode of operation [4,5].

Nanolithography is another STM application that allows the manipulation of atoms. For example, in Figure 7.5a and b, a surface oxygen atom is transferred from the original adatom to the neighbor, producing a displaced dark site. In Figure 7.5c and d, the removal of the surface oxygen can be observed [6].

7.3 SCANNING ELECTROCHEMICAL MICROSCOPY

Scanning electrochemical microscopy (SECM) is part of the SPM family. SECM uses an ultramicroelectrode (UME) to scan very close to the sample surface and measures the current through the UME when it is moving in a solution closely to a substrate which can be biological material, polymers, glass, metals, or liquids. It is used to scan interfaces obtaining topographical, chemical or physical characterization. The SECM has four basic elements, as shown in Figure 7.6a: UME tip, biopotentiostat,

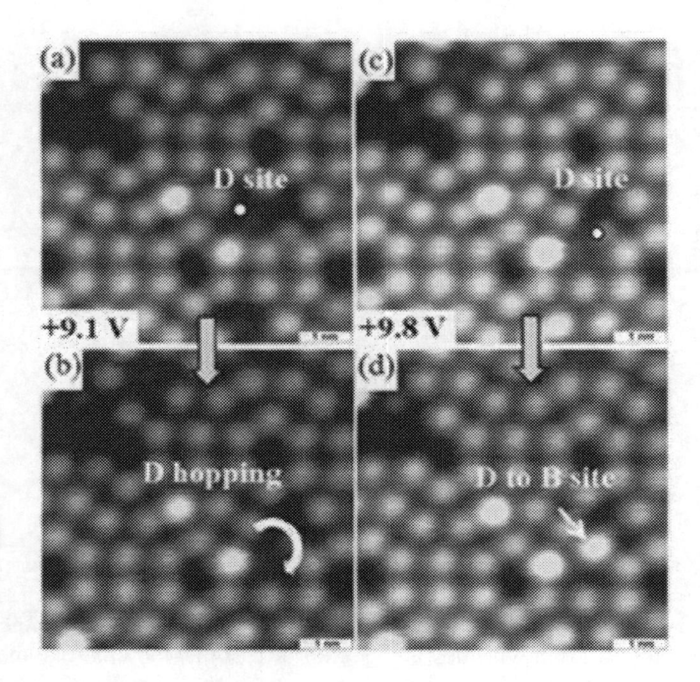

FIGURE 7.5 Manipulation of atoms. (a) and (b) Dark site displaced to a neighboring adatom (+9.1 V). (c) and (d) Site transformation into a bright site (+9.8 V). (Adapted from: D. Kaya, R. J. Cobley, and R. E. Palmer, "Combining scanning tunneling microscope (STM) imaging and local manipulation to probe the high dose oxidation structure of the Si(111)-7 × 7 surface," Nano Res., vol. 13, no. 1, pp. 145–150, 2020) [6].

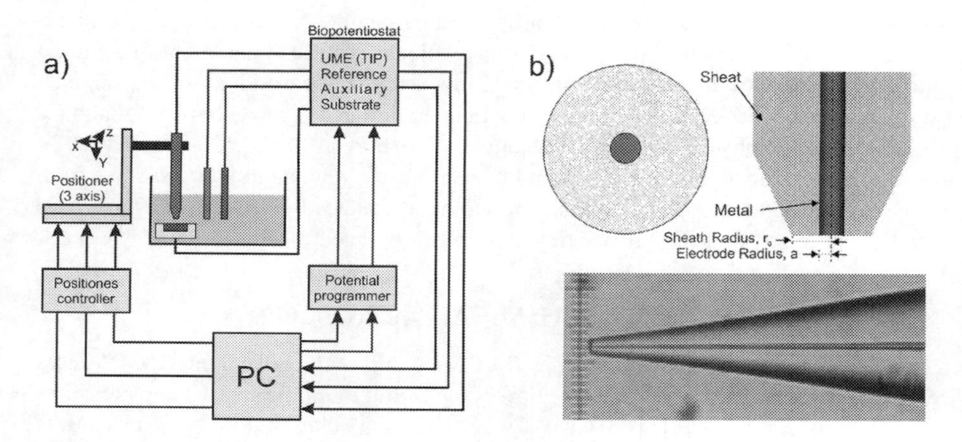

FIGURE 7.6 (a) Schematic of SECM. (b) Schematic (top) and optical micrograph (bottom) of a Pt tip. (L. Huang, Z. Li, Y. Lou, F. Cao, D. Zhang, and X. Li, "Recent advances in scanning electrochemical microscopy for biological applications," Materials (Basel)., vol. 11, no. 8, 2018) [9].

three-axis positioner and the data acquisition system. The tip is mounted in the positioner that provides *X*, *Y*, and *Z* movement. The biopotentiostat is used to control the voltage bias between tip and the sample [7]. The UME are commonly fabricated with carbon, platinum, or gold; Figure 7.6b shows a UME schematic; the resolution of the image depends on its size. Commercially, there are tips with some micrometer of diameter; however, it's possible to obtain finer tips with methods such as electrochemical etching [8].

In SECM, the UME (tip) and substrate are parts of an electrochemical cell with reference and auxiliary electrodes, as shown in Figure 7.6. The electrolyte solution, with an oxidized species Ox, allows the diffusive transport of O to the electrode. When applying a negative voltage to the UME, the reduction of Ox occurs at the UME and a cathode current flow through it.

At the same time, a reaction occurs at the auxiliary electrode. If the sample is a conductor, the current at the tip will increase if the reducing species can be oxidized to Ox (Figure 7.7a). On the other hand, if the sample is an insulator, the diffusion layer will be blocked and the current decreases (Figure 7.7b) [10].

With a stable reference electrode (Ag or Ag/Cl), the potential of the tip is monitored. Measuring the UME current as the tip scans the sample, images of the surface that contain chemical and electrochemical information can be obtained. The current flows vs. the UME potential plot is named a voltammogram.

When the tip-sample separation is bigger than the tip diameter, the steady-state current ($i_{T,\infty}$) in the tip is given by

$$i_{T,\infty} = 4nFCDa \tag{7.7}$$

where *F* is the Faraday constant, *D* is the diffusion coefficient of species Ox, *C* is the concentration constant, *n* is the number of electrons in the electrode reaction, and *a* is the tip radius. Figure 7.8 shows the typical curves for a conductive and insulating sample; the plot does not depend on the concentration or diffusion of Ox. If a conductive sample and the tip are close *R* can back to Ox, then the Ox of the substrate flows to the tip and leads to an increase in current as *L* decreases. For an insulating sample,

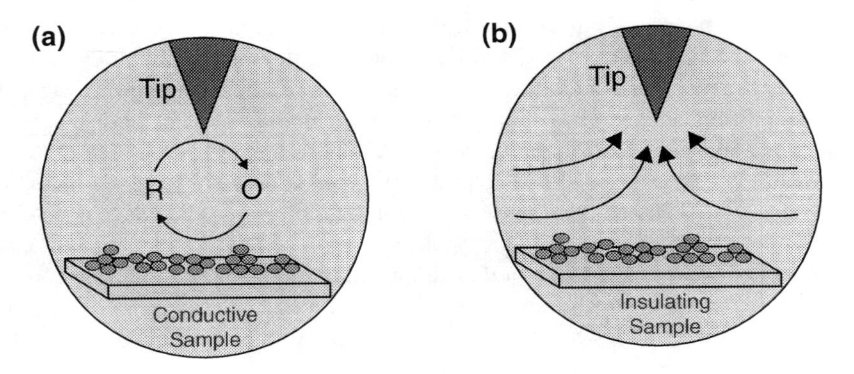

FIGURE 7.7 Basic principles of STM. (a) Conductive sample. (b) Insulating sample.

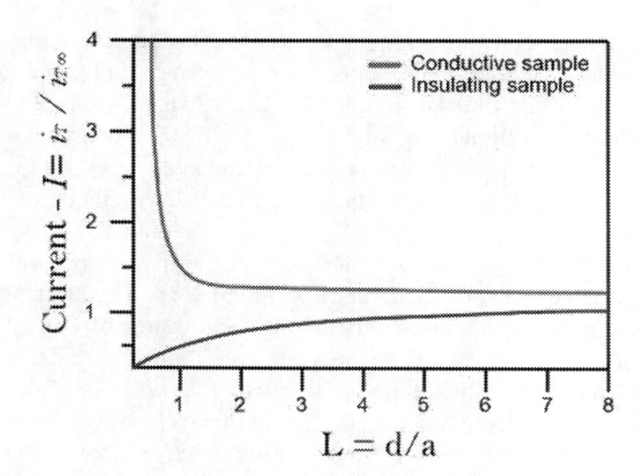

FIGURE 7.8 Plot of current as a function of tip-substrate separation.

the diffusion of Ox is difficult. The closer tip to the sample, the smaller will be the current; this behavior is called *negative feedback* [11].

SECM has been used for multiple applications, including analysis of surface reactivity, as shown in Figure 7.9a, where the evolution of the electrically insulating character of a Si electrode surface was studied. To study the corrosion processes on metallic surfaces, as shown in Figure 7.9b. To analyze conductive and topographic properties of oligomeric films, as depicted in Figure 7.9c [12].

7.4 ATOMIC FORCE MICROSCOPY

As mentioned earlier, a disadvantage of STM is that it is limited to conductive samples. In 1985, Binning, Quate, and Rohrer proposed a new instrument, the atomic force microscope (AFM), based on the principles of STM and of the profilometer that also made it possible to study the topography of insulating materials [16]. The atomic force microscope is also known as scanning force microscope (SFM), in analogy to the scanning tunnel microscope, but unlike the STM in which the tunnel effect is the measured magnitude, in the AFM, the interaction forces between a tip and the sample are measured [17]. An AFM allows knowing the topography of a surface in three dimensions (3D), reaching nanometric and even atomic resolution; AFM can be used with all types of materials, hard or soft, both artificial and biological samples regardless of their opacity or conductivity [18].

Essentially, an AFM consists of a sharp tip mounted on a cantilever spring which can be soft or hard, a method to sense the cantilever deflection, a piezoelectric system to move the sample and the tip, and a display system to convert the data into an image [19]. A schematic representation of the equipment is shown in Figure 7.10.

AFM is a scanning technique; the tip moves along the *X–Y* axis, causing the cantilever to flex up or down in the process; this operation is like that of a turntable,

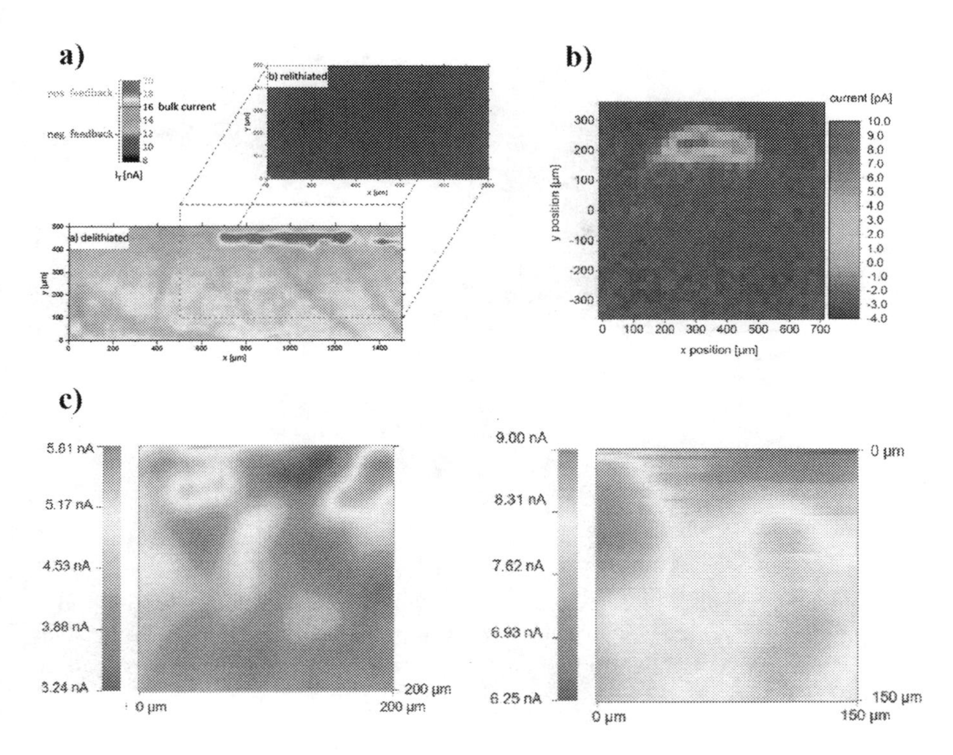

FIGURE 7.9 SECM mapping. (a) Si electrode after delithiation and after second lithiation [13]. (Published by The Royal Society of Chemistry [E. Ventosa, P. Wilde, A.-H. Zinn, M. Trautmann, A. Ludwig, and W. Schuhmann, "Understanding surface reactivity of Si electrodes in Li-ion batteries by in-operando scanning electrochemical microscopy," Chem. Commun, vol. 52, p. 6825, 2016]. (b) Signals of corrosion on a stainless-steel electrode after injection of H_2O_2. (Adapted from [M. Hampel, M. Schenderlein, C. Schary, M. Dimper, and O. Ozcan, "Efficient detection of localized corrosion processes on stainless steel by means of scanning electrochemical microscopy (SECM) using a multi-electrode approach," Electrochem. commun., vol. 101, no. February, pp. 52–55, 2019]) [14]. (c) Bi-dimensional scans of oligo-(S)-BT_2T_4-Au with tip biased at 0 V substrate unbiased (left) and biased at 1 V (right). (Adapted from: M. Hampel, M. Schenderlein, C. Schary, M. Dimper, and O. Ozcan, "Efficient detection of localized corrosion processes on stainless steel by means of scanning electrochemical microscopy (SECM) using a multi-electrode approach," Electrochem. commun., vol. 101, no. February, pp. 52–55, 2019. M. Donnici et al., "Characterization of Inherently Chiral Electrosynthesized Oligomeric Films by Voltammetry and Scanning Electrochemical Microscopy (SECM)," Molecules, vol. 25, no. 5368, 2020) [15].

with the needle traversing the grooves etched into the vinyl. In principle, any type of force can be measured if the tip on a cantilever beam (which acts as a spring) can be sensed. Measuring the deflection force and maintaining it constantly by varying the vertical position of the tip produces a constant force image, which,

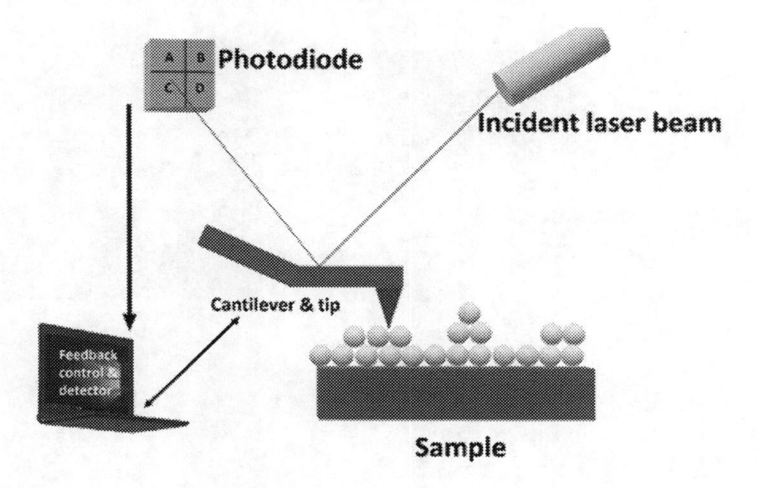

FIGURE 7.10 AFM schematic representation.

when done sufficiently close to the surface where van der Waals forces dominate, represents the force. After applying the appropriate image processing, visual information as well as quantitative data such as grain size and roughness can be obtained [20].

7.4.1 Operating Principle

AFM is based on the detection of forces between the tip and the sample, which depend on the distance among them as presented in Figure 7.11. We can model this situation using the Lennard–Jones potential; this potential is the combination of the attractive forces of van der Waals [21] that predominate at large distances (greater than 1 nm), and repulsive forces that occur at small distances (less than 0.5 nm) due to the overlapping of the electron layers of the tip and the surface of the material (Pauli's principle of exclusion).

Since interatomic repulsive forces are short-range forces that are confined to an extremely small area, they can be used to track surface topography with atomic resolution. In addition to these short-range forces, far-reaching forces (e.g., electrical forces, magnetic forces, van der Waals forces) can be observed, which can be attractive or repulsive. These forces interact over larger areas and are therefore not suitable for atomic resolution imaging [22]. But long-range detection of these forces offers various modalities that can be implemented in an AFM, such as magnetic force microscopy (MFM) and scanning area microscopy (SSPM).

Because the forces acting on the tip are of the order of nN, the tip is mounted on a cantilever, which works like a spring converter, depending on the nature of the forces it will bend upward, if it is attractive, or downwards if it is repulsive, the deflection of the cantilever can be expressed as

$$\Delta z = F / K_0 \tag{7.8}$$

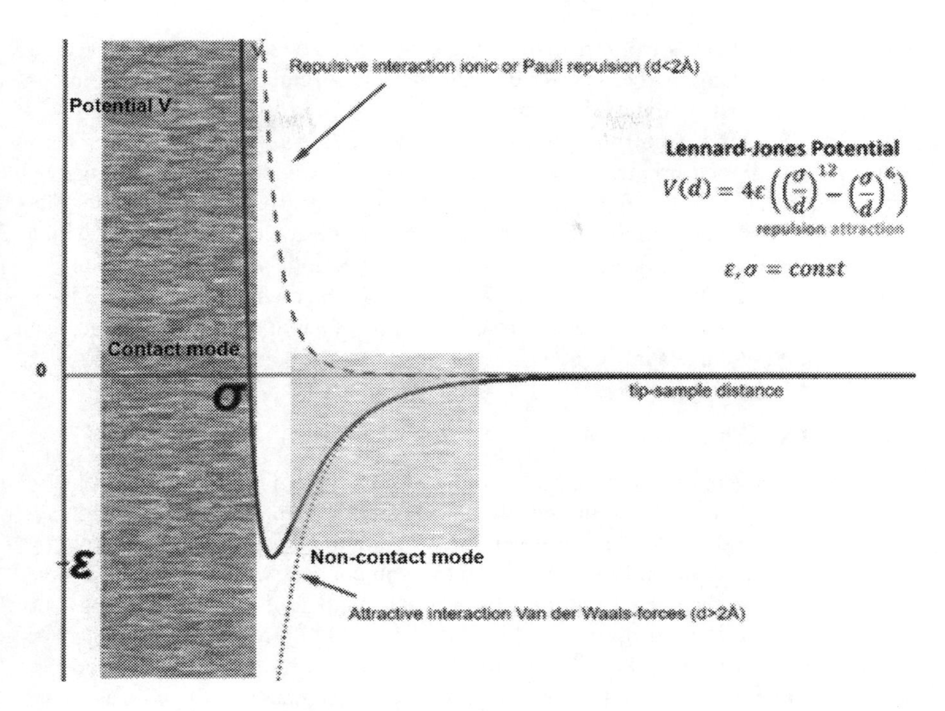

FIGURE 7.11 Representation of Lennard–Jones potential.

where:

K_0: It is the hardness constant of the cantilever, and the hardness constant can be expressed as a function of the modulus of elasticity (E), width (w), length (L), and thickness (t) of the cantilever.

$$K = \frac{wEt^3}{4L^3} \tag{7.9}$$

Hence:

$$\Delta z = \frac{4L^3F}{wEt^3} \tag{7.10}$$

Today, in the market, there are tips of various geometries with hardness from 0.01 to 50 N m^{-1}, which allows measuring a wide range of forces; these two parameters, the hardness constant and the shape of the tip, are of great relevance. Since the quality of the images to be obtained depends on them [23], naturally these parameters must be adapted according to what is to be measured.

To convert cantilever deflections into an electrical signal, a detection system is used in the AFM. Different techniques can be used: interferometric, optical, piezoelectric, capacitive, piezoresistive. The most commonly used technique, however, remains the optical lever method [24]. This technique uses a laser diode whose beam is focused and reflected from the rear of the cantilever. The signal is then projected onto a four-quadrant photodiode, which analyzes its movements by detecting intensity fluctuations between the quadrants.

Due to the highly localized nature of the interaction forces involved, the AFM performs a local measurement. Therefore, the instrument integrates a scanning system that generates a relative displacement of the tip with respect to the sample. The movements are often carried out through a system of piezoelectric actuators which ensure, under the action of voltages applied to the electrodes, nanometric displacements in the three directions of space (X, Y, Z). The objective is to probe the local properties of the sample by means of the constitution of a regular mesh of points that will constitute the pixels of the image.

7.4.2 AFM Operation Modes

AFM can operate in different modes according to the type of application that can be divided in two groups: the static mode (contact) and the dynamic mode (no contact and intermittent contact or tapping). In the contact mode, information is acquired by monitoring the interaction forces while the tip is in contact with the surface sample [25]. Whereas for the study of atomic, electric, or magnetic forces on a sample, the noncontact mode is used by moving the cantilever slightly away from the surface of the sample (~10 nm) and oscillating the cantilever near its natural resonance frequency. By mounting the cantilever on a piezoelectric element (PZT) and measuring the change in its natural resonance frequency due to attractive sample interactions, topographic information can be extracted [26].

Alternatively, the intermittent contact mode of operation combines the qualities of contact and noncontact modes by collecting sample data and oscillating the cantilever tip at or near its natural resonant frequency, while the cantilever tip briefly impacts on the sample surface during a short time [27].

Figure 7.12 shows a scheme of the AFM operation in the three modes as well as the surface topography that would be generated in each mode. The intermittent contact mode can reveal the surface topography of the sample to a greater extent than the contact mode. Operating the AFM in contact mode can damage the sample surface due to the side drag forces exerted by the probe tip but usually provides a higher resolution. A brief description and operational characteristics of each mode can be found below.

7.4.2.1 Contact Mode

In the contact mode, the tip is in close contact with the sample, typically the distance from the tip to the surface is smaller than 1 nm [28]. At these distances, the repulsion forces are very intense, as can be seen in the Lennard–Jones potential shown in Figure 7.11, because the electron clouds in the atoms of the tip and the sample repel each other; therefore, these repulsion forces predominate over any other attractive force that can be generated. In this mode, the topography of the sample can be obtained in two ways, keeping the height constant or the force constant.

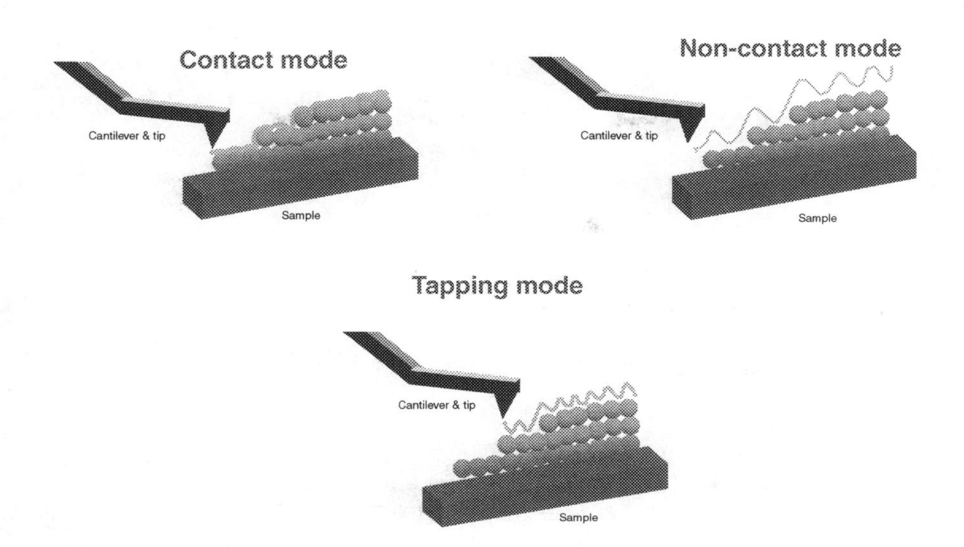

FIGURE 7.12 AFM operation modes.

7.4.2.2 Constant Force Mode

According to Equation (7.8), the deflection of the cantilever depends only on the force between the atoms of the tip and those of the sample since the hardness constant K is the same throughout the measurement; therefore, in the mode constant force, the deflection is kept constant by means of the feedback system. Once the deflection value is set, the AFM begins to sweep the sample in the x–y axis, and as the detector detects variation in the deflection, the tip displaces in the z-axis, in order to keep constant the deflection in the cantilever and therefore the force.

So, as mentioned above, since interatomic repulsive forces are short-range forces that are confined to an extremely small distances, when obtaining an image of the forces, the image of the sample topography is obtained.

The main problem with contact-mode AFM measurement is that shear forces, due to side sweep of the tip, generally tend to damage soft samples and distort the properties of the produced image. Therefore, this mode is not suitable for screening soft and polymeric biological surfaces, as it causes significant degradation of the sample.

7.4.2.3 Height Force Mode

In this mode, it is considered that the piezoelectric scanner is holding the tip while scanning the sample in the x–y axis, but without moving in the z-direction (height). So, if the height does not change, the distance between the tip and the surface change and then the force and the deflection caused in the cantilever, so we can use this information to estimate the surface topography.

So far, we have assumed that the sample surface is aligned to the scanner, that is, there is no inclination between the scanning plane and the sample surface (no scanning slope); therefore, when probing in the x–y axes the changes in the deflection are

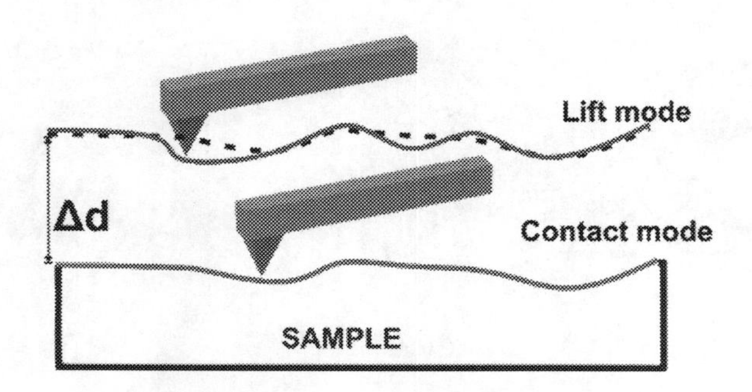

FIGURE 7.13 A schematic diagram of AFM's lift mode.

measured obtaining an image of the topography of the samples; in this case, feedback was not used and the measurement is very fast.

However, in practice, the procedure is different from the one previously described; as an example, the measurement of the magnetic interaction with a ferromagnetic tip is described; in order to obtain topography information without variations due to the scanning slope, two measurements must be made. The first in contact mode and the second one in noncontact mode, for which a relative distance is set, so that at each moment the sample and the tip are at the same distance; the measured topography is followed with a distance offset relative to the previous scan line, as shown in Figure 7.13 by the dashed line. In the second line, long-range magnetic interactions are detected by a corresponding offset of the cantilever shown as a solid line. The difference between the two signals (the dashed and the solid lines) corresponds to the magnetic signal. This type of constant height mode is also called lift mode [29].

7.4.3 Noncontact Mode

When operating in this mode, the tip is not in contact with the surface; instead, in this configuration, the tip is held around 10 nm above the surface to be sensed, and the cantilever is not static but oscillates at or slightly above its resonance frequency (frequency modulation) [30]. Because the attractive forces in this mode are so small, a small oscillation is applied to the tip of the cantilever during the measurement. By detecting changes in the amplitude, phase, or frequency of the oscillation of the cantilever during the measurement, it is possible to detect these slight variations in strength.

It is possible to construct topographic images of samples by scanning them across their surfaces while maintaining a constant frequency change during scanning. When there is a distance of the order of 10 nm between the atoms of the surface of the sample and those on the tip, attractive van der Waals forces are dominant, and these interactions induce a variation in the natural frequency of the cantilever; the topographic image is constructed by scanning the sample while maintaining a constant frequency change during scanning.

7.4.4 Intermittent-Contact Mode (Tapping)

Using a PZT element, the cantilever is forced to oscillate at or near its natural resonance frequency. This is performed by applying a force to the cantilever base, causing the cantilever tip to vibrate at amplitudes that are typically in the range of 20–100 nm when the cantilever is not yet in contact with the surface. This is accomplished by bringing the vibrating tip as close to the surface as possible until it begins to lightly tap it. High-frequency oscillations (usually between 100 kHz and 300 kHz) are used to take the measurement when the tip hits the surface and rises off alternately. To compensate for the energy losses produced by intermittent touching between the tip and sample surface, the amplitude of the vibration is changed in response to the surface topography. When operating in the intermittent AFM mode, there are various advantages that make this a widely used operating mode. The vertical and lateral resolutions are excellent, and there is less contact with the sample when compared to the C-AFM mode. Furthermore, measurements in a liquid environment are possible.

It is possible to image soft samples at high resolution with this measurement method, which is more difficult to do with the contact AFM mode. In this way, problems such as friction and adhesion, which were typically encountered with conventional AFM imaging systems, are eliminated. The intermittent contact mode, also known as tapping AFM mode as shown in Figure 7.12c, it was developed by a company acquired by Bruker Analytics and was a significant advancement in AFM technology.

7.4.5 Applications

The AFM is a very versatile tool, and it has become indispensable in research within many disciplines, such as materials science, biology, tribology, surface chemistry, energy storage, molecular biology, among others. Topography is the most common application of AFM; however, the identification of morphological characteristics, grain sizes, and roughness is among the most performed analyses. Some specific applications of the AFM are briefly mentioned below.

7.4.5.1 Imaging

The study of the topography of a sample is important because many times it is required to know if, for example, a film was deposited uniformly on a substrate or the influence of the topography on different properties such as adhesion, friction, corrosion, and so on. To illustrate the example, 2D and 3D images of a CdTe film deposited on copper are shown in Figure 7.14. Information such as grain size and roughness can be extracted from image analysis.

In the next example, a soft, loosely adhered material is shown. Blood cells onto a glass slide were imaged with the intermittent contact mode (Figure 7.15).

The intermittent contact mode was chosen to avoid scratching the cells from the glass slide. The red cells are clearly observed after carefully tuning the PID control gains and the amplitude of the oscillating tip.

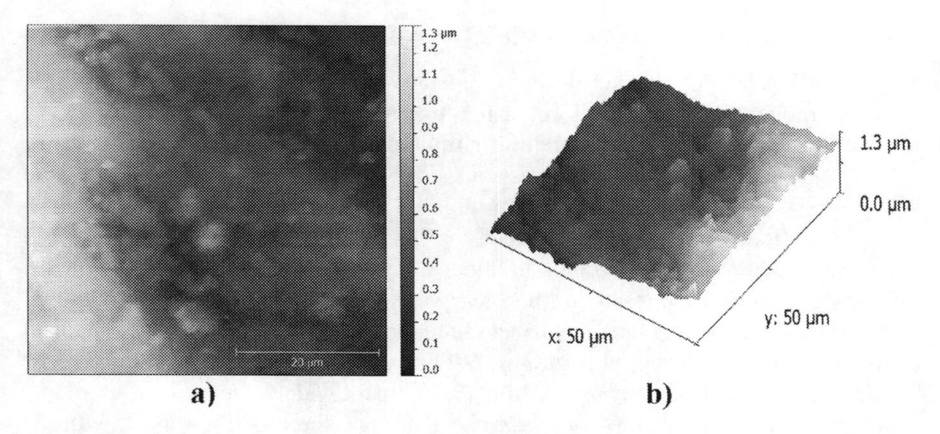

FIGURE 7.14 50 × 50 µm² topography of a CdTe film, 2D view (a) and 3D view (b).

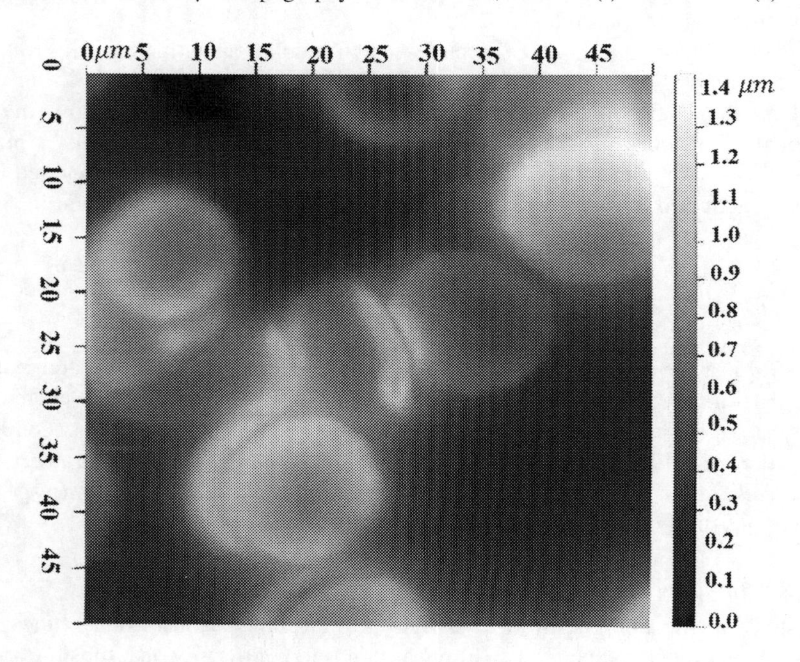

FIGURE 7.15 50 × 50 µm² AFM image of blood red cells, obtained in the intermittent contact mode (Image credits J. A. Baron-Miranda).

7.4.5.2 Determining the Film Thickness Using AFM

It is possible to measure the thickness of a thin film deposited by some "line of sight" deposition technique, such as sputtering or ion beam, where the atoms are coming from the same direction and the diffusion rate is minimal at the surface. In this type of depositing techniques, it is possible to create a kind of mechanical step on the edge of the film using tape or a piece of material that prevents the growth of the film in the substrate (Figure 7.16). Once this step is created, the AFM can differentiate the

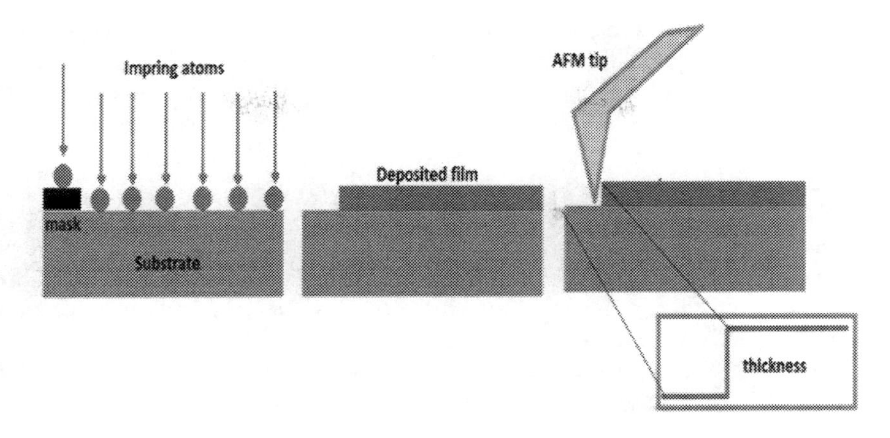

FIGURE 7.16 Scheme of the thickness determination on a sample using the AFM.

height between the film and the non-deposited area and thus obtain an estimate of the thickness of the film.

7.4.5.3 Correlation of the Sample Topography with Different Properties

As discussed, the SPM techniques involve scanning a surface with a probe that can detect physical or chemical properties with high spatial resolution. Particularly, based on an AFM instrument, several approaches referred to as AXM techniques, where the *X* symbolizes some form of interaction between the tip and the surface of the object being studied, can be done. A description of these modes can be found below.

Conductive AFM. Conductive AFM can perform simultaneous topography and current mapping by applying a DC or AC voltage between the tip and the sample. In conductive samples, the current is recorded in a 2D map. Also, applying a voltage ramp, an I–V curve is obtained, where the information on the contact such as sheet resistance, as well as barrier height and other diode parameters, can be extracted. For an insulating layer onto a conductive sample such as SiO_2/Si, the tunneling current through the SiO_2 layer can also detect thickness variations in the sub-Å regime and image surface potentials and contact potential differences between the tip and the sample.

Magnetic AFM. Magnetic AFM (M-AFM) is a technique in which magnetic forces are quantified by measuring the interaction of magnetic domains on the surface sample with a magnetized tip. The measurement is done by lifting the tip over the surface, so no correlation between topography and magnetic dominion is observed, the orientation and dominion size can be observed, and an estimation of the magnetic strength is possible too.

AFM Lithography. AFM technology evolved from STM as a tool for manipulating or modifying surfaces; the precise control of the probe's motion across the surface makes even a typical AFM a versatile tool for manipulating nanometric size surfaces. Making programmed patterns by displacing atoms as well as by indentation onto the surfaces is possible. Also, the mechanical properties at the nanoscale can be studied.

Electrochemical AFM. Electrochemical AFM (E-AFM) is a simple technique for studying the behavior of a surface as a function of applied voltage. Electrochemical reactions cause changes in the surface topography as a function of applied voltage in an electrochemical cell formed by the tip, the surface and a water drop. More controlled experiments can be done by using an actual EC cell in a three-electrode configuration. Biological, sensing and corrosion studies can be performed this way.

7.5 RAMAN SPECTROSCOPY

Raman scattering is a spectroscopic technique used to obtain structural information and molecular composition of solid, liquid, or gas samples. In 1928, Sir Chandrasekhar Venkata Raman, observed, for the first time, the inelastic scattering of light, phenomenon named after him, Raman scattering. When light (photon beam) interacts with a sample, various phenomena can occur. Light can be dispersed by atoms or molecules, most of scattered photons have the same energy and therefore the same wavelength as the incident photons (Rayleigh scattering). However, a tiny fraction of light (only one of 10^7 photons) is scattered at different frequencies from the incident light; this is the Raman scattering. Raman scattering can occur in two different ways, in the first one the radiation has less energy than the incident beam (Stokes) and in the second it has more energy (anti-Stokes) [31]. When the excitation beam, with a defined wavelength, said for example 532 nm, strikes the material being analyzed, the vibrational modes of the material scatter the incident light by gaining or losing a quantum of vibration, i.e., a phonon; the scattered light is registered as a peak (Figure 7.17). The wavelength is registered respective to the excitation source

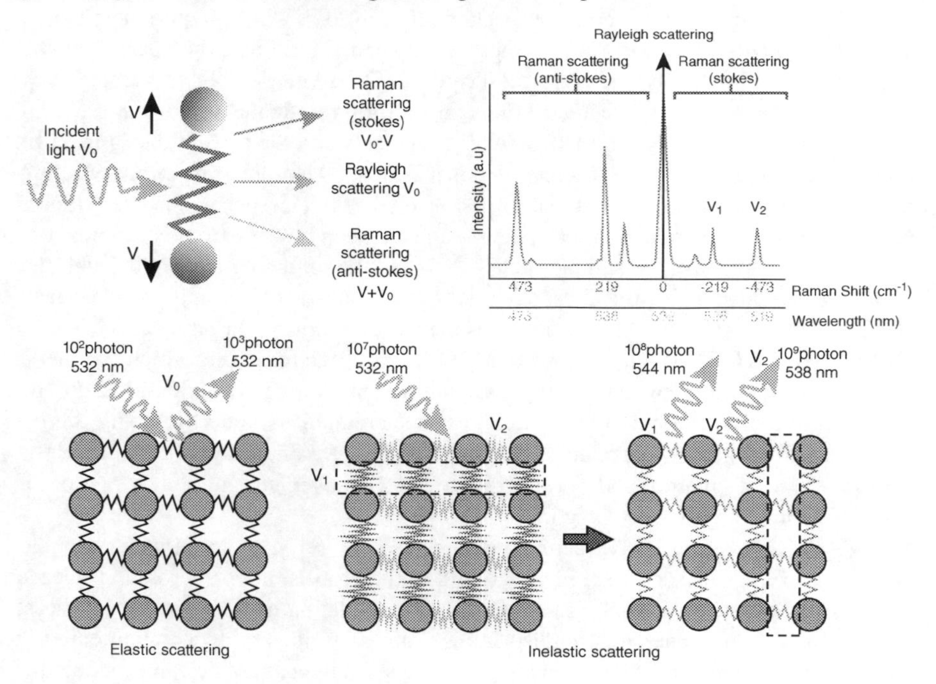

FIGURE 7.17 Raman effect, Rayleigh scattering, and Stokes scattering to produce Raman spectrum.

that can be higher (Stokes scattering) or lower (anti-Stokes scattering), as mentioned before, and therefore it is said that the Raman shift is toward positive or negative frequencies. The inelastically scattered Rayleigh line appears at a Raman shift of zero, i.e., at the same wavelength of the incident beam [32].

7.5.1 Instrumentation

There are four basic components in a Raman spectrometer, shown in Figure 7.18, which are:

1. Excitation source (photon beam), generally a laser. Lasers are highly monochromatic and provide many photons to outcome the intrinsic low yield of the Raman effect.
2. Filters. Notch filters are used to block the laser line (Rayleigh scattering) to allow the system to record the weak Raman photons.

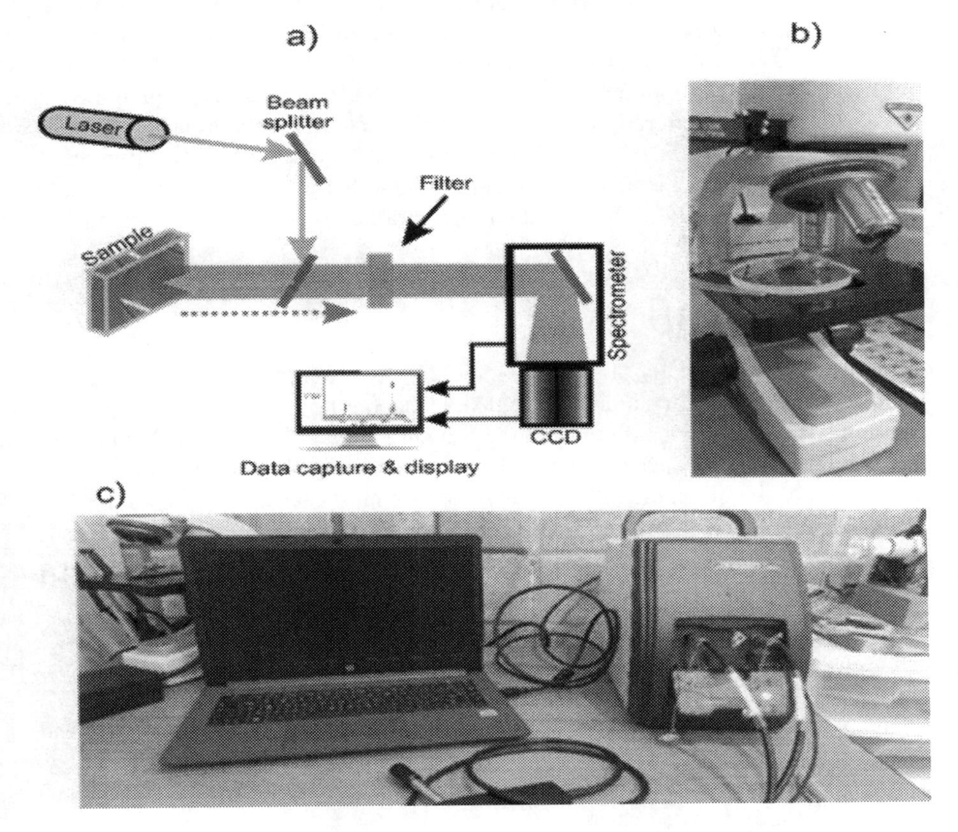

FIGURE 7.18 (a) Scheme of the typical setup of the Raman instrument; (b) Raman box, optical fiber for acquisition in solids and liquids, and laptop for processing; (c) Raman microscope with the optical fiber inserted in the microscope head; images from the Raman i-Plus (BWTek) with 532 nm laser, at GESMAT Lab.

3. Spectrometer. The spectrometers are composed of an entrance slit, a diffraction grating, and a detector, commonly a charge coupled device (CCD). Raman-dedicated spectrometers are normally configured in the ultraviolet-visible-near-infra light range, depending on the excitation wavelength, to be extremely sensitive to the low level of Raman photons. CCDs are usually thermoelectrically cooled to increase sensitivity and reduce noise levels.
4. A computer for data storage and deployment.

   Other Raman systems have variants such as the beam is focused through a microscope lens, typically 20–100×, then the technique is called micro-Raman and allows spatial resolution onto the sample, that is interestingly when the sample is not homogenous, such as a paint or another art object. Other case includes the use of a confocal microscope that allows for obtaining in-depth information of the sample. An *x*–*y* stage is used to move the sample and it is possible to map the intensity of a given Raman line to produce a 2D image of the surface, thus called Raman mapping. Other setups are the focusing and acquisition of the Raman signal onto a sharp tip such as in a scanning probe microscope, to have a simultaneous Raman and topographic characterization. Many instruments have an optical fiber to guide the laser and the Raman signal in a backscattered geometry to a CCD detector; others have a laser focused onto the sample, either solid or liquid and collect the Raman signal through mechanical monochromators, that increase the sensibility and the range where the signal can be obtained down to some few cm^{-1} close to the Rayleigh line.

7.5.2 Raman Spectra

Raman spectra are usually plotted using the detector counts of the intensity of the Raman signal in arbitrary units (arb. units) versus the Raman shift (cm^{-1}) and less commonly vs. the wavelength (nm) of the Raman photons. The unit of Raman shift is the reciprocal of the wavelength, which is proportional to the energy of the electromagnetic wave based on the following equation for photon energy:

$$E = h\nu = \frac{hc}{\lambda} \tag{7.11}$$

where h is the Planck constant, ν the frequency of the wave, c the light speed and, λ is the wavelength. The Raman shift is the result of subtracting the wavenumber of the scattered light from the wavenumber of the incident beam, as shown in Figure 7.17. Therefore, the Raman shift is the difference between energy of incident beam and the inelastic scattered light, which is proportional to a molecular vibrational energy. The Raman scattering intensity is proportional to the square of the molecular polarizability a resulting from a normal mode q:

$$I_{Raman} \propto \left(\frac{\partial a}{\partial q}\right)^2 \tag{7.12}$$

In a Raman spectrum, the bands correspond to a vibration of chemical bonds or functional groups into the molecule. The position of each band is characteristic of the bond or functional group, so the Raman spectrum can be considered as its fingerprint [33,34]. RS is a complementary technique of IR spectroscopy, which has been extensively documented in the literature. In IR spectroscopy, strong bands correspond to polar and heteronuclear functional groups. On the other hand, nonpolar functional groups produce strong bands at Raman spectra. Some functional groups have a characteristic vibration frequency, which does not change regardless of which molecule the group is in [35]. Figure 7.19 shows that the vibrational spectrum is divided into typical regions and some examples from each one.

RS has some advantages over IR, as an example:

1. RS can be used in aqueous medium or into glass matrixes, while IR does not because of the strong signal from OH in water and many glassy systems.
2. In IR spectroscopy, samples must be dispersed in a transparent matrix or measured in a total attenuated reflectance accessory. This is not necessary in RS, where powders, films, solids, and liquids can be measured without special treatment.

In Figure 7.20, the features of a typical Raman spectra are shown, related with the physical properties of the sample. Therefore, the effect of the temperature or stress, the structural defects, degree of crystallinity, the phase or molecular entity as well as the surface roughness, can be assessed by recording the peak position, width, peak shoulders, and intensity, in addition to the fingerprint of a given material.

Figure 7.21 shows the complementary nature of infrared and Raman. In Figure 7.21 left, the spectra of a graphene oxide (GO) samples are presented. GO is composed of a quasi-2D graphene layer with carboxyl, hydroxyl, carbonyl, and epoxy groups located in the borders and onto the graphene planes. These groups have very defined IR bands, from which the relative composition can be inferred. On the other hand, the C=C group has a very low infrared absorbance, but in the graphite structure, it

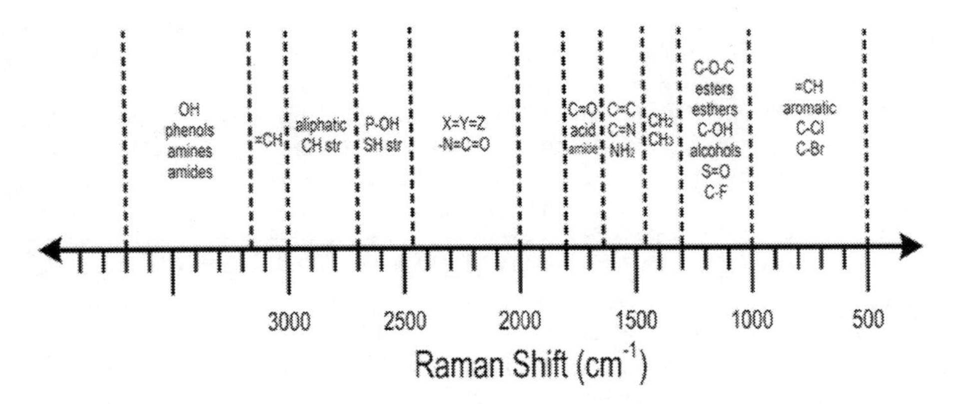

FIGURE 7.19 Spectral ranges at the mid-infrared where some functional groups can be observed either by Raman or by infrared spectroscopy.

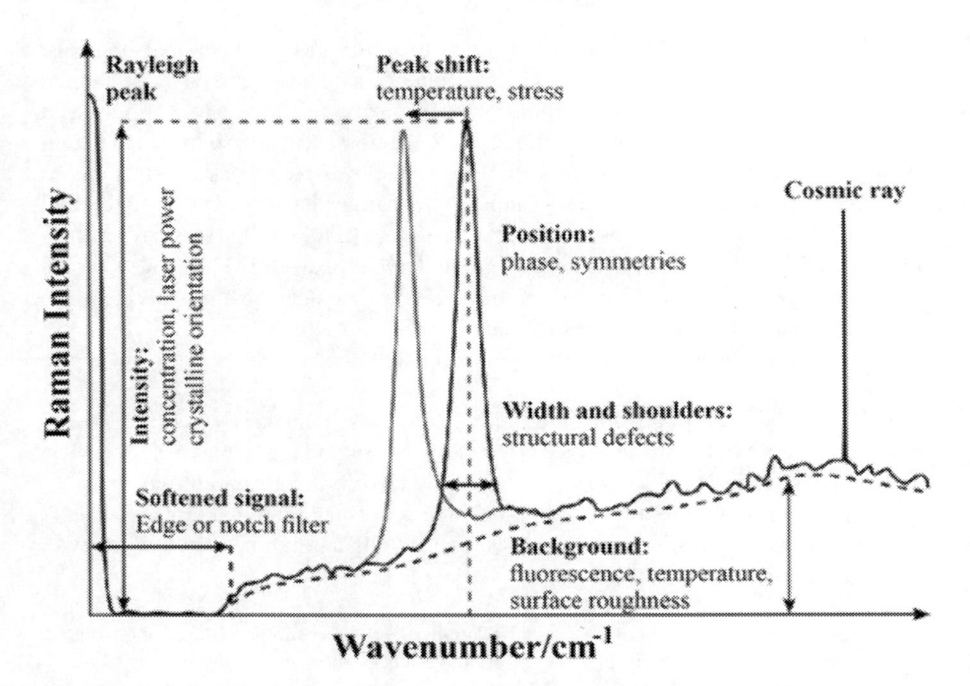

FIGURE 7.20 Features of a typical Raman spectrum and their relationship with structural and chemical parameters. (Reproduced from: Frédéric Foucher, Guillaume Guimbretière, Nicolas Bost and Frances Westall (February 15, 2017). Petrographical and Mineralogical Applications of Raman Mapping, Raman Spectroscopy and Applications, Khan Maaz, IntechOpen) [36].

presents a very well-defined Raman mode, the G band. Finally, the Raman feature named as D-band does not have a correspondence in the IR spectrum, but its origin arises from structural defects that activate this mode in the Raman spectrum, such as sp3 hybridization and reduction in the crystallite size. Figure 7.21 right, presents the corresponding IR and Raman spectra of a sample of reduced graphene oxide (rGO) decorated with ZnO nanoparticles. The IR spectrum is practically featureless and informs of a deep reduction degree; only an inflection at the point of the C=C bonding and a semi-defined peak at about 700 cm^{-1}, related to vibrations along the Zn–OH bond, are observed. The mid-infrared range (400–4000 cm^{-1} or 2.5–25 μm) is not well suited for inorganic materials; however, Raman presents information related to different vibrational modes of the nanoparticles, i.e., the fundamental Eg mode as well as the rare TA+TO and LA+TO modes usually present in nanoparticulate systems. Another Raman band is observed at ca. 2650 cm^{-1}, the 2D band, that contains information of crystalline order and number of graphene layers, an information unavailable in the infrared spectrum.

In the case of solid samples, the Raman spectrum contains information of the structure with a resolution higher than X-ray diffraction (XRD), where only the long-range order is visible. In Raman spectra, the short-range ordering, i.e., the ordering at molecular level is achievable, making it very useful to characterize the structure

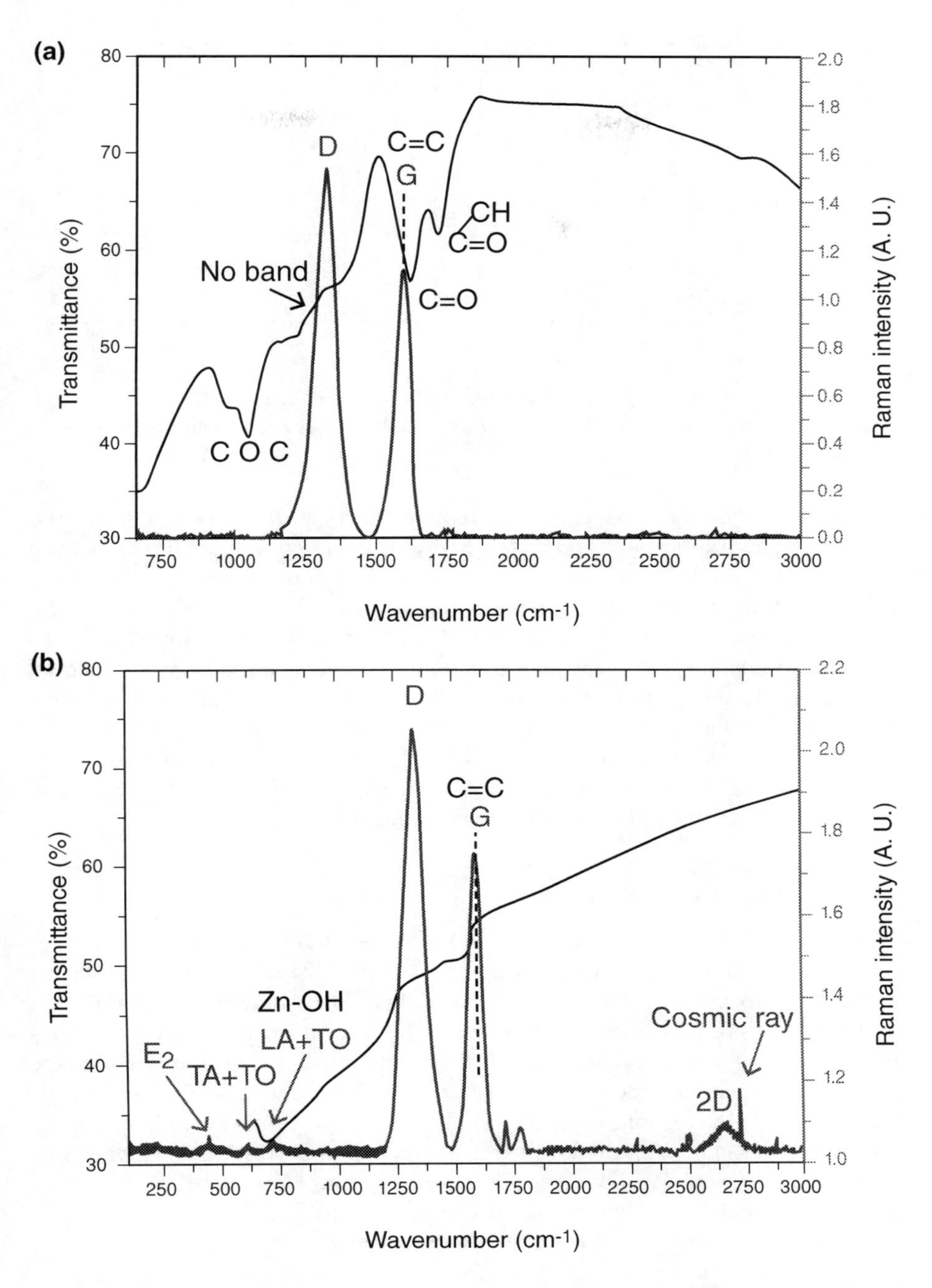

FIGURE 7.21 Comparison of Raman and infrared spectra of graphene oxide (left) and ZnO-decorated reduced graphene oxide (right). (Credits, R. V. Tolentino-Hernandez, F. Caballero-Briones).

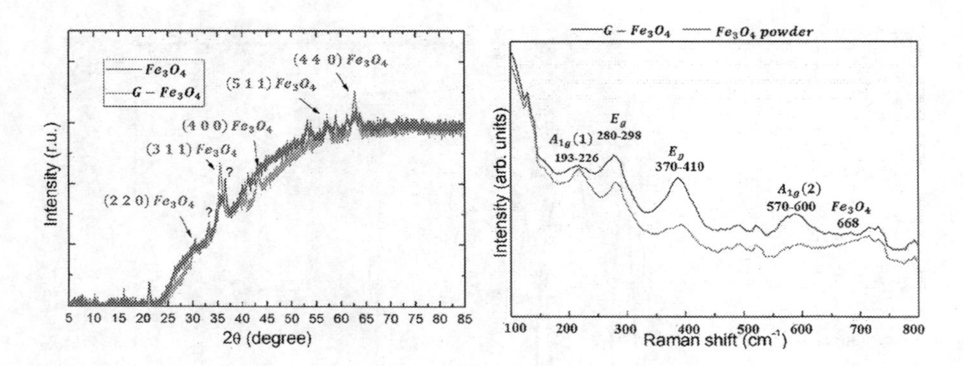

FIGURE 7.22 Comparison of X-ray diffraction (left) and Raman spectra (right) of Fe_3O_4 nanoparticles and GO decorated with Fe_3O_4 nanoparticles. (Credits J. Guerrero-Contreras, F. Caballero-Briones).

of noncrystalline solids and nanostructured materials, where the bond type as well as polymorph identification is possible. For example, in Figure 7.22, the comparison between the X-ray diffractograms and the Raman spectra of a magnetite powder and a GO sample decorated with magnetite is presented. The X-ray diffractograms present strong fluorescence due to the use of a Cu anode onto a Fe-containing sample; the peaks corresponding to the magnetite sample are well defined and indexed; however, the decorated sample presents unidentified peaks with a poor signal-to-noise ratio, made worst by the fluorescence. On the other hand, the Raman spectra show the presence of the characteristic vibrations of magnetite, but other set of peaks at ca. 290 and 390 cm^{-1} related to the ordered vacancy phase maghemite, which is hard to differentiate by XRD as it is isomorph with magnetite.

7.5.3 Other Applications

Figure 7.23 left presents the Raman spectra of ammonium nitrate solutions with different concentrations. A main peak at ca. 1050 cm^{-1} is characteristic of the nitrate ion and its intensity is proportional to the concentration. The nitrate peak at 1050 cm^{-1} is proposed as a kinetic marker to follow the evolution of the deposition of CdS films by Chemical Bath Deposition (CBD). The CBD process for CdS films includes the formation of an amino–cadmium complex, which slowly decomposes to release the Cd^{2+} ion, while the S^{2-} ion is released by the hydrolysis of thiourea in basic media. Previous to the addition of thiourea to the bath containing the amino–cadmium ion, the substrate is inserted in the bath to allow the formation of $CdCO_3$ nuclei. After thiourea addition and S^{2-} and the complex decomposition, the film is grown. Figure 7.23 right shows the evolution of the film thickness compared with the concentration of the nitrate ion, obtained after in-operando acquisition of Raman spectra of aliquots of the bath. The thickness follows a classic logistic curve, with an induction period, an exponentially decaying growth, and a steady-state limit. The nitrate ion concentration, on its side, reduces abruptly after a perceptible film thickness can be observed. This has been interpreted in terms of an arrangement of the amino–cadmium complex

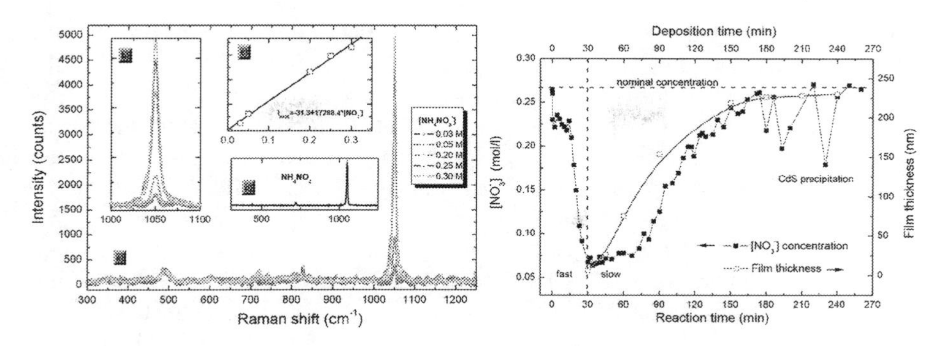

FIGURE 7.23 Left: Raman spectra of ammonium nitrate solutions indicating the linear response of the 1050 cm^{-1} peak with the concentration; Right: evolution of the NO_3^- concentration and CdS film thickness with the deposition time during a CBD process [7].

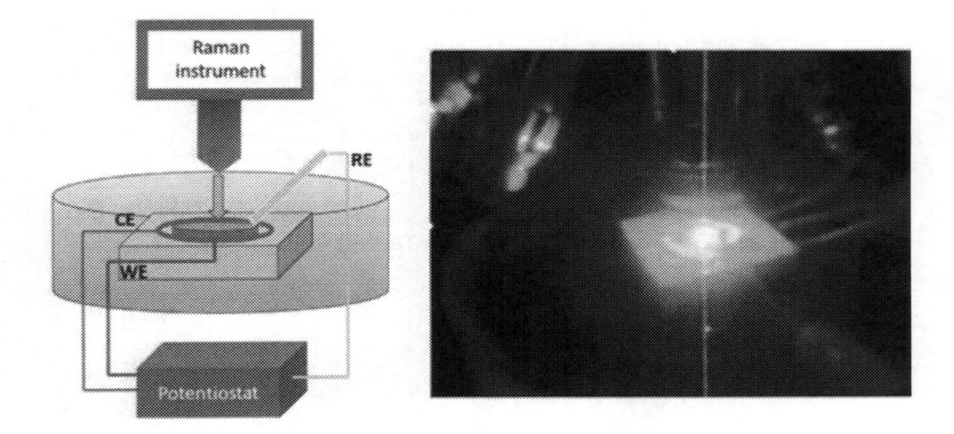

FIGURE 7.24 Left: Eectrochemical setup for the acquisition of Raman spectra during the *in-situ* oxidation of copper; Right: still from the Raman acquisition with a 532 nm laser, where the Teflon cell with the electrodes and the Cu disk can be observed. (Credits. F Caballero-Briones).

with the nitrate ions before a critical complex decomposition. The nitrate ion increases up to the initial concentration, and after the steady-state is achieved, its concentration varies abruptly, suggesting CdS particle precipitation [37].

Figure 7.24 presents an experiment for studying by micro-RS the *in-situ* growth of copper oxides in 0.1 M NaOH by copper anodization. In the experiment, a copper disk was placed into a Teflon® cell, connected as working electrode to a potentiostat. Also, a mini-Ag/AgCl electrode was used as reference and a Pt wire around the Cu disk as counter electrode. The setup was placed below the microscope to focus the 532 nm laser onto the Cu surface covered with a shallow electrolyte layer.

The electrochemical potential, the microscope objective, the laser power onto the surface and the substrate roughness were varied to observe the formation of Cu_2O

and CuO as described in Figure 7.24. The first point was to observe the effect of the objective on the signal in the spectra in Figure 7.24a. Using the 100× objective, the peaks became defined; this is because the laser is focused onto a smaller area, ca 1 μm diameter, thus increasing the power density in the surface (fluence) and then increasing the yield of Raman photons. However, despite, in this case, the signal became better; in other cases, the increase in the fluence can damage or chemically alter the material under study; thus, a careful evaluation must be done for each case. In Figure 7.25b, the effect of the surface roughness on the signal intensity was studied, by chemically oxidizing and reducing the Cu electrode before electrochemical oxidation, thus increasing its roughness and causing the so-called surface-enhanced effect (SERS) on the Raman spectra, which briefly consists in an increase of the electric field caused by metal plasmons (the Cu nano peaks at the rough surface in this example). It is noticeable that the SERS effect leads to an increased signal-to-noise ratio. In Figure 7.25c, neutral density filters (NDFs) were used to attenuate the laser power onto the surface, to avoid as possible the modification by the laser onto the SERS-modified electrode. The power onto the surface was estimated from the attenuation power of each filter vs. the nominal laser intensity. The best results were obtained with the maximum power of 14.5 mW onto the surface. Finally, under the best acquisition conditions, Raman spectra of the Cu substrate, a Cu_2O film obtained after exposing the substrate at −265 mV vs. SSC during 60 minutes, and a CuO film obtained at ca. +0.5 mV vs. SSC after another 60 minutes are shown in Figure 7.25d.

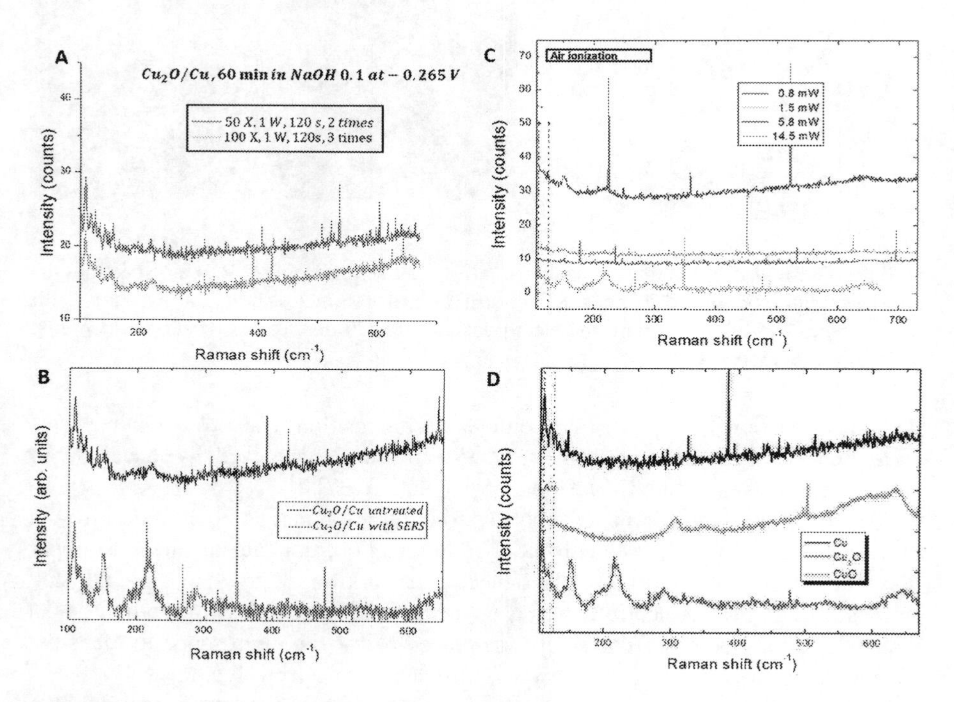

FIGURE 7.25 Different experimental conditions to acquire in situ Raman spectra of the potentiostatic oxidation of a Cu electrode. Figures described in the text. (Credits. F. Caballero-Briones).

7.6 X-RAY PHOTOELECTRON SPECTROSCOPY

The XPS, also known as Electron Spectroscopy for Chemical Analysis (ESCA), is a technique used in materials science for surface characterization, due to its versatility to assess surface chemistry, bonding structure, and composition of surfaces and interface in materials and is the most used electron spectroscopy technique for defining the elemental composition and/or speciation of the outer 1–10 nm below the surface in solid substrates.

A simple way to describe photoelectron production is seen as the process in which an electron is bounded to an atom, and this is ejected by the absorption of a photon. Photons are massless and chargeless energy packages; the photon energy can be transferred to the electron during the photon–electron interaction, if this energy is sufficient, the emission of the electron from atom as well as the solid can be achieved. The emitted electron has a discrete kinetic energy (KE) that can be measured, which is related to the binding energy (BE), which is specific of each element and its chemical environment. When a photoelectron is emitted from an atom, a vacancy is left in its inner orbital and becomes unstable. Auger transition is the process by which the unstable excited atom relaxes. In the Auger process, an outer orbital electron decays to the vacancy, and the excess energy is released through the emission of another outer orbital electron (KLL Auger electron). XPS measures both photoelectrons and Auger electrons. XPS and Auger processes are schematized in Figure 7.26.

7.6.1 INSTRUMENTATION

A schematic of the photoelectron emission process (XPS instrument) is shown in Figure 7.27 along with the most common means of acquiring the data:

- Spectral energy distributions of electron emissions within a predefined energy range
- Spatial distributions of specific electron emissions across the surface (for elemental or speciation distributions mapping)
- In-depth distributions of specific electron emissions to predefined depth (this can be from less than 10 nm to several micrometers)

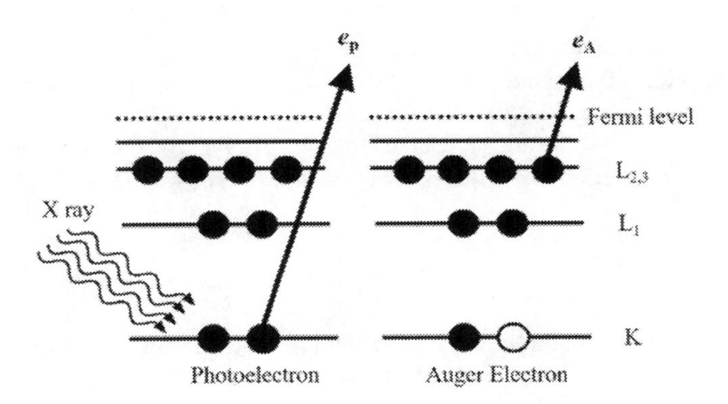

FIGURE 7.26 Photoelectron (e_P) and Auger electron (e_A) emission processes induced by X-rays. (Image credits R. V. Tolentino-Hernandez, M. S. Ovando-Rocha).

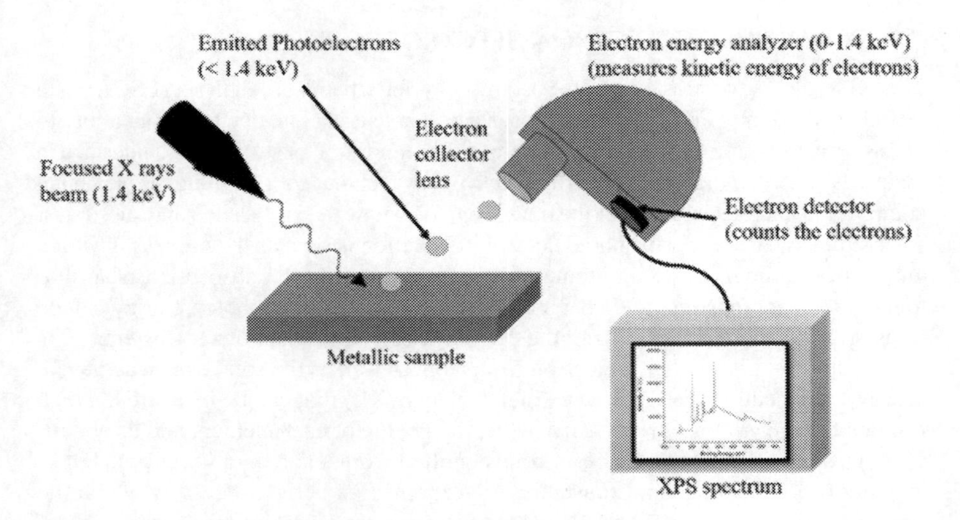

FIGURE 7.27 Basic components of XPS instrument with energy spectrum obtained. (Image credits R. V. Tolentino-Hernandez, M. S. Ovando-Rocha).

During analysis performance, the sample is irradiated with X-ray photons of known energy, which gives rise to the photoelectric effect. The electrons generated close to the surface are ejected from the sample into vacuum and projected into the analyzer slit of the spectrometer, which can measure the electron current as a function of their energy.

XPS analysis is usually performed by first collecting a survey energy spectrum over all accessible energies BE < 1400 eV and then collecting a high-resolution spectrum of photoelectron signals. The survey spectrum allows quantifying all the present elements, while the high-resolution spectra are for identifying the chemical environment for each atom.

The most common X-ray sources employed in XPS analysis are the characteristic Kα lines from Al and Mg anodes; these primary lines have energies of 1253.6 eV and 1486.6 eV, respectively, and are high enough to penetrate in the core level electrons from most of the elements, except H and He [38,39].

Although KE is the quantity measured in the XPS instrument, it is the derived BE that is used to construct the energy spectrum, since the KE is dependent on the X-ray energy, whereas the BE is not. The values for KE, BE, and the incident photon energy (hv) are related through the expression proposed by Einstein for the photoelectric effect [40,41].

$$KE = hv - BE - \phi \tag{7.13}$$

where ϕ is the work function of the instrument, not of the sample. This is included due to the energy required to bring the electron from its zero-BE (or Fermi) level to that of the instrument, in the assumption that a conductive sample in physical contact with the instrument is analyzed.

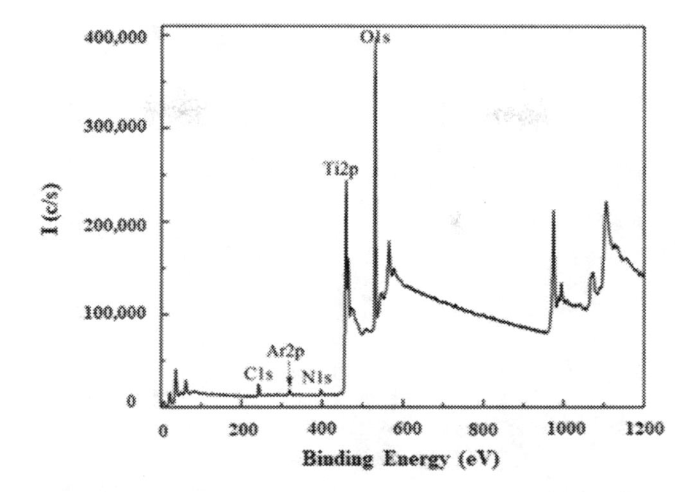

FIGURE 7.28 Survey scan of XPS spectrum of the as-grown TiN film. (Reused from: M. Maarouf, M. B. Haider, Q. A. Drmosh, M. B. Mekki, X-Ray Photoelectron Spectroscopy Depth Profiling of As-Grown and Annealed Titanium Nitride Thin Films. Crystals 2021, 11, 239) [42].

7.6.2 XPS Spectra

A typical example of XPS survey scan is shown in Figure 7.28; the spectra belong to a sample of TiN film in which Ti 2p, N 1s, C 1s, O 2p and Ar 2p peaks are observed. Notwithstanding that the sample only must contain Ti and N, peaks corresponding to C and Ar are also present in the spectra, the presence of C 1s peak is due to organic contamination of the sample when exposed to air after the growth, the so-called adventitious carbon, which is used as a reference to correct the peaks shift due to charging effect, the position of the adventitious carbon peak is 284.9 eV, and the explanation for the Ar 2p peak is due to ion implantation during the film etching, which is made with a Ar^+ gun during the XPS acquisition.

High-resolution XPS spectra allow the elucidation of the different chemical species present in the sample for a specific element, as well as their quantification. This is a key point in the development of new materials with specific properties that are related to the surface chemistry of the materials. In the following example, an XPS high-resolution spectrum in the region of the Au 4f doublets from Au nanoparticles is presented in Figure 7.29; in the same energy region, a spectral fitting was performed: the main peaks belong to the metallic Au (0) (red), while a little double peak (blue) shows that sample contains a little amount of Au (I).

7.6.3 Good Practices in Data Processing

During the development of the XPS technique and with the increasing number of equipment available to perform surface characterizations on materials, the community of researchers that use this technique has grown exponentially; however, the treatment of the data obtained by XPS is not a simple task to do for nonspecialized

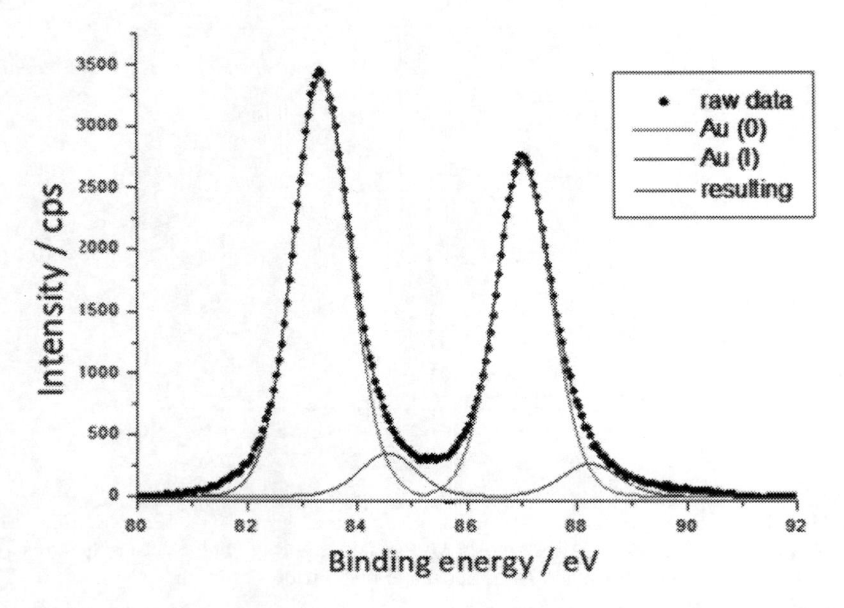

FIGURE 7.29 Au 4f doublet XPS spectral region of deposited Au nanoparticles. (Reused from: S. Caporali, F. Muniz-Miranda, A. Pedone, M. Muniz-Miranda, SERS, XPS and DFT Study of Xanthine Adsorbed on Citrate-Stabilized Gold Nanoparticles. Sensors 2019, 19, 2700) [42].

people with this technique. Despite the breakthrough in computing power to fit data on overlapping high-resolution XPS spectra, there are only a few studies that rely on developing chemically and physically meaningful approaches to curve fitting, and this entails a serious problem regarding the information published in scientific articles, in which erroneous results are presented regarding surface chemistry [43].

A basic approach to XPS spectra peak fitting is defined in terms of component peaks and background. These component peaks are integrated by their line shapes (which are mathematical functions), and the fitting parameters that allow a component peak to vary in diverse ways, which include position, full width at half maximum (FWHM), area, Lorentzian character, degree of asymmetry, and Gaussian character. All present components are summed and with the addition of background from the original data. Before starting with the data adjustment, it is necessary to carry out the removal of the background in the spectrum, with the specialized software packages for the treatment of XPS spectra; there are background remotion models such as linear, Shirley or Tougaard; the background model and the selection of the end points are of great importance to obtain the area under the curve, and therefore the relationships of the individual peaks that are used in the spectrum curve fit.

Another point to consider is the function chosen to perform the adjustment of the peaks; Lorentzian or Gaussian functions are generally used indiscriminately, since the contributions that are the origin of the spectrum obtained are not considered, for example, it is known that when X-rays are incident on a surface, these promote the movement of an electron, from an initial occupied energy level to a final one; the virtual energy of the promoted electron is greater than the energy of the Fermi level of the studied surface. This allows the electrons to escape from the solid at a specific energy, and then this energy is measured by the spectrometer analyzer. When dealing with subatomic particles, the uncertainty principle gives the initial energy level a Lorentz energy distribution. However, the basic shapes of the peaks recorded in an XPS spectrometer due to the emission phenomenon itself are modified due to instrumental factors, which gives a Gaussian contribution. Therefore, using a Gaussian or Lorentzian function omits some of the contributions of the spectrum obtained, to make an adequate fitting of the XPS peaks, a function that is a convolution between a Gaussian and a Lorentzian function must be used, the so-called Voigt function [44], which gives a better data fit, because it considers the different distributions of the process.

The use of XPS in many areas of materials science and technology has become increasingly important due to its versatility for the characterization on the surface chemistry, speciation, and bonding structure. But also the complexity of new advanced materials has increased, the extraction of important physicochemical information increasingly requires a proper fitting of XPS data. The good practices and information presented in this section are provided as a guide to improving the quality of the XPS curve fitting reported and results in the reported literature [45].

7.7 X-RAY ABSORPTION SPECTROSCOPY

XAS is a local atomic structure and unoccupied electron probe technique that provides unique information to unveil structural and electronic information at the local atomic scale. We briefly review the basis of XAS, experimental requirements, and analytical procedures for *in-situ* and *ex-situ* experimentation.

7.7.1 Extended and Local Atomic Structure of Complex Materials

The periodic atomic structure of the vast majority of materials was discovered by von Laue and further developed by Bragg [46,47]. This evolved into the development of modern crystallography defining every atom position in a space group and atom positions within it. Nowadays, structural information of crystalline materials is routinely obtained from standard laboratory XRD in both, research laboratories and industry. Small deviations from perfect periodicity are commonly described in terms of point, line, or boundary defects. However, it can be overlooked, specifically in complex materials, that the basic assumption of perfect periodicity could be only partially fulfilled. Heterogeneous materials [48], dynamic lattice instabilities [49], local atomic distortions [50], and amorphous components could be difficult, if possible, at all, to detect in a standard XRD experiment. The basis for this limitation is the reduced amount of momentum transfer of the scattering in a typical Cu anode laboratory-based XRD experiment, with values below $Q = 8$ A^{-1} where $Q = \dfrac{4\pi \sin(\theta)}{\lambda}$, with θ being the scattering angle and λ the wavelength. We note that the large coherence length in XRD provides high precision in the lattice parameter determination [51]. Deviations from perfect periodicity could be observed in a synchrotron-based (or neutron source) diffraction experiment performed with Q values of 30 A^{-1} (or above) with high-energy X-rays (or neutrons) [51]. By applying the PDF technique, which involves the Fourier transformation of the spectra, a radial distribution function in real space is revealed [52]. In this way, all atomic pairs of the material are described, including local atomic distortions within them.

7.7.2 X-Ray Absorption Spectroscopy

XAS provides information on the local atomic structure around a specific atomic species that does not rely on periodicity, so it can be applied to solids, liquids, or gases. Historically, the discovery of X-ray absorption bands by Maurice de Broglie could be taken as the beginning of XAS [53]. At the beginning of the 20th century, the fine structure oscillations of the X-ray absorption of atoms in solids, even hundreds of eV after the X-ray absorption edge, were observed, but not completely understood [54]. Although some advances were performed with laboratory-based X-ray sources, it was until the development of synchrotron facilities, particularly in the 1970s, and the concomitant development of the theoretical understanding of the complex phenomena of X-ray absorption in solids that took place in the last decades of the 20th century, where it became quantitative and evolved into a powerful local atomic structural technique [55,56].

The complexity of X-ray absorption in solids is based on the involved phenomena. The Beer-Lambert law describes the absorption in a material, where the incident intensity I_0 exponentially decays with thickness and energy by $I = I_0 e^{-\mu(E)t}$, where t is the sample thickness and $\mu(E)$ the absorption coefficient. The absorption coefficient decreases with energy; however, at ionization energies, when a core electron absorbs an X-ray leaving a core hole, it creates a step-like increase in the absorption. The atom then relaxes and fills that hole, emitting an X-ray (X-ray fluorescence) or an Auger electron. Either the X-ray fluorescence or the Auger electron yield can be used to indirectly measure the X-ray absorption. Figure 7.30, black line, shows a simulation of the X-ray absorption of an isolated atom of Fe using the *ab-initio* code FEFF9 [56]. The spectrum is separated into three regions: the pre-edge region where background absorption monotonically decreases until the energy reaches the BE of a core electron (Fe 1s or "K shell" electron). Second, the large increase in the absorption is called the absorption edge, and third the postedge region with a monotonic decrease in the absorption. The probability for an X-ray to be absorbed by a core electron depends on the initial and the final state, mediated by a (mainly) dipole transition, i.e., Fermi's golden rule: $\mu(E) \propto |i|\epsilon \cdot \boldsymbol{r}|f|^2$. The initial state is an atom with a core electron and an X-ray, while the final state, for the case of an isolated atom, is a free electron and a core hole in the absorbing atom. In the case of a solid, at the absorption edge and energies slightly above it, the photoelectron reaches unoccupied electron states, modulated by the unoccupied electron density of states (Figure 7.31a). This is called the X-ray absorption near-edge region or "XANES" region (Figure 7.30, red solid line). At higher energies, in the EXAFS region (Figure 7.30, blue solid line), the final state of the photoelectron can be described, in a Green's function formalism, as a wave that propagates and scatters from neighbors in single and multiple scattering paths, inducing constructive and destructive interference (see Figure 7.31b).

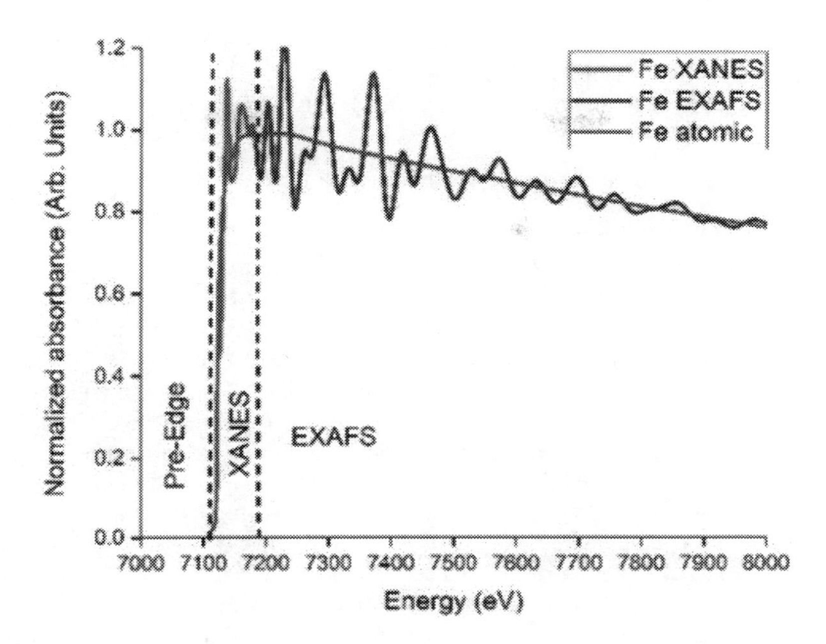

FIGURE 7.30 Simulation of the X-ray absorption of iron. Blackline: an isolated iron atom (with no neighbors). Redline: X-ray near-edge absorption spectra and blue line: extended X-ray absorption spectra of α-Fe.

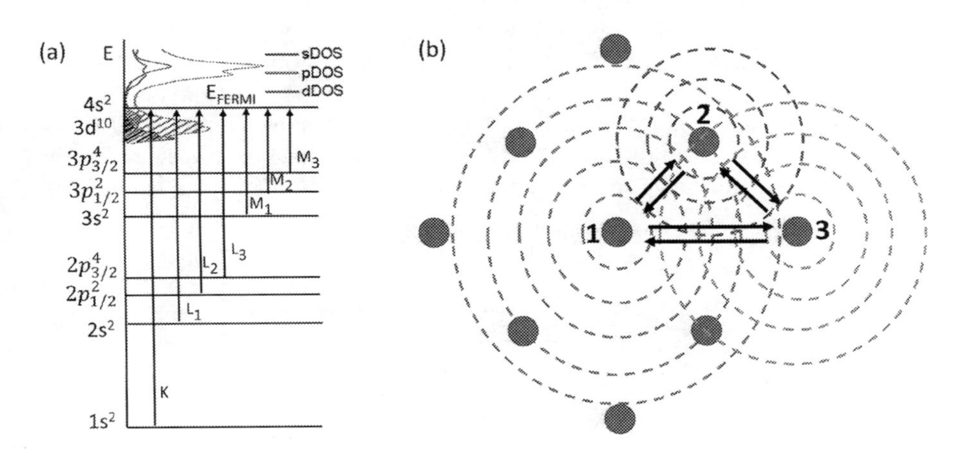

FIGURE 7.31 (a) Near-edge X-ray absorption showing the *K*, *L*, and *M*, X-ray absorption edges, and the corresponding excited core electron of a specific atomic species to unoccupied electron states above the Fermi level. (b) Photoelectron scattering in the extended X-ray absorption regime, with single scattering paths (1-2-1 and 1-3-1) and multiple scattering paths (1-2-3-1 and 1-3-2-1). Figure 7.31a is based on Figure 3 from [56]. (Reused from: Rehr, J. J., & Albers, R. C. (2000). Theoretical approaches to X-ray absorption fine structure. Reviews of Modern Physics, 72(3), 621–654).

This interference modulates the absorption coefficient generating oscillations in the absorption spectra. It is important to note that in the EXAFS region, the oscillations in the spectra are generally dominated by single scattering paths.

7.7.3 Experimental and Analytical Procedures

7.7.3.1 Synchrotron Radiation and Beamline Instrumentation

Synchrotron radiation (SR) is generated when a charged particle (i.e., an electron or a positron) at relativistic speed changes its trajectory due to the Lorentz force produced by, for example, a magnet. This electromagnetic radiation, also known as magneto bremsstrahlung radiation, can be produced from a bending magnet in a circularly shaped storage ring or from an "insertion device" such as a wiggler or an undulator inserted before a beamline. Historically, SR facilities evolved from being an interesting "by-product" of high-energy physics laboratories operating in "parasitic mode" to modern dedicated user facilities with a large community of users worldwide. The key benefits of SR are the continuous and highly intense X-ray spectra. Depending on the characteristics of the storage ring, we distinguish between soft and hard X-rays. Among techniques, photoelectron spectroscopy has benefited enormously from soft X-rays, while EXAFS and XRD benefited from hard X-rays. Modern third-generation facilities concentrate not only on the ray flux of photons but also on the brilliance, i.e., the flux per unit area of the source per unit solid angle of the radiation cone per unit spectral bandwidth, since some experiments get important benefits from it.

A typical beamline in a synchrotron facility is depicted in Figure 7.32. The ribbon-shaped X-ray is delimited by rectangular slits. The highest energy resolution from the monochromator is obtained with a narrow horizontal line-shaped beam. Thus, to obtain high photon fluxes, the brightness of the source becomes relevant. From the white spectra of the synchrotron, the liquid nitrogen cooled monochromator selects specific energy governed by Bragg's diffraction law. Typical monochromator crystals are Si (200) and Si (111). The selection of crystal depends on the energy and the "glitch map" of each crystal. It is important to note that not only the

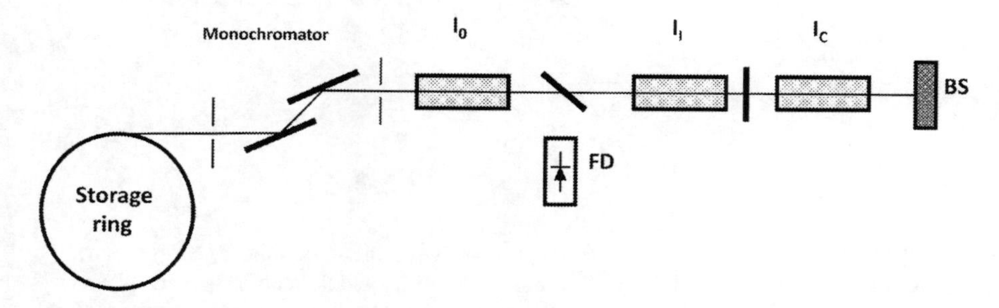

FIGURE 7.32 Experimental setup for X-ray absorption spectroscopy. Ion chambers I_0, I_I, and I_c measure beam intensity as it passes through sample (I_0/I_c) and reference foil (I_I/I_c). Fluorescence detector (FD) at a right angle from the beam with the sample at 45° for fluorescence measurements.

fundamental reflection passes through the double-crystal monochromator but also harmonics. To get rid of harmonics, two techniques are commonly used: the first is to finely "detune" the second monochromator crystal to position it at an angle slightly shifted from the exact Bragg angle (which the first crystal is at). The "detuning" is fixed when the intensity of the X-ray beam falls to 50%. This gets rid of the harmonics since they are narrower than the fundamental reflection. The second technique is using a mirror typically made of Pt or Rh. This works based on the total external reflection of X-rays, so carefully modifying the angle of the mirror changes the cutoff energy. Photons below the cutoff reflect from the mirror, while higher energy photons get absorbed.

Inside the user experimental station or "hutch," a table with all instrumentation is aligned and tracks the beam coming from the double-crystal monochromator making the beam pass through the center of the first ionization chamber (I_0), made of parallel plates. Ionization chambers are filled with either nitrogen, for lower energy hard X-rays, or argon, for higher energy. The measured current, product of the gas ionization, and the high voltage between the plates is proportional to the photon flux and is thus taken as the incoming beam intensity monitor. The next object in the beam path is the sample. For X-ray absorption measurements in transmission mode, the optimal sample thickness is around one absorption length. In the case of a powder sample, it could be compacted in the form of a pellet with homogeneous thickness. Thick or thin-film samples of proper thickness could also be used. In fluorescence mode, a thin sample should be prepared to minimize self-absorption. Among the advantages of fluorescence measurements is that the sample does not need to be homogeneous so even powder impregnated in an adhesive tape could be measured. To minimize thermal broadening, measurements at low temperature are, in general, preferred using liquid nitrogen or a He cryostat. Fluorescence from the sample is generally measured using an energy discriminating detector (Si, Ge). Photons coming from the main fluorescence line of the atom of interest are isolated and counted in a multichannel analyzer to obtain the absorption spectra of that specific atom without interference from other incoming X-rays from the sample such as elastic scattering, Compton scattering, or fluorescence from other elements. Diffraction peaks also must be avoided. Modern detectors are multielement packed in their measuring head from a few to a 100 X-ray detector elements. Each element is "windowed" to the main fluorescence line of the element of interest and the average spectra divided by the intensity of the X-ray beam; I_0, is taken as the fluorescence measurement, $I_f = F/I_0$. Following the beam path, ion chamber I_1 measures the transmitted X-rays. The logarithm of the ratio between I_0 and I_1 is taken as the absorption spectra in transmission mode. Finally, a metallic or known composition foil of the element of interest is mounted in front of I_2 for calibration purposes during data analysis.

7.7.4 Data Reduction and Analysis

Data reduction includes correction for deadtime in solid-state detectors (fluorescence detector) and the adjustment of the absorption spectra for fluctuations in the bean intensity by performing the ratio of the fluorescence signal and the incident intensity (I_F/I_0), or the $\log(I_0/I_1)$ for transmission mode. Normalization of the X-ray absorption spectra involves the least-squares fitting, a second-order polynomial over the pre-edge, and a third-order polynomial over the postedge, making the difference at the edge energy (E_0) unity. Edge energy is commonly selected at the energy where the maximum of the first derivative of the absorption edge is located.

With the normalized spectra, comparisons and simulations of XANES are evaluated. For the extraction of the EXAFS, the difference between the data, $\mu(E)$, and a spline function (that resembles the atomic absorption spectra, $\mu_0(E)$), is calculated by optimizing the position of the knots of the spline function to reduce low-frequency contributions minimizing the amplitude of the Fourier Transform of the EXAFS signal (the radial distribution function) at distances below 1 Å, where no atomic contribution is expected. This process is probably the more complex part of EXAFS data reduction and should be performed systematically for a given set of data. Finally, we obtain the experimental EXAFS signal $\chi(E)$.

$$\chi(E) = \frac{\mu(E) - \mu_o(E)}{\mu_0(E)} \tag{7.14}$$

The EXAFS oscillations are finally transformed from energy to photoelectron wave-vector such that $k = \sqrt{(2m / \hbar^2)(E - E_0)}$.

In practice, theoretical developments, as described before, lead to the EXAFS equation:

$$\chi(k) = \sum_j \frac{S_0^2 N_j f_j(k) e^{-2k^2\sigma_j^2}}{kR_j^2} \sin\left[2kR_j + \delta_j(k)\right] \tag{7.15}$$

This describes the amplitude and phase of the EXAFS oscillations separated as "shells" of neighbors (with index "*j*"). To extract the neighbor environment around the absorbing species and measure deviations from the crystallographic positions, theoretical scattering amplitudes $f_j(k)$ and phases $\delta_j(k)$ are calculated with the *ab-initio* code FEFF [57]. Theoretical parameters are then included in the EXAFS equation together with the estimated values of distance (R), pair-wise Debye–Waller factor (σ), number of Neighbors (N), and the many-body amplitude reduction factor (S_0^2), to be fitted in the least-squares refinement procedure.

7.7.5 In-situ/Operando XAS Experimentation

To increase the efficiency, safety, and finally cost of energy storage devices including energy density and cycling life, structural and electronic information at the molecular level is required. These studies could benefit by analyzing the evolution of the atomic structure of energy storage materials as they operate. Electrode dynamics include charge transfer, chemical bonding, diffusion, phase changes, etc. The combination of characterization tools like electrochemical methods with *in-situ* XAS provides valuable information. *In-operando* characterization depends on timeframe: data collection times need to be at least an order of magnitude faster than charging/discharging time. XANES spectra can comply with required times in a standard beamline; however, a quick-XAS beamline/setup is required for fast phenomena [58–60]. Experimentally, cells for *in-situ* experiments are specifically designed and build for XAS [60–65]. Design parameters include X-ray energy range, measurement type (transmission/fluorescence), pressure, temperature, and electrodes. Body material can be stainless steel or a thermoplastic like a polyether ether ketone (PEEK) (with a glass transition temperature of ~143°C and melting point at ~343°C) or PTFE (Teflon®) (glass transition temperature of ~114°C, and melting point at 327°C) or any other material like polycarbonate or acrylic with the mechanical and chemical resistance. Screws to seal the body of the cell should be tightened with care to sustain an even pressure. The same screws could be used as electrical contacts. Transmission mode measurements require an entrance and exit window of a low-Z material. Beryllium is a typical window material for X-rays, although safety concerns could prevent its use. For room and moderate temperatures (<400°C), polyimide (Kapton) of a few tenths of a millimeter is an option, avoiding safety concerns and the possible reaction of BE with cell electrolyte/materials. Kapton is highly resistant to humidity, strong acids, most solvents, and a little less resistant to strong bases. It could also be glued to the stainless-steel body material with an epoxy adhesive to form a vacuum-tight window. However, O_2 and N_2 sensitive experiments should take into account diffusion rates. If a rigid X-ray window is required, graphite or a glassy carbon in combination with polyamide or thin quartz glass could be used. If electrical conductivity or better isolation from outside gasses is required, a thin film of a suitable low-Z metal (aluminum) could be deposited (sputtered) into the polyimide. Electrical connection can be derived from the assembly screws or passing through the body material in ceramic or plastic tubes for insulation with epoxy adhesive or thermoplastic filling to leak-tight seal the feedthroughs. Liquid or gas inlet/outlets can also be integrated into the body of the cell. PTFE jackets and spacers can also be used to build a stack within the cell. Finally, silicone or perfluoro elastomer O-rings are used seal the cell or any subsection of the stack within it. Care should be taken in the assembly to avoid bubbles of gas to remain in the cell.

Electrical connections in a compact cell for *in-situ/operando* experimentation depend on the type of experiment. For a two-electrode anode–cathode connection, with X-rays passing through both the anode and cathode, simple feedthroughs are required with proper materials thicknesses to optimize the X-ray absorption experiment. For three-electrode designs, the situation is more complex. In a typical cell, the working and counter electrode are in front of each other and relatively far apart, so the electric field is close to being constant at regions close to the working electrode (i.e., as in a parallel plate capacitor). However, for XAS measurements, the counter electrode should not block the X-rays, so a ring of platinum wire could be used in a configuration with the working electrode close to the entrance X-ray window, a micro reference electrode in the middle with sealing O-rings, and the counter electrode close to X-ray exit window [58,66,67].

7.7.6 *Ex-situ* XAS Experimentation

In *ex-situ* experiments, the electrochemical process is stopped at regular intervals, and samples are measured. This has the potential disadvantage of contamination or

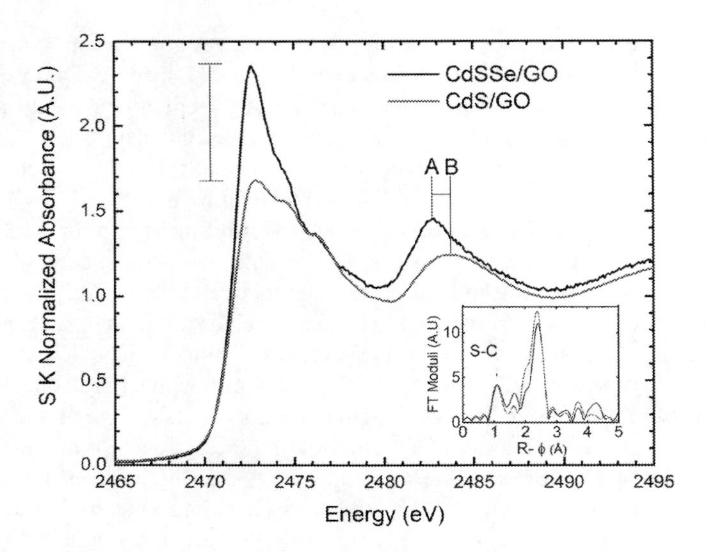

FIGURE 7.33 Normalized XANES spectra of CdS/GO (Red) and CdSSe/GO (black). Inlet: Fourier Transform Magnitude comparison of the EXAFS of both samples showing the S–C bond. (Reprinted (adapted) with permission from [68]. © 2019 American Chemical Society).

unwanted oxidation/reduction, but the advantage of being very useful for acquiring EXAFS data to measure changes in the local atomic structure. Materials can also be specifically synthesized/grown for EXAFS experiments to obtain the best possible signal-to-noise ratios. In a recent study by members of our group [68], multilayer $CdS_{1-x}Se_x$/Graphene oxide films with $x=0.5$ and $x=0$ and 10 deposition cycles were analyzed by XAS. Although several phenomenological band diagrams have been proposed to explain graphene effects, there is still scarce experimental evidence on the actual mechanisms of graphene interaction with semiconductors. Experimental data were measured at the Stanford Synchrotron Radiation Light source. Figure 7.33 shows the XANES spectra with two distinct features. First, a difference in the amplitude of the main peak, ascribed to an increase of unoccupied electron states, and second a shift of the second peak, ascribed to the difference in atomic radius between S and Se. Figure 7.34 shows the fit to the EXAFS by multiple shells of neighbors. From this information, it was concluded that a chemical bond between S and C is the link between CdS(Se) and GO. Band diagrams were proposed by combining optical and photocurrent data. We found that films present intercalated nanoparticles within the GO matrix and defined Raman transitions related to crystalline nanoparticles, as well as S–C bonding. From chemical data, it was found that CdS/GO film is S-deficient, while the CdSSe/GO film is Cd deficient, which caused positive charge accumulation in the CdS/GO interface and a negative one in the CdSSe case, respectively. The charging behavior leads to unexpected stability of the photocurrent upon the electric potential increase, due to recombination through surface states originated by the Cd excess. On the other hand, Se accumulation leads to an increase of the observed photocurrent, thus leading to the possibility of quantum switches controlled by the composition and probably by the particle size/

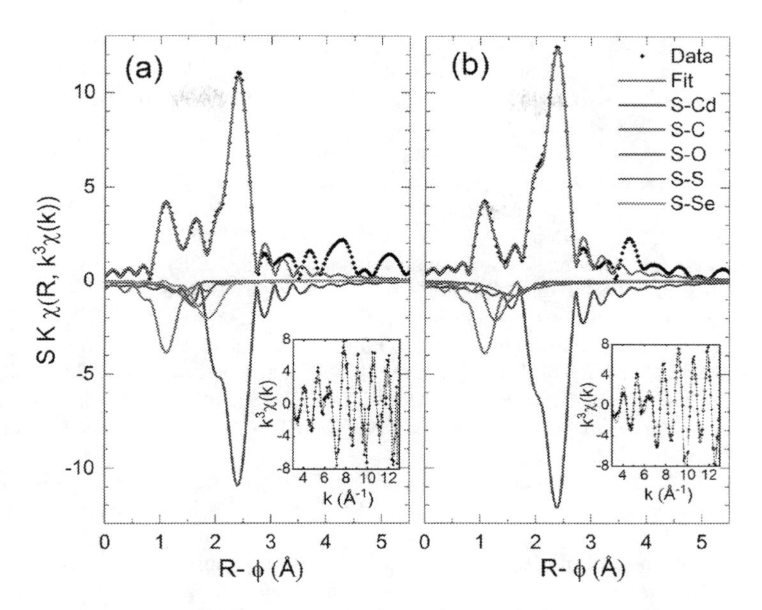

FIGURE 7.34 Sulfur EXAFS data (black) and fit (red) in k-space (insets) and real space of (a) CdS/GO and (b) CdSSe/GO. The lower part (inverted) are the individual contributions of each shell of atoms to the spectra. (Reprinted (adapted) with permission from, Colina-Ruiz, Roberto A., Tolentino-Hernandez, R. V., Guarneros-Aguilar, C., Mustre de León, J., Espinosa-Faller, F. J., & Caballero-Briones, F. (2019). Chemical Bonding and Electronic Structure in CdS/GO and CdSSe/GO Multilayer Films. The Journal of Physical Chemistry C, acs.jpcc.9b03328) [68]. © 2019 American Chemical Society.

semiconductor thickness. As mentioned before, an insightful approach to study such systems is the coupling of electrochemical, spectroscopic, and structural methods to obtain a realistic atomic structure of materials and be able to properly simulate and optimize them.

REFERENCES

[1] J. E. Griffith and G. P. Kochanski, "Scanning tunneling microscopy," *Annu. Rev. Mater. Sci.*, vol. 20, no. 1, pp. 219–244, 1990, doi: 10.1146/annurev.ms.20.080190.001251.

[2] Y. Jiang et al., "Inducing Kondo screening of vacancy magnetic moments in graphene with gating and local curvature," *Nat. Commun.*, vol. 9:2349, no. 1, pp. 1–7, 2018, doi: 10.1038/s41467-018-04812-6.

[3] B. Voigtländer, *Scanning Probe Microscopy–Atomic Force Microscopy and Scanning Tunneling Microscopy.* Springer, Berlin, 2015, doi: https://doi.org/10.1007/978-3-662-45240-0

[4] F. Fang and B. Ju, "Scanning tunneling microscope," in *CIRP Encyclopedia of Production Engineering*, 1st., L. Laperriére and G. Reinhart, Eds. Springer, Berlin, Heidelberg, 2019, pp. 1090–1092.

[5] Y. He, S. Chen, and Q. Yu, "Scanning tunneling microscope (STM)," in *Encyclopedia of Tribology*, vol. 150, no. 1–2, Q. J. Wang and Y.-W. Chung, Eds. Springer, Boston, MA, 1991, pp. 2983–2988.

[6] D. Kaya, R. J. Cobley, and R. E. Palmer, "Combining scanning tunneling microscope (STM) imaging and local manipulation to probe the high dose oxidation structure of the Si(111)-7×7 surface," *Nano Res.*, vol. 13, no. 1, pp. 145–150, 2020, doi: 10.1007/s12274-019-2587-1.

[7] A. J. Bard, G. Denuault, C. Lee, D. Mandler, and D. O. Wipf, "Scanning electrochemical microscopy: A new technique for the characterization and modification of surfaces," *Acc. Chem. Res.*, vol. 23, no. 11, pp. 257–263, 1990, doi: 10.1021/ar00179a002.

[8] L. Huang, Z. Li, Y. Lou, F. Cao, D. Zhang, and X. Li, "Recent advances in scanning electrochemical microscopy for biological applications," *Materials (Basel).*, vol. 11, no. 8, 2018, doi: 10.3390/ma11081389.

[9] Wikipedia Commons Contributors, "Fig2_SECM_Modded.jpg." Wikimedia Commons, the free media repository, [Online]. Available: https://commons.wikimedia.org/w/index.php?title=File:Fig2_SECM_Modded.jpg&oldid=465401528.

[10] F.-R. F. Fan, J. Fernandez, B. Liu, and J. Mauzeroll, "Scanning electrochemical microscopy," in *Handbook of Electrochemistry*, C. G. Zoski, Ed. Elsevier Science, Amsterdam, 2007, pp. 471–540.

[11] A. J. Bard, "Introduction and principles," in *Scanning Electrochemical Microscopy*, 2nd., A. J. Bard and M. V. Mirkin, Eds. CRC Press Taylor & Francis, Boca Raton, FL, 2012, pp. 1–15.

[12] M. V. Mirkin and B. R. Horrocks, "Fundamentals of scanning electrochemical microscopy," in *Electrochemical Microsystem Technologies*, J. W. Schultze, T. Osaka, and M. Datta, Eds. Taylor & Francis, London, 2002, pp. 66–103.

[13] E. Ventosa, P. Wilde, A.-H. Zinn, M. Trautmann, A. Ludwig, and W. Schuhmann, "Understanding surface reactivity of Si electrodes in Li-ion batteries by in operando scanning electrochemical microscopy," *Chem. Commun*, vol. 52, p. 6825, 2016, doi: 10.1039/c6cc02493a. This article is licensed under a Creative Commons Attribution 3.0 Unported Licence. You can use material from this article in other publications without requesting further permissions from the RSC, provided that the correct acknowledgement is given.

[14] M. Hampel, M. Schenderlein, C. Schary, M. Dimper, and O. Ozcan, "Efficient detection of localized corrosion processes on stainless steel by means of scanning electrochemical microscopy (SECM) using a multi-electrode approach," *Electrochem. Commun.*, vol. 101, no. February, pp. 52–55, 2019, doi: 10.1016/j.elecom.2019.02.019. This is an open access article distributed under the terms of the Creative Commons CC-BY license, which permits unrestricted use, distribution, and reproduction in any medium, provided the original work is properly cited. You are not required to obtain permission to reuse this article.

[15] M. Donnici et al., "Characterization of Inherently Chiral Electrosynthesized Oligomeric Films by Voltammetry and Scanning Electrochemical Microscopy (SECM)," *Molecules*, vol. 25, no. 5368, 2020, doi: 10.3390/molecules25225368. This is an open access article distributed under the Creative Commons Attribution License which permits unrestricted use, distribution, and reproduction in any medium, provided the original work is properly cited.

[16] Binnig, G., Quate, C. F., & Gerber, C, "Atomic force microscope," *Phys. Rev. Lett.*, vol. 56, no. 9, pp. 930–933, 1986.

[17] B. Voigtländer, *Scanning Probe Microscopy—Atomic Force Microscopy and Scanning Tunneling Microscopy*. Springer, Berlin (2015).

[18] V. Bellitto, ed. *Atomic Force Microscopy: Imaging, Measuring and Manipulating Surfaces at the Atomic Scale*. IntechOpen, London (2012).

[19] R. M. Feenstra, Scanning tunneling spectroscopy, *Surf. Sci.*, vol. 299–300, pp. 965–979, 1994.

[20] F. Caballero-Briones, *Semiconducting Thin Films Prepared by Electrochemical Modulation for Technological Devices*. Diss. PhD thesis. Barcelona, Spain: Universidad de Barcelona, (2009). Y. Boukellal, Contribution à la mise en place d'un microscope à force Atomique métrologique (mAFM): Conception d'une tête AFM métrologique et caractérisation métrologique de l'instrument (Doctoral dissertation, École normale supérieure de Cachan-ENS Cachan). (2015)

[21] H. Zhang, J. Huang, Y. Wang, R. Liu, X. Huai, J. Jiang, C. Anfuso, Atomic force microscopy for two-dimensional materials: A tutorial review, *Optics Commun.*, vol. 406, pp. 3–17, 2018.

[22] F. T. Limpoco. *How to Choose the Right Afm Probe*. [Webinar]. Oxford Instruments. (2017).

[23] S. Alexander, L. Hellemans, O. Marti, J. Schneir, V. Elings, P. Hansma, M. Longmire et J. Gurley. An atomic resolution atomic force microscope implemented using an optical lever, *J. Appl. Phys.*, vol. 65, pp. 164–167, 1988.

[24] R. Fung, S. Huang, Dynamic modeling and vibration force microscope, *ASME J. Vib. Acoust*, vol. 123, pp. 502–509, 2001.

[25] M. Basso, L. Giarre, M. Dahleh, I. Mezic. Numerical analysis of complex dynamics in atomic force microscopes. *Proc IEEE Conf Control Appl. Trieste, Italy*, vol.2 pp. 1026–1030, 1998.

[26] A. Sebastian, M. Salapaka, D. Chen, J. Cleveland, Harmonic analysis based modeling of tapping-mode AFM. *Proc Am Control Conf. San Diego CA*, vol. 1, pp. 232–236, 1999.

[27] R. Wiesendanger, *Scanning Probe Microscopy and Spectroscopy: Methods and Applications*, Cambridge University Press, Cambridge, UK, 1994.

[28] B. Voigtländer, Static atomic force microscopy. In: *Atomic Force Microscopy. Nano Science and Technology*. Springer, Cham. (2019) https://doi.org/10.1007/978-3-030-13654-3_12

[29] A. Jawad, *AFM Handbook; Theoretical Principles and Experimental Parameters: An Introduction to Theoretical principles and Experimental Parameters of Atomic Force Microscopy*. Scholars´ Press, Mauritius, (2020).

[30] JPK Instruments AG. Nanowizard ® *AFM Handbook*. Germany. (2012).

[31] Smith E, Dent G. *Modern Raman Spectroscopy–A Practical Approach*. 2nd ed. Wiley, Chichester (2019).

[32] Ferraro JR, Nakamoto K, Brown CW. Introductory raman spectroscopy, Elsevier, Amsterdam, 2003.

[33] Wartewig S. *IR and Raman Spectroscopy–Fundamental Processing*. Wiley, Weinheim. (2003).

[34] Nafie LA. Theory of raman scattering. In: Lewis IR, Edwards HGM, editors. *Hand Book. Raman Spectroscopy*, Marcel Dekker, New York. (2001).

[35] McCreery RL. *Raman Spectroscopy for Chemical Analysis*. John Wiley & Sons, Inc., New York. (2001).

[36] Frédéric Foucher, Guillaume Guimbretière, Nicolas Bost and Frances Westall. Petrographical and mineralogical applications of raman mapping, raman spectroscopy and applications, IntechOpen, London (2017) DOI: 10.5772/65112.

[37] F. Caballero-Briones, O. Calzadilla, F.J. Espinosa-Faller, J.O. Arias, M. Vidales, personal communication.

[38] Reed SJB. *Electron Probe Microanalysis*. 2nd ed. Cambridge, UK: Cambridge University Press; 1993.

[39] Potts AW. *X-ray Science and Technology*. Bristol, UK: Institute of Physics Publications; 1993, 48.

[40] Hertz H. *Ann Phys U Chem (Wied. Ann)*, vol. 31, p. 421, 1887.

[41] Einstein A. On a heuristic point of view about the creation and conversion of light. *Ann Phys*, vol. 17, pp. 132–48, 1905.
[42] M. Maarouf, M. B. Haider, Q. A. Drmosh, M. B. Mekki, X-ray photoelectron spectroscopy depth profiling of as-grown and annealed titanium nitride thin films. *Crystals*, vol. 11, p. 239, 2021.
[43] S. Caporali, F. Muniz-Miranda, A. Pedone, M. Muniz-Miranda, SERS, XPS and DFT study of xanthine adsorbed on citrate-stabilized gold nanoparticles. *Sensors*, vol. 19, p. 2700, 2019.
[44] P. H. Citrin, D. R. Hamann, *Phys. Rev. B,* vol. 15, p. 2923, 1977.
[45] W. Voigt, *Akad. Wiss. München, Math. Phys.,* vol. 216, p. 603, 1912.
[46] G. H. Major, N. Fairley, P. M. A. Sherwood, M. R. Linford, J. Terry, V. Fernandez, K. Artyushkova, Practical guide for curve fitting in X-ray photoelectron spectroscopy, *J. Vac. Sci. Technol. A* vol. 38, p. 061203, 2020.
[47] Laue, M. Eine quantitative Prüfung der Theorie für die Interferenzerscheinungen bei Röntgenstrahlen. *Annalen Der Physik*, vol. 346, no. 10, pp. 989–1002, 1913. https://doi.org/10.1002/andp.19133461005
[48] Bragg, W. H. X-rays and crystals [1]. *Nature,* vol. 90, no. 2243, p. 219, 1912). Nature Publishing Group. https://doi.org/10.1038/090219a0
[49] Conradson, S. D. et al., Intrinsic nanoscience of δ Pu-Ga alloys: Local structure and speciation, collective behavior, nanoscale heterogeneity, and aging mechanisms. *J. Phys. Chem. C*, vol. 118, no. 16, pp. 8541–8563, 2014. https://doi.org/10.1021/jp5004038
[50] Bianconi, A. Superconductivity in quantum complex matter: The superstripes landscape. *J. Supercon. Novel Magn.,* vol. 33, no. 8, pp. 2269–2277, 2020). Springer. https://doi.org/10.1007/s10948-020-05602-2
[51] Colina-Ruiz, R. A., Mustre de León, J., Lezama-Pacheco, J. S., Caballero-Briones, F., Acosta-Alejandro, M., & Espinosa-Faller, F. J. Local atomic structure and analysis of secondary phases in non-stoichiometric Cu_2ZnSnS_4 using X-ray absorption fine structure spectroscopy. *J. Alloys Comp*, vol. 714, pp. 381–389, 2017. https://doi.org/10.1016/j.jallcom.2017.04.191
[52] Egami, T., & Billinge, S. J. *Underneath the Bragg Peaks* (2nd ed.). Pergamon Press, Oxford, (2012) https://doi.org/10.1016/s1369-7021(03)00635-7
[53] Proffen, T., Billinge, S. J. L., Egami, T., & Louca, D. Structural analysis of complex materials using the atomic pair distribution function–A practical guide. *Zeitschrift fur Kristallographie,* vol. 218, no. 2, pp. 132–143, 2003). https://doi.org/10.1524/zkri.218.2.132.20664
[54] de Broglie, M. La spectrographie des rayons de Röntgen. *J. Phys. Appl.*, vol. 4, no. 1, pp. 101–116, 1914. https://doi.org/10.1051/jphystap:019140040010100ï
[55] Koningsberger, D. C., & Prins, R. X-ray absorption: principles, applications, techniques of EXAFS, SEXAFS and XANES, Wiley, New York, 1988.
[56] Rehr, J. J., & Albers, R. C. Theoretical approaches to X-ray absorption fine structure. *Rev. Modern Phys.*, vol. 72, no. 3, pp. 621–654, 2000. https://doi.org/10.1103/RevModPhys.72.621
[57] Rehr, John J, Kas, J. J., Prange, M. P., Sorini, A. P., Takimoto, Y., & Vila, F. Ab initio theory and calculations of X-ray spectra. *C. R. Physique*, vol. 10, pp. 548–559, 2009. https://doi.org/10.1016/j.crhy.2008.08.004
[58] Rehr, John J., Kas, J. J., Vila, F. D., Prange, M. P., & Jorissen, K. Parameter-free calculations of X-ray spectra with FEFF9. *Phys. Chem. Chem. Phys.,* vol. 12, no. 21, pp. 5503–5513, 2010). The Royal Society of Chemistry. https://doi.org/10.1039/b926434e
[59] Goubard-Bretesché, N., Crosnier, O., Douard, C., Iadecola, A., Retoux, R., Payen, C., Doublet, M. L., Kisu, K., Iwama, E., Naoi, K., Favier, F., & Brousse, T. Unveiling pseudocapacitive charge storage behavior in $FeWO_4$ electrode material by operando X-ray absorption spectroscopy. *Small*, vol. 16, no. 33, p. 2002855, 2020. https://doi.org/10.1002/smll.202002855

[60] Giorgetti, M., & Stievano, L. (2017). X-Ray absorption spectroscopy study of battery materials. *X-ray Charac. Nano. Ener. Mater. Sync. Radi.*, InTech. https://doi.org/10.5772/66868

[61] N. A., Nayak, C., Halankar, K. K., Jha, S. N., & Bhattacharyya, D. First results of in-situ X-ray absorption spectroscopy study on charging-discharging cycles of lithium ion battery at Indus-2 SRS. *Nuclear Instr. Methods Phys. Res., Sec. A: Accel. Spectro. Detect. Assoc. Equip.*, vol. 972, p. 164032, 2020. https://doi.org/10.1016/j.nima.2020.164032

[62] Borkiewicz, O. J., Shyam, B., Wiaderek, K. M., Kurtz, C., Chupas, P. J., & Chapman, K. W. The AMPIX electrochemical cell: A versatile apparatus for in situ X-ray scattering and spectroscopic measurements. *J. Appl. Crystall.*, vol. 45, no. 6, pp. 1261–1269, 2012. https://doi.org/10.1107/S0021889812042720

[63] Bak, S. M., Shadike, Z., Lin, R., Yu, X., & Yang, X. Q. In situ/operando synchrotron-based X-ray techniques for lithium-ion battery research. *NPG Asia Mater.*, vol. 10, no. 7, pp. 563–580, 2018). Nature Publishing Group. https://doi.org/10.1038/s41427-018-0056-z

[64] Bare, S. R., Mickelson, G. E., Modica, F. S., Ringwelski, A. Z., & Yang, N. Simple flow through reaction cells for in situ transmission and fluorescence x-ray-absorption spectroscopy of heterogeneous catalysts. *Rev. Sci. Instr.*, vol. 77, no. 2, p. 023105, 2006. https://doi.org/10.1063/1.2168685

[65] Bare, S. R., Yang, N., Kelly, S. D., Mickelson, G. E., & Modica, F. S. Design and operation of a high pressure reaction cell for in situ X-ray absorption spectroscopy. *Catalysis Today*, vol. 126, no. 1–2, pp. 18–26, 2007. https://doi.org/10.1016/j.cattod.2006.10.007

[66] Jentoft, R. E., Deutsch, S. E., & Gates, B. C. Low-cost, heated, and/or cooled flow-through cell for transmission x-ray absorption spectroscopy. *Rev. Sci. Instr.*, vol. 67, no. 6, pp. 2111–2112, 1996. https://doi.org/10.1063/1.1147023

[67] Kao, L. C., Feng, X., Ha, Y., Yang, F., Liu, Y. S., Hahn, N. T., MacDougall, J., Chao, W., Yang, W., Zavadil, K. R., & Guo, J. In-situ/operando X-ray absorption spectroscopic investigation of the electrode/electrolyte interface on the molecular scale. *Surf. Sci.*, vol. 702, p. 121720, 2020. https://doi.org/10.1016/j.susc.2020.121720

[68] Colina-Ruiz, Roberto A., Tolentino-Hernandez, R. V., Guarneros-Aguilar, C., Mustre de León, J., Espinosa-Faller, F. J., & Caballero-Briones, F. Chemical bonding and electronic structure in CdS/GO and CdSSe/GO multilayer films. *J. Phys. Chem. C*, 2019. Acs.jpcc.9b03328. https://doi.org/10.1021/acs.jpcc.9b03328

8 The Fundamental Idea of Electrochemical Devices

N. Gnanaseelan, S.K. Kamaraj, and F. Caballero-Briones

CONTENTS

8.1 INTRODUCTION

The foundation of electrochemistry was laid down by Galvani and Volta during the 18th century. After that, there has been an impressive growth in electrochemical research, resulting in plethora of electrochemical devices such as batteries, supercapacitors, fuel cells, electrochemical sensors, etc. Electrochemical devices became indispensable in human life due to the arrival of numerous portable electronic devices and electric automobiles (even traditional automobiles require batteries to power lights and stereo). This growth has skyrocketed the market value of electrochemical devices. The global market size of battery was USD 108.4 billion in 2019 and expected to grow 14.1% until 2027, whereas the supercapacitor had market of USD3.27 billion in 2019

DOI: 10.1201/9781003216308-8

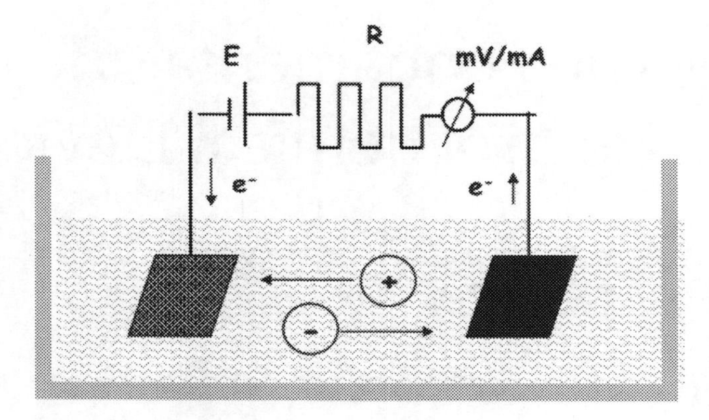

FIGURE 8.1 Schematic diagram of an electrochemical cell. (Reproduced from reference [4] with permission of Springer Nature.)

and anticipated to grow 23.3% until 2027 [1,2]. Fuel cell owned a market value of USD 4.1 billion and sought the growth of 23.2% until 2028 [3].

Electrochemistry refers to science of interfacial reaction between phase containing electrons and phases-containing ions. The phase containing the ions that conduct electrical charge is an electrolyte. The phase is usually the solid, which conducts electrons called electrodes. An electrochemical cell consists of two electrodes immersed in the electrolyte that contains positive and negative ions (Figure 8.1). If the electrons travel through the external circuit, the electrons enter where the reduction happens and leave where the oxidation takes place. The electrons transfer occurs between an interface, an electrode (solid) and electrolyte (liquid or solid), responsible for the development of various fascinating electrochemical devices as some of them are batteries, fuel cell, supercapacitor, and electrochemical sensors.

This book chapter starts with the historical evolution of electrochemical devices. Then, it outlines the basic working principle of different electrochemical devices, which are useful for energy storage, energy generation, and sensing. The chapter also discusses the recent advancement made in those devices.

8.2 HISTORICAL EVOLUTION OF ELECTROCHEMICAL DEVICES

The history of electrochemistry predates to the end of the 17th century after an observation made by Luigi Galvani in 1791. He observed that frog legs hanging in the iron rail were twitched convulsively when they contact with copper wire, and it made him believe that animals are capable of producing electricity. After many experiments, Alessandro Volta found that frog's movements caused by current was produced due to the contact established between a bimetallic conductor and frog's body fluid. Later, Volta constructed the first battery called voltaic pile in around 1800, announced at the Royal Society in London. The battery constructed of staking dissimilar metal discs (zinc and silver) separated by the moistened cloth of sodium chloride. In the span of less than six weeks, Nicholson and Carlisle demonstrated the concept of electrolysis by connecting a voltaic pile to two platinum wires immersed in salt water (conversion of electrical energy into chemical energy). Hydrogen and oxygen accumulated at the end of wires were proportional to the amount of current utilized.

The proposal of Michael Faraday's two quantitative laws of electrolysis and Sir William Grove's invention of fuel cell happened in 1834. Grove showed the generation of electricity using hydrogen and oxygen kept in a separate cell, which were connected in series externally. He used platinum electrodes, which were immersed in sulfuric acid. The issue of overpotential was overcome with the usage of more single fuel cell to produce enough voltage to undergo electrolysis of water. In 1836, a British chemist named John Frederic Daniell designed the first successful battery for telegraphs, which consisted of copper and zinc electrodes in the copper sulfate solution. Edmund Becquerel illuminated light on silver chloride in the acidic solution connected to platinum electrodes, which resulted in photovoltaic (PV) effect of power generation. This was the first solar cell demonstrated in 1839 when Becquerel was just 19.

In 1866, Georges Leclanché developed a battery presently known as dry cell, which consists of zinc rod as anode and manganese oxide-carbon mixture as cathode immersed in an aqueous solution of ammonium chloride. Gaston Planté came up with the discovery of lead acid battery in 1859 and Waldmar Jungner invented the rechargeable nickel-cadmium battery in 1901. The evolution of electrochemical energy storage devices dramatically changed after the arrival of portable devices in 1960 that require high energy density, high power density and long-lasting reliable energy sources. Lithium batteries arrived in the 1970s which powered electronic watches, toys and cameras. Lithium-ion batteries have undergone tremendous modifications as their application widened from tiny portable electronic devices to electric vehicles [5–9].

8.3 ELECTROCHEMICAL ENERGY STORAGE DEVICES

Figure 8.2 shows the Ragone plot for different electrochemical devices, which plotted the power density with respect to energy density. Supercapacitor can deliver higher power density than batteries and fuel cell. Life cycle of supercapacitor is higher than batteries and fuel cells, which don't involve any faradaic reaction for energy storage. Batteries have higher energy density than a supercapacitor, which means batteries can deliver electric power for long time than a supercapacitor. Fuel cell is energy conversion device that has the highest energy density than the rest of the devices. Each device was explained in the preceding sections.

8.3.1 SUPERCAPACITORS

Supercapacitors or electrochemical capacitor accumulate electrical charges at electric double layer forms at the interface between an electrolyte and a conductor [11]. Otherwise, electrochemical pseudocapacitance where charge storage happens at the faradic redox reaction or intercalation occurs at the electrode [12]. Supercapacitors offer high power density, fast charging, wide operating temperature, and long cycle

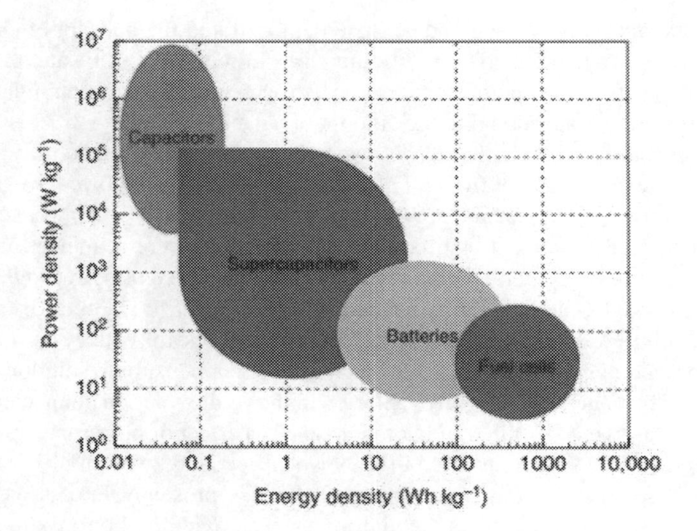

FIGURE 8.2 Ragone plot for electrochemical storage and conversion devices. (Reproduced from reference [10] with permission of John Wiley and sons.)

stability (>100,000) due to electrostatic storage of charges, which do not allow any reversible chemical reactions. Despite these merits, wide usage of supercapacitors is hindered due to its lower energy density. The energy density of commercially available supercapacitor (~5 Wh/kg) is much lower than batteries (~200 Wh/kg) and fuel cell (~350 Wh/kg) [13]. Another disadvantage with a supercapacitor is that the operating voltage of a supercapacitor should be maintained low to prevent chemical decomposition of electrolytes. A supercapacitor consists of two electrodes and a separator placed between them. The same electrodes are configured for symmetric cells, whereas different electrodes involved with asymmetric cells. The separator should possess high ion permeability, high electrical resistivity, and low thickness to attain high performance. In general, polymer or paper separator is used as a separator for an organic electrolyte and ceramic or glass fiber is utilized for an aqueous electrolyte [14].

There are three types of supercapacitors based on the storage mechanism or cell configuration:

1. Electric double layer capacitor (EDLC)
2. Pseudocapacitors
3. Hybrid capacitor

8.3.1.1 Electric Double Layer Capacitors

An EDLC, as its name suggests, stores energy by formation of electric double layer. As voltage is applied between the electrodes, it attracts oppositely charged ions in the electrolyte where it diffuses at the pores of the electrode. The electrode materials are

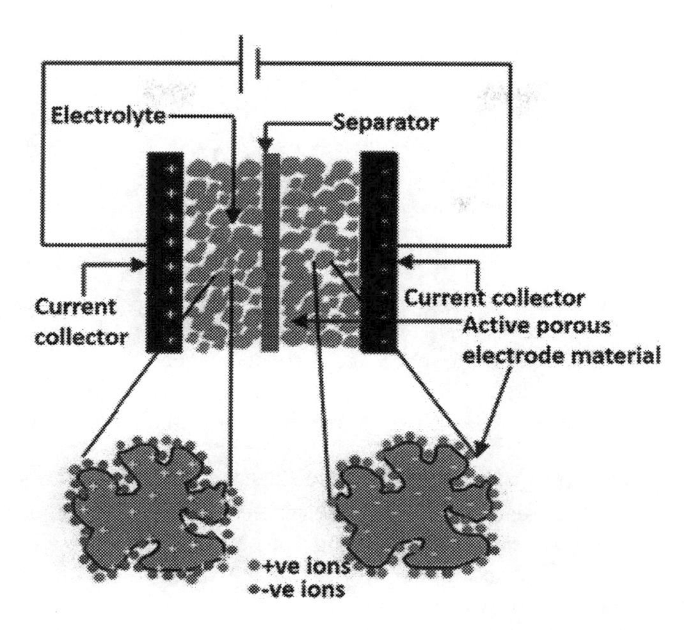

FIGURE 8.3 Schematic representation of the working of EDLC. (Reproduced from reference [15] with permission of Royal Society of Chemistry.)

engineered to prevent the recombination. It leads to the formation of electric double layer and stores energy [16]. Also, it is having much higher surface area and much thinner dielectric medium contributes more energy storage in the supercapacitor, as shown in Figure 8.3.

EDLC stores energy electrostatically through adsorption and desorption of ions at the interface of an electrolyte and conductor. It does not involve any chemical reaction or charge transfer and non-faradaic reactions are primary reason for EDLC, which is highly reversible and stable over more than 10^6 cycles. Its performance could also be altered by using different electrolytes. The capacitance for an acidic electrolyte is higher than the basic electrolyte. Carbon materials are generally used as electrode material for EDLC.

8.3.1.2 Pseudocapacitors

Pseudocapacitor stores energy faradaically through the charge transfer between the electrode and electrolyte. It occurs by the following process: electrosorption, reduction-oxidation reactions and intercalation. It allows a capacitor with higher energy density and capacitance when compared to EDLC. But it is having lower reversibility and cycle stability than EDLC. Conducting polymers (polyaniline, polypyrrole, polythiophene) and metal oxides (RuO_2, MnO_2, CO_3O_4, Co_3O_4 etc.) are generally used as an electrode material for in the case of a pseudocapacitor. RuO_2 exhibits a specific capacitance of ~ 600 F/g, but its higher cost limits its application in a supercapacitor.

8.3.1.3 Hybrid Capacitors

Hybrid capacitors are designed to incorporate the behavior of both EDLC and pseudocapacitance. One electrode material is made with EDLC behavior and another with pseudocapacitance behavior. Hence, a hybrid capacitor can achieve higher energy density and stability.

There are two types of capacitors:

1. Symmetrical capacitors
2. Asymmetrical capacitors

In symmetrical capacitors, both anode and cathode are made up of the same material. In asymmetrical capacitors, both anode and cathode are having different materials in the thrust of achieving large cell potential window and huge energy density.

8.3.1.4 Supercapacitor Devices–Notable Research Developments

Deng et al. fabricated wearable asymmetric supercapacitor, which consists of VO_x used as positive electrode and MnO_x nanoporous structure developed on the conducting paper as negative electrode. The electrolyte was a combination of polyvinyl alcohol and acetamide and $LiClO_4$. The device exhibited specific capacity of 100 F/g, good cyclic stability up to 6000 cycles and maximum energy density/power density about 245 Wh/kg / 95.3 kW/kg [17]. Zhu et al. made coin cell and pouch cell supercapacitor devices using defect-engineered graphene electrodes with polymer electrodes. Nitrogen-doped graphene oxide showed fivefold increase in capacitance than pristine graphene, whereas plasma-processed graphene induced the enhancement in performance [18]. Amit Sanger et al. reported a symmetrical supercapacitor device using silicon carbide as electrode materials with 1M Na_2SO_4 aqueous solution as electrolyte. Device exhibited specific capacity of 300 F/g at 5 mV/s, cycle life of 5000 cycles, high energy density of 31.43 Wh/kg and power density of 18.8 kW/kg at 17.76 Wh/kg at a voltage range of 1.8 V. Unique 3D structure of electrode material leads formation of a conductive network which reduced internal resistance of working electrode, which contributed enhanced performance [19]. Ramaden et al. made an asymmetric supercapacitor consists of activated carbon as negative electrode and binder-free MnO incorporated in carbon as negative electrode. The device was stable for 7000 cycles with an energy density of 35.5 Wh/kg at 1 Ag^{-1} and a power density of 1000 W/kg [20]. Yongmin Ko et al. adopted a ligand-mediated layer-by-layer approach to fabricate a paper-based supercapacitor. This approach was adopted to nullify the problem of low packing density, high contact density, etc. The supercapacitor exhibited the power density of 15.1 mW/cm^2 and an energy density of 267.3 $\mu Wh/cm^2$ with maintenance of 91.8% (with loading of 1.56 mg/cm^2) of initial capacitance for 5000 cycles [21]. Archana et al. synthesized Ni/NiO/nitrogen-doped carbon through the pyrolysis process, which used as positive electrode and reduce graphene oxide as negative electrode to fabricate supercapattery device. The device had a high specific energy of 37 Wh/kg at a specific power of 18 Wh/kg and a high specific power of 2750 W/kg at an energy density of 18 Wh/kg. The device was reportedly stable for 5000 cycles with 100% columbic efficiency. Large surface area and shortened diffusion path of ion/electron transport in conductive carbon network are the reasons for attractive performance [22]. Junais et al. made a laboratory prototype of supercapattery using CeO_2 nanoparticle as positive electrode and reduced graphene oxide as negative electrode. The device showed the specific energy of 20 Wh/kg and a high specific power of 1475 W/kg. The device retained 82% of cycle specific capacity with coulombic efficiency of 100% at 1000th cycle [23].

8.3.2 Sodium Ion Capacitor

Li-ion capacitors are explored for its high energy density, and it lacks of power density and cycle stability. While non-faradic capacitors could deliver high power density and cycle stability with limited energy density when compared to Li-based materials. Ever growing demand for portable electronic devices and EV, which require Li-based storage devices, leads to the depletion of Li reserves on the earth. With practically inexhaustible and less cost, sodium could replace Li as charge carriers in electrochemical storage devices. Na-ion capacitor (NIC) is one type of hybrid storage device which delivers high energy density, high power density and long-life span by performing high bulk intercalation (faradic) at one electrode and high rate surface adsorption (non-faradic) at another electrode [24,25,26]

Figure 8.4 illustrates the mechanism and configurations of NIC, which consists of anode, cathode, electrolyte, separator and current collector. Anions and cations are transported between the anode and cathode during charging or discharging, respectively. As shown in Figure 8.4a, battery-type faradaic redox reactions occur at anode while the capacitive behavior occurs at the cathode. In some cases, capacitive behavior takes place at cathode and redox reactions proceed at the anode as shown in Figure 8.4b. Due to the presence of both capacitive and redox electrodes, NIC could deliver high energy and power density. The electrolytes such as aqueous, organic and gel polymer can be incorporated in the NIC [27].

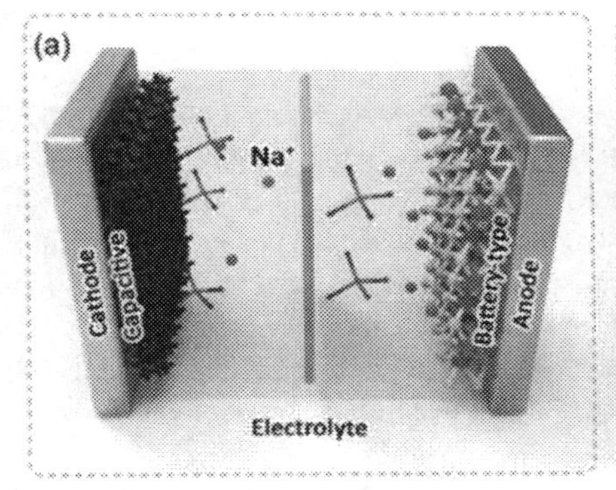

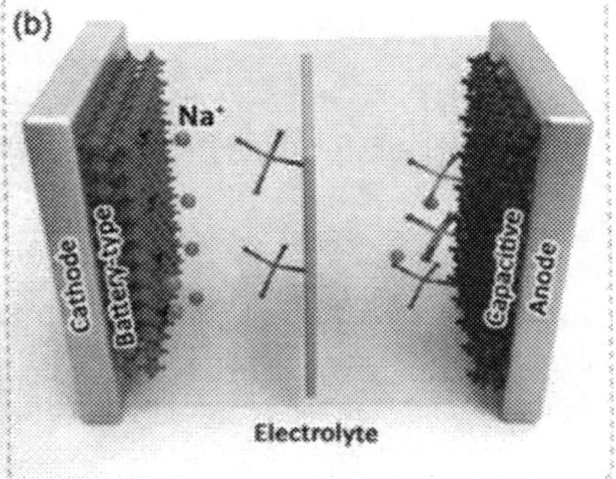

FIGURE 8.4 Charge–discharge mechanism of NIC. (a) Cathode: capacitive, anode: battery-type. (b) Cathode: battery-type, anode: capacitive. (Reproduced from reference [28] with permission of John Wiley and Sons.)

8.3.2.1 Notable Reports on a Sodium Ion Capacitor

Biomass-based carbons act as attractive electrode material for NIC due to efficient, low cost, higher surface area, large ion diffusion length in which some reports were briefed below. Chen et al. developed hierarchically porous carbon using peanut shell, wheat straw, rice straw, corn stalk, cotton stalk and soybean stalk, which is shown in Figure 8.2 [29]. The carbon derived from the cotton stalk exhibited a high reversible capacitance of 160 F/g at 0.2 A/g due to a high surface area of 1994 m^2/g and a large pore volume of 1.107 cm^3/g. A hybrid NIC assembled using cotton stalk-derived carbon and Nb-doped $Na_2Ti_3O_7$ as negative and positive electrodes delivered a high specific energy of 169.4 Wh/kg at 120.5 W/kg.

Wang et al. reported carbon microspheres featured with regular and sharp-edge-free geometry, and high stability. Surface functionalities contained in the material facilitated Na^+ accommodation by dilating the lattice distance. Specific capacity of the carbon microsphere anodes was 183.9 mA/h/g at 50 mA/g with 99% capacity retention for 10,000 cycles [30]. Liu et al. developed a garlic-derived porous carbon after KOH pretreatment and different high-temperature calcination. These carbons were utilized in making the NIC as electrode materials. Charge storage was accomplished in the cell through an ion adsorption mechanism, which resulted in excellent initial coulombic efficiency, high specific capacity and superior cycling stability. The NIC served the energy density of 156 W/h/kg and a power density of 355 W/kg [31]

Zhang et al. reported a moss-covered rock-like hybrid porous carbons for high specific capacitance, excellent rate capacity and exceptional cycle stability using rice husk and coal tar. The reported carbon material had an ultrahigh specific surface area of 3619 m^2/g and graphitic carbon layers. It exhibited the specific capacity of 205 F/g at 0.2 A/g with stability of 1000 cycles [32]. Sujithkrishnan et al. derived biomass from carbon as electrode material for NIC from *Acacia auriculiformis*, which was carbonized and chemically activated using KOH. The assembled pouch cell device exhibited a high specific capacity of 150 Wh/kg at a specific power of 1495 W/kg. The device maintained excellent stability of 10,000 cycles with columbic efficiency of 100% [33].

8.3.3 Li-Ion Battery

Sony first introduced the recharge lithium-ion battery in 1991 and then joint venture of Asahi Kaesei and Toshiba in the following year. The Li-ion battery consisted of carbon as negative electrode material and lithium cobalt oxide as positive electrode material. Those commercialized batteries had energy density that was twice of nickel-cadmium or nickel-metal hydride batteries in terms of volume as well as weight. Due to its low weight long life and high energy density, the application of Li-ion battery widened to cell phones, laptops, electric bicycles, electric scooters, hybrid trucks and buses and electric buses [34,35]. Li-ion battery is also used to buffer the intermittent and fluctuating energy supply from renewable energy technologies such as solar PVs and wind energy turbines. Additional power produced by the solar PV during the sunny times stored in the Li-ion battery could be utilized during the nighttime. Compatibility, absence of memory effect, high gravimetric and volumetric voltage, reliability, and very low self-discharge rate are merits, which make Li-ion batteries a unformidable dominant position in the market. Li-ion cells are connected either in series (to increment current) or parallel (to increment voltage) to construct a Li-ion battery. Seven thousand one hundred four cells are connected in such a way to make the 85 k/wh battery pack in a typical Tesla car [36]. Environmental impact, safety, cycle life, energy and power are the critical areas, which have to be considered before Li-ion battery adoption for various applications [37].

A Li-ion cell is fabricated using anode and cathode, isolated by the separator (microporous polymer membrane) which is sunk in an electrolyte to establish a contact between them, as shown in Figure 8.5. Anode (negative electrode) is mainly made from insertion compound materials, which are graphite and amorphous carbon compounds. Cathodes are possibly made from three structural choices, namely, layered $LiMO_2$ (M=Mn, Co and Ni), spinal $LiMn_2O_4$ and olivine $LiFePO_4$ [37]. Lithium containing an organic electrolyte (For example, $LiPF_6$ salt dissolved into organic liquids, such as ethylene carbonate (EC), dimethyl carbonate (DMC), diethyl carbonate (DEC), ethyl methyl carbonate (EMC) and others) is generally employed. The lithium ions are intercalated and deintercalated during charging and discharging.

When charging happens, lithium-ion is deintercalated from the anode material and reaches the cathode material through a separator and electrolyte. Simultaneously, electrons travel from a negative electrode to a positive electrode via an external circuit. The process is reversed during the charging, where the lithium-ion shuttles from the cathode and intercalated into the anode material through a separator and electrolyte.

The Li-ion-based batteries are classified as the following: [39]

1. Lithium-ion battery
2. Lithium-air/O_2 battery
3. Lithium-sulfur battery
4. Lithium polymer battery

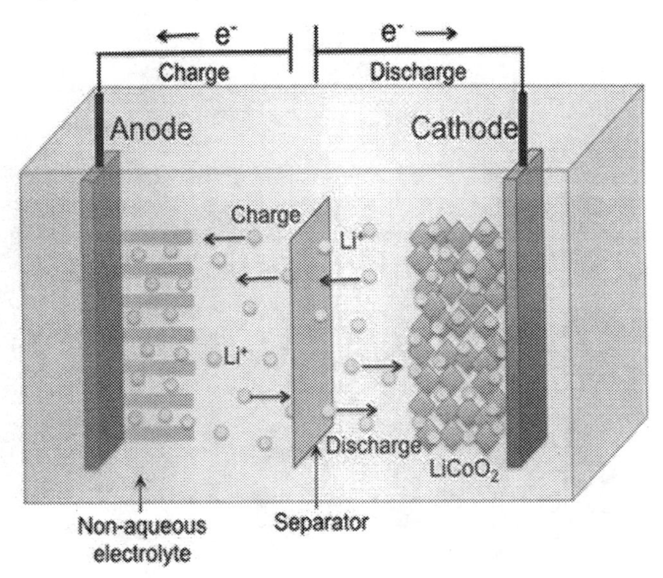

FIGURE 8.5 Working of Li-ion battery. (Reproduced from reference [38] with permission of Royal Society of Chemistry.)

As the lithium-ion battery was discussed in the previous section, rest of the Li-ion-based batteries will be discussed in the following sections.

8.3.3.1 Lithium-Sulfur Battery (LSBY)

The history of Li–sulfur battery started in the 1960s with the usage of sulfur and lithium metal as cathode and anode, respectively. The interest in LSBY gained in recent years due to its theoretical capacity of 1672 mA/h/g, volumetric energy density of 2835 W/h/L and gravimetric energy density of 2572 W/h/kg. Merits of engaging sulfur in energy storage devices include nontoxic, cheap due to abundant reserve of sulfur, and environmentally benign. LSBY consists of lithium metal anode, organic electrolyte and sulfur composite cathode, which is shown in Figure 8.6. The cell operation begun with discharge as sulfur is in charged state. Lithium metal is get oxidized at the negative electrode to generate lithium ions and electrons during the charging.

Lithium ions migrate to a positive electrode through an electrolyte, whereas electrons reach the positive electrode through an external circuit, thus generating the electric current.

$$2\text{Li} \rightarrow 2\text{Li}^{+} + 2\text{e}^{-}$$

After accepting lithium ions and electrons, sulfur is reduced to form to lithium sulfide at the positive electrode.

$$\text{S} + 2\text{Li}^{+} + 2\text{e}^{-} \rightarrow \text{Li S}$$

The overall cell discharge reaction

$$2\text{Li} + \text{S} \rightarrow \text{Li}_2\text{S}$$

During charging, backward reaction will be happening in the cell.

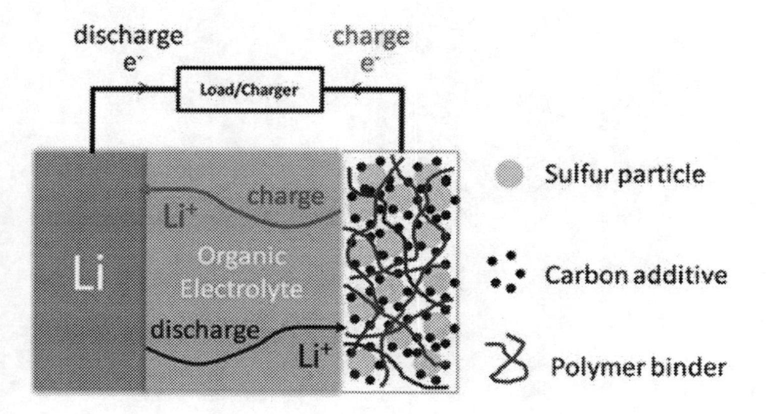

FIGURE 8.6 Schematic diagram of a Li–S battery cell. (Reproduced from reference [40] with permission of ACS Publications.)

The following are potential drawbacks, which should be overcome.

1. Lower electrical conductivity of sulfur,
2. Pulverization of cathode due to large volume expansion of sulfur during lithiation and shuttle effect.
3. Capacity decay of electrochemical cell due to shuttling of polysulfides between an anode and cathode [40–43].

8.3.3.2 Lithium Air/O_2 Battery (LABY)

Abraham and Jiang first reported LABY in 1996, through a fabricated prototype with a polymer electrolyte [44]. The theoretical capacity of the LABY is 3500 Wh/kg, which is more suitable for long-range electric vehicles (>500 km). LABYs are categorized into the following types based on the electrolytes: (a) non-aqueous (aprotic organic solvents and ionic liquids), (b) aqueous, (c) hybrid and (d) solid electrolytes. Among the above electrolytes, non-aqueous has the highest energy density with a simple battery structure. A typical non-aqueous lithium-air battery consists of lithium metal anode, a porous carbon-based cathode exposed to atmospheric oxygen and non-aqueous electrolyte soaked in a separator (Figure 8.7). Carbon act as a host for active material at cathode and picked for its good electrical conductivity, cheap, lightweight, and high chemical stability. During discharge (Figure 8.7a), dimolecule oxygen from air oxidizes lithium metal to form lithium ions and electrons at anode. Meanwhile, Li^+ reacts with oxygen and electron to form Li_2O_2 on the air cathode.

$$\text{Anode}: 2\text{Li} \leftrightarrow 2\text{Li}^+ + 2\text{e}^-$$

$$\text{Cathode}: 2\text{Li}^+ + \text{O}_2 + 2\text{e}^- \leftrightarrow \text{Li}_2\text{O}_2$$

$$\text{Total}: 2\text{Li} + \text{O}_2 \rightarrow \text{Li}_2\text{O}_2 \quad \left(E_{ocv} = 2.96\ V \text{ vs. Li / Li}^+\right)$$

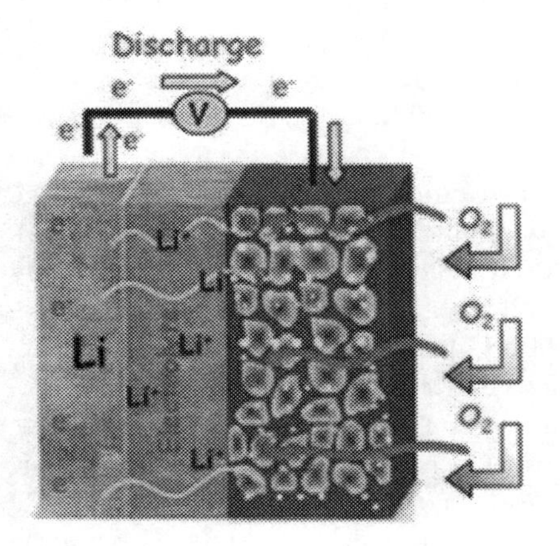

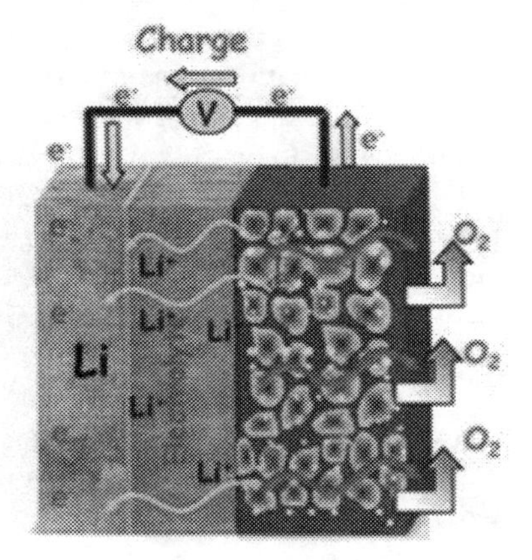

FIGURE 8.7 Schematic diagram of (a) charge and (b) discharge processes of Li–air battery. (Reproduced from reference [45] with permission of ACS Publications.)

The reactions are reversed during charging, Li_2O_2 is reduced to Li and air released back to atmosphere as shown in Figure 8.7b. Pending problems with LABYs are short cycle life, low trip efficiency, poor rate capability and poor capacity [46–49].

8.3.3.3 Lithium Polymer Battery (LPBY)

Lithium polymer batteries have been explored since 1973, which could fulfill the requirement of portable electronic devices, mobile phones and wearable electronics. LBPY offers high energy density, long cyclability, being reliable, flexible and safety. Here, polymer-based electrolytes are used to have plastic nature, different shapes and thin configuration. For example, poly (ethylene oxide) (PEO) with lithium salt, LiX (where *X* is mostly large soft anion). Li ions are coiled around in the PEO-LiX with loosely coordinated anion, which offers fast Li^+ ion transport and high conductivity. This could occur only when PEO is in amorphous state, which occurs at the temperature of 70°C. This kind of arrangement is more suitable for EV traction. LPBY is generally constructed by laminating a lithium metal (otherwise a composite carbon) anode, a lithium-ion conducting membrane (polymer electrolyte) and composite cathode. Other types of polyelectrolytes are based on poly(acrylo-nitrile) (PAN), poly(methyl methacrylate (PMMA), poly (vinyl chloride) and poly(vinylidene fluoride) (PVDF). The charging and discharging operation would be like intercalation and deintercalation mechanism of Li-ion battery, which is explained in the previous section. Polymer electrolytes overcome the drawbacks of liquid electrolytes such as electrolyte leakage, poor electrochemical window, accumulation of extra Li ions on anode surface (dendrite aggregation) and flammability. However, it suffers from issues pertaining to high interfacial resistance of electrode/electrolyte, low ionic conductivity and lithium-ion transference number [50–53].

8.3.4 Recent Trends in Li-Based Batteries

Sui et al. reported organic-inorganic composite by impregnating electrochemically active 4,8-dihydrobenzo [1,2-b:4,5-b'] dithiophene-4,8-dione (BDT) into the pores of 3D cross-linked graphene-based honeycomb carbon (3D graphene) as cathode material for Li-ion battery. The reported composite material offered reversible capacity of 210 mA/h/g at 0.1 C, good rate capability and stable cycle life of 80% capacity retention at 0.5 C for 200 cycles. Shorten ion diffusion distance due to the presence of abundant interconnected pores in the composite, high conductivity of 3D graphene and high electrochemical activity of BDT behind the improved performance [54]. Namsar et al. demonstrated $Sn(SnO_2)–SiO_2/rGO$ as anode material for high energy density LIB. The capacity was retained at 1205.2 mA/h/g after 200 cycles with a columbic efficiency of 99%. Better distribution of fine nanoparticle and buffering the volume change of the nanoparticle enhanced electrochemical performance [55]. Bhargav et al. used polyethylene hexasulfide-carbon nanotube (PEHS–CNT) composite as cathode material for Lithium-sulfur battery. Prevention of long-chain intermediates at cathodes leads to high material level capacity of 774 mA/h/g at 1 C rate and stability performance over 350 cycles at a capacity fade rate of 0.083% per cycle [56]. Huang et al. made MWCNTs@S-PPy for cathode material in lithium-sulfur battery. The composite showed an initial capacity of 1304.9 mA/h/g at 0.2 C and retained the capacity of 986.3 mA/h/g after 300 cycles. Excellent electronic property of the composite and halting of shuttle effect of lithium polysulfide were attributed to higher performance [57]. Pakseresht et al. utilized flower extract of Matricaria chamomilla to biosynthesize TiO_2 as cathode material for lithium-air battery. The material exhibited initial specific discharge capacity of 2000 mA/h/g at full discharge along with stable for 30 cycles at a limited capacity of 500 mA/h/g [58]. Peng et al. optimized Co_3O_4/graphene (48.2:51.8. w/w)) as a catalyst for hybrid electrolyte lithium-air battery. The optimized material displayed high stability and best columbic efficiency under ambient air of 70%–100% humidity [59].

8.4 FUEL CELL

Fuel cell converts the chemical energy of fuel into electrical energy by undergoing an electrochemical reaction [60]. The merits of the fuel cell are high electrical efficiency ($\geq$ 40%), long operational cycles noiseless operation, low emission of pollutants and modular construction [61]. Hydrogen, metal hydrides, methanol, formic acid and ethanol are some of the fuels used for the fuel cell.

The schematic diagram of fuel cell is shown in Figure 8.8, where fuel (for example, hydrogen) fed into the anode compartment and an oxidant (generally atmospheric air) fed into cathode compartment continuously. The electrochemical reaction occurs at the electrodes to generate the electrical energy through the electrolyte. At anode, the hydrogen splits into electrons and hydrogen ion.

$$H_2 \rightarrow 2H^+ + 2e^- \tag{8.1}$$

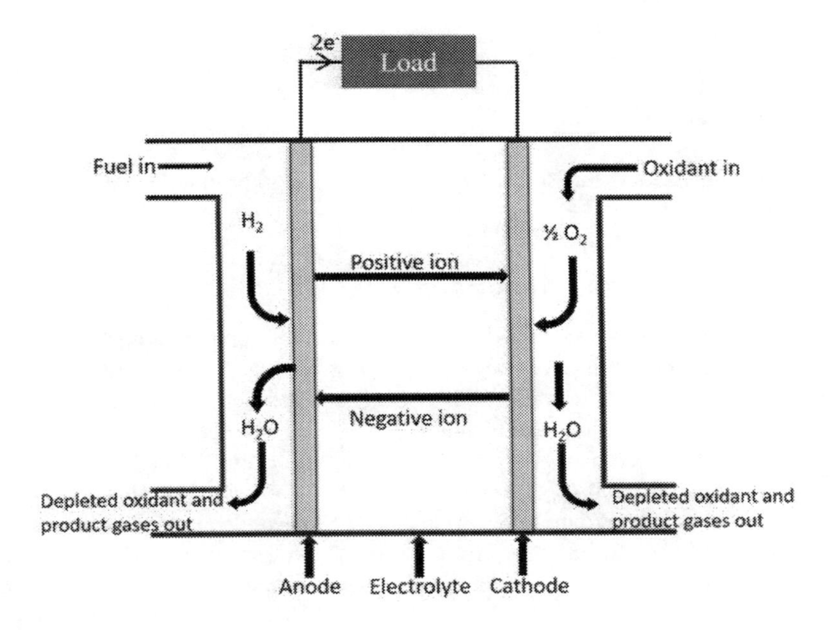

FIGURE 8.8 Basic structure of a fuel cell.

Electrons are driven through an external circuit to cathode, and hydrogen ions migrate via an electrolyte. Hydrogen ion, electrons and oxygen react to produce water.

$$H_2 + \frac{1}{2}O_2 + 2e^- \rightarrow H_2O \tag{8.2}$$

The overall reaction of fuel cell generates electrical work (W_{EW}), heat energy (Q_{HE}) and water

$$H_2 + \frac{1}{2}O_2^- \rightarrow H_2O + W_{EW} + Q_{HE} \tag{8.3}$$

Seemingly, fuel cells are identical to batteries by considering that both deliver the power through an electrochemical reaction. However, batteries store the electrical energy through electrochemical reactions and cease to deliver power once chemicals are consumed. Fuel cell involves energy conversion of fuel and oxidant through electrochemical reactions and never cease to deliver electrical energy as long as fuel is fed. Fuel cells are staked in series and parallel for the required electrical power [62,63,64].

The fuel cells are categorized based on its construction, electrolyte, fuel and operating temperature, viz.:

1. Proton exchange membrane or polymer electrolyte fuel cell (PEMFC),
2. Alkaline fuel cell (AFC),
3. Phosphoric acid fuel cell (PAFC),
4. Molten carbonate fuel cell (MCFC)
5. Solid-state fuel cell (SOFC)
6. Direct methanol fuel cell (DMFC)
7. Direct ammonia fuel cell (DAFC)
8. Direct carbon fuel cell (DCFC)

Hydrogen is utilized as fuel for all types of fuel cells other than DMFC, DAFC and DCFC [64].

8.4.1 Recent Trends in Fuel Cell

Hydroxide exchange fuel cells (HEMFCs) are gaining popularity due to the possibility to use non-noble electrocatalyst unlike proton exchange membrane fuel cells, which are built with expensive Pt-based catalysts and perfluorinated membranes. Xue et al. realized low cost and high-active HOR (hydrogen oxidation reaction) catalysts for HEMFCs through Ru_7Ni_3/C. The reported catalyst showed exceptional mass activity of 9.4 $A/mg/_{Ru}$, specific activity of 23.4 mA/cm^2 and high PPD (peak power density) of 2.03 W/cm^2 at 5.0 A/cm^2. The catalyst also exhibited great stability in the HEMFC durability test with less than 5% voltage decay after 100 hours at 500 mA/cm^2 [65]. Li et al. modified membrane electrode assembly (MEA) using a sulfonated poly(ionic liquid) block copolymer (SPILBCP) to improve kinetics and mass transported limited regions. SPILBCP enhanced the kinetic performance for both lower loading (I/C = 0.3) and higher loading (I/C = 0.8) (I/C denotes the ionomer and carbon ratio), where the improvement in H_2/air was 100% and 30% relative humidity (RH). Later, SPILBCP and Nafion ionomers introduced in catalyst layer for more optical electrode structure. MEA with a combination of SPILBCP and Nafion actually improved the proton accessibility at low RH due to sulfonated block and copolymer [66].

Alkaline polymer electrolyte fuel cell (APEFCs) has advantage of non-usage of precious catalyst, which inhouse an alkaline polymer electrolytes (APEs). Many reports suggested that APEs have high ionic conductivity and excellent stability. Flooding of water is a serious problem in the anode of APEFCs, particularly when it operates at high power density. Hu et al. adopted fluorine-containing ionomer, quaternary ammonia poly (arylene perflouroalkylene) (QAPAF), which actively curtailed the problem of flooding and negatively enhanced the hydrophobicity of anode. Improvement of cell performance, gain in voltage up to 140 V, sealing of anode emission and ability to fed H_2 gas without humidification are other benefits [67]. Ishak et al. undergone comparative study on single-step synthesis of Pt nano particles (NPs), which sourced from pineapple peel, sugarcane extract and banana peel. Among them, biogenic Pt NPs from sugarcane demonstrated superior electro-catalytic activity, improved utilization efficiency of Pt and good stability toward methanol oxidation. The nanomaterial exhibited active specific surface area (ECSA) of 94.58 m^2/g, mass activity of 398.20 mA/mg and specific activity of 0.8471 mA/cm^2 [68].

Li et al. fabricated a flexible solid-state direct ethanol fuel cell using AlPdNiCuMo as anode and $(AlMnCO)_3O_4$, as shown in Figure 4.2a. The anode material possesses high ECSA of 88.53m^2/g, which confirms formation of nanoscale quinary alloy influence in enhancing ECSA and the utilization of efficiency of Pd.

The cathode material showed promising results in an ethanol-tolerance Oxygen reduction reaction (ORR) catalyst. The energy density of solid-state DEFC was 13.63 m/Wh/cm^2 and durable discharge time up to 120 h for just 3 mL of ethanol applied on the anode side of the cell, which is shown in Figure 4.2b [69].

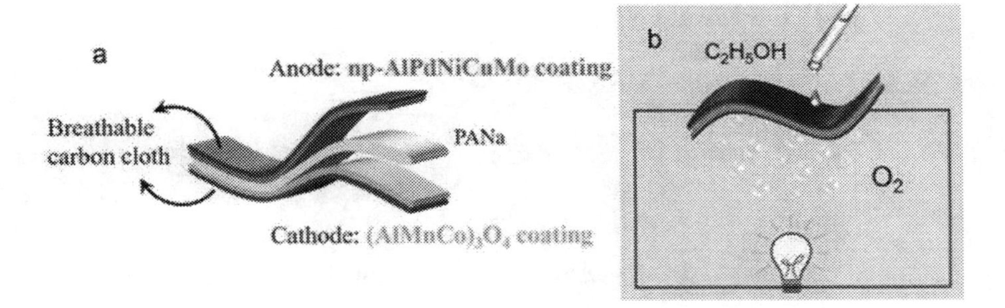

FIGURE 8.9 (a) Flexible solid-state direct ethanol fuel cell. (b) Way of applying ethanol on a fuel cell. (Reproduced from reference [69] with permission of John Wiley and Sons.)

8.5 ELECTROCHEMICAL SENSORS

Electrochemical sensors (ESs) came into existence in the 1950s for monitoring industrial oxygen. Labor laws were brought to ensure the safety of workers against toxic gases and fuels in confined space, and propelled the research on electrochemical research with accuracy and selectivity of different gases. Initial discoveries of ES suffered the problem of unstable electric current signal, which hampered its application. Later, the discovery of advanced instruments such as scanning tunnelling microscope (STM) and atomic force microscope (AFM) in the 1980s, facilitated to analyze and visualize the material in nanoscale. Advent development of nanomaterials led astounding improvements in electrochemical sensors and their application in various fields [70]. ESs transform electrochemical output (which produced due to a reaction between an electrode and analyte) into quantifiable signal. The merits of the ES are fast, high sensitivity, requires little power, reliable, greater resolution, economical and stable [71].

A modern-day sensor includes:

Receptor: Works on three main principles with an analyte, such as physical (not involving any chemical reaction, e.g., measurement of mass, temperature, pressure, etc.), chemical (involves chemical reaction, e.g., chemical sensors), and biochemical (where analytical signals stem from biochemical process, e.g., immunosensing devices).

Transducer: Transforms receptor output into an electrical signal.

Data processing unit: Process the transducer signal and convert into understandable practical data.

The stages of EC operation are shown in Figure 5.1 [72]. ECs are basically consists of a metallic anode and a metallic cathode, surrounded by the electrolyte solution. During its operation, an electric current is passed on the sensing electrode due to the electrochemical reaction that takes place at the surface of the sensing electrode which is coated with the catalyst. Later, the generated electric current due to electrochemical reaction is processed and displayed in the understandable way. The merits of the EC are fast, high sensitivity, requires little power, reliable, greater resolution, economical and stable [71]. The demerits of the EC are loss of sensitivity due to catalyst degradation, and easily being contaminated during the operation [73].

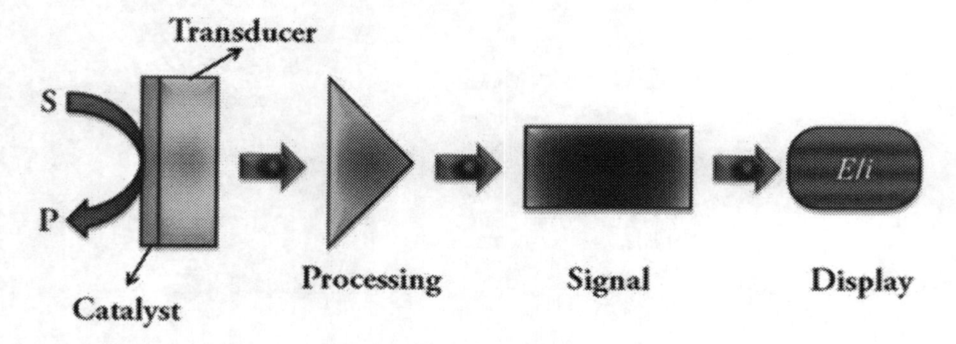

FIGURE 8.10 Stages of sensor operation. (Reproduced from reference [72] with permission of Elsevier.)

The electrochemical sensors are classified into potentiometric, and amperometric sensors. In amperometric, applied voltage is kept at constant and the current is taken as a sensor signal. The sensor signal in potentiometric is potential difference between the sensing electrode and a reference electrode, where the current is kept at zero.

8.6 CONCLUSION

The chapter discussed various aspects of electrochemical devices including historical evolution, working principle, and recent reports on their various improvements. Undeniably, the electrochemical devices have become a vital part of life from storing energy for mobile phones to life-saving medical devices. Efforts are being put to enhance the various aspects of electrochemical devices including various enhancements in operation durability, reliability, reduction of cost and safety. At the same time, the focus should be on the recovery of critical raw material (for example, platinum in fuel cell) and recycling of electrochemical devices after their lifetime. The care should be taken to properly dispose the devices without creating any environmental hazard.

REFERENCES

[1] Battery Market Size & Share | Industry Report, 2020–2027, 2020. https://www.grandviewresearch.com/industry-analysis/battery-market. (Accessed: August 19, 2021).
[2] Supercapacitor Market Size, Share | Future Analysis and Trends by 2027, 2020. https://www.alliedmarketresearch.com/supercapacitor-market. (Accessed: August 19, 2021).
[3] Fuel Cell Market Size, Share & Trends, Industry Forecast Report, 2018. https://www.grandviewresearch.com/industry-analysis/fuel-cell-market. (Accessed: August 19, 2021).
[4] S. Petrovic, History of electrochemistry, *Electrochemistry Crash Course for Engineers*. Springer, Cham, 2021, https://doi.org/10.1007/978-3-030-61562-8_1.
[5] Petrovic, S. (2021). Basic electrochemistry concepts, *Electrochemistry Crash Course for Engineers*. Springer, Cham. https://doi.org/10.1007/978-3-030-61562-8_2
[6] D.B. Hibbert, Introduction to electrochemistry, *Macmillan Physical Science Series*, Palgrave, London. 1993, https://doi.org/10.1007/978-1-349-22721-1_1.
[7] B. Scrosati, History of lithium batteries, *J. Solid State Electrochem.* 15, 2011, 1623–1630. https://doi.org/10.1007/s10008-011-1386-8.
[8] R. Cecchini and G. Pelosi, Alessandro Volta and his battery. *IEEE Antennas Propag. Mag.*, 34(2), 30–37, April 1992, doi: 10.1109/74.134307
[9] S. Srinivasan, Evolution of electrochemistry, *Fuel Cells*. Springer, Boston, MA 2006, https://doi.org/10.1007/0-387-35402-6_1.
[10] B.K. Kim, S. Sy, A. Yu, J. Zhang, Electrochemical supercapacitors for energy storage and conversion, *Handbook of Clean Energy Systems*. John Wiley & Sons, Hoboken, NJ, 2015, https://doi.org/10.1002/9781118991978.hces112.
[11] M. Winter, R.J. Brodd, What are batteries, fuel cells, and supercapacitors? *Chem. Rev.* 104, 2004, 4245–4269. https://doi.org/10.1021/cr020730k.
[12] S. Najib, E. Erdem, Current progress achieved in novel materials for supercapacitor electrodes: Mini review, *Nanoscale Adv.* 1, 2019, 2817–2827. https://doi.org/10.1039/c9na00345b.
[13] Y. Shao, et al., Design and mechanisms of asymmetric supercapacitors, *Chem. Rev.* 118, 2018, 9233–9280. https://doi.org/10.1021/acs.chemrev.8b00252.

[14] A. González, E. Goikolea, J.A. Barrena, R. Mysyk, Review on supercapacitors: Technologies and materials, *Renew. Sustain. Energy Rev.* 58, 2016, 1189–1206. https://doi.org/10.1016/j.rser.2015.12.249.

[15] L. Zhang, X.S. Zhao, Carbon-based materials as supercapacitor electrodes, *Chem. Soc. Rev.* 38, 2009, 2520–2531. https://doi.org/10.1039/b813846j.

[16] L. Zhou, C. Li, X. Liu, Y. Zhu, Y. Wu, T. van Ree, *Metal Oxides in Supercapacitors*, Elsevier Inc., 2018. Amsterdam. https://doi.org/10.1016/b978-0-12-811167-3.00007-9.

[17] M.J. Deng, T.H. Chou, L.H. Yeh, J.M. Chen, K.T. Lu, 4.2 V wearable asymmetric supercapacitor devices based on a VOx//MnOx paper electrode and an eco-friendly deep eutectic solvent-based gel electrolyte, *J. Mater. Chem. A*. 6, 2018, 20686–20694. https://doi.org/10.1039/c8ta06018e.

[18] J. Zhu, et al., Defect-engineered graphene for high-energy- and high-power-density supercapacitor devices, *Adv. Mater.* 28, 2016, 7185–7192. https://doi.org/10.1002/adma.201602028.

[19] A. Sanger, A. Kumar, A. Kumar, P.K. Jain, Y.K. Mishra, R. Chandra, Silicon carbide nanocauliflowers for symmetric supercapacitor devices, *Ind. Eng. Chem. Res.* 55, 2016, 9452–9458. https://doi.org/10.1021/acs.iecr.6b02243.

[20] M. Ramadan, A.M. Abdellah, S.G. Mohamed, N.K. Allam, 3D interconnected binder-free electrospun MnO@C nanofibers for supercapacitor devices, *Sci. Rep.* 8, 2018, 1–8. https://doi.org/10.1038/s41598-018-26370-z.

[21] Y. Ko, M. Kwon, W.K. Bae, B. Lee, S.W. Lee, J. Cho, Flexible supercapacitor electrodes based on real metal-like cellulose papers/639/4077/4079/4105/639/301/299/1013 article, *Nat. Commun.* 8, 2017, 1–10. https://doi.org/10.1038/s41467-017-00550-3.

[22] S. Archana, M. Athika, P. Elumalai, Supercapattery and full-cell lithium-ion battery performances of a [Ni(Schiff base)]-derived Ni/NiO/nitrogen-doped carbon heterostructure, *New J. Chem.* 44, 2020, 12452–12464. https://doi.org/10.1039/d0nj01602k.

[23] P.M. Junais, M. Athika, G. Govindaraj, P. Elumalai, Supercapattery performances of nanostructured cerium oxide synthesized using polymer soft-template, *J. Energy Storage*. 28, 2020, 101241. https://doi.org/10.1016/j.est.2020.101241.

[24] F. Wang, et al., A quasi-solid-state sodium-ion capacitor with high energy density, *Adv. Mater.* 27, 2015, 6962–6968. https://doi.org/10.1002/adma.201503097.

[25] K. Kuratani, M. Yao, H. Senoh, N. Takeichi, T. Sakai, T. Kiyobayashi, Na-ion capacitor using sodium pre-doped hard carbon and activated carbon, *Electrochim. Acta*. 76, 2012, 320–325. https://doi.org/10.1016/j.electacta.2012.05.040.

[26] X. Zhao, Y. Zhang, Y. Wang, H. Wei, Battery-type electrode materials for sodium-ion capacitors, *Batter. Supercaps*. 2, 2019, 899–917. https://doi.org/10.1002/batt.201900082.

[27] H. Zhang, M. Hu, Q. Lv, Z. Huang, F. Kang, R. Lv, Advanced materials for sodium-ion capacitors with superior energy–power properties: Progress and perspectives, *Small*. 1902843, 2019, 1–22. https://doi.org/10.1002/smll.201902843.

[28] H. Zhang, M. Hu, Q. Lv, Z.H. Huang, F. Kang, R. Lv, Advanced materials for sodium-ion capacitors with superior energy–power properties: Progress and perspectives, *Small*. 16, 2020, 1–22. https://doi.org/10.1002/smll.201902843.

[29] J. Chen, X. Zhou, C. Mei, J. Xu, S. Zhou, C.P. Wong, Evaluating biomass-derived hierarchically porous carbon as the positive electrode material for hybrid Na-ion capacitors, *J. Power Sources*. 342, 2017, 48–55. https://doi.org/10.1016/j.jpowsour.2016.12.034.

[30] S. Wang, R. Wang, Y. Zhang, D. Jin, L. Zhang, Scalable and sustainable synthesis of carbon microspheres via a puri fi cation-free strategy for sodium-ion capacitors, *J. Power Sources*. 379, 2018, 33–40. https://doi.org/10.1016/j.jpowsour.2018.01.019.

[31] H. Liu, et al., High-performance sodium-ion capacitor constructed by well-matched dual-carbon electrodes from a single biomass, *ACS Sustainable Chem. Eng.* 2019, 7, 14, 12188–12199. https://doi.org/10.1021/acssuschemeng.9b01370.

[32] H. Zhang, X. He, F. Wei, S. Dong, N. Xiao, J. Qiu, Moss-covered rock-like hybrid porous carbons with enhanced electrochemical properties. *ACS Sustainable Chem. Eng.*2020, 8, 8, 3065–3071 https://doi.org/10.1021/acssuschemeng.9b05075.

[33] E. Sujithkrishnan, A. Prasath, M. Govindasamy, R.A. Alshgari, P. Elumalai, Pyrrolic-nitrogen-containing hierarchical porous biocarbon for enhanced sodium-ion energy storage, *Energy and Fuels.* 35, 2021, 5320–5332. https://doi.org/10.1021/acs.energyfuels.0c04226.

[34] R. Korthauer, Lithium-ion batteries: Basics and applications, *Lithium-Ion Batteries: Basics and Applications.* Springer Nature, Berlin 2018. https://doi.org/10.1007/978-3-662-53071-9.

[35] A. Yoshino, The birth of the lithium-ion battery, *Angew. Chemie–Int. Ed.* 51, 2012, 5798–5800. https://doi.org/10.1002/anie.201105006.

[36] D. Deng, Li-ion batteries: Basics, progress, and challenges, *Energy Sci. Eng.* 3, 2015, 385–418. https://doi.org/10.1002/ese3.95.

[37] A. Manthiram, An outlook on lithium ion battery technology, *ACS Cent. Sci.* 3, 2017, 1063–1069. https://doi.org/10.1021/acscentsci.7b00288.

[38] P. Roy, S.K. Srivastava, Nanostructured anode materials for lithium ion batteries, *J. Mater. Chem. A.* 3, 2015, 2454–2484. https://doi.org/10.1039/C4TA04980B.

[39] A. Saxena, N. Gnanaseelan, S.K. Kamaraj, F. Caballero-Briones, Polymer electrolytes for lithium ion batteries, *Rechargeable Lithium-ion Batteries: Trends and Progress in Electric Vehicles*, CRC Press, 2020, 260–288. https://doi.org/10.1201/9781351052702-8.

[40] A. Manthiram, Y. Fu, S.H. Chung, C. Zu, Y.S. Su, Rechargeable lithium-sulfur batteries, *Chem. Rev.* 114, 2014, 11751–11787. https://doi.org/10.1021/cr500062v.

[41] T. Li, et al., A comprehensive understanding of lithium–sulfur battery technology, *Adv. Funct. Mater.* 29, 2019, 1–56. https://doi.org/10.1002/adfm.201901730.

[42] P. Kurzweil, *Post-Lithium-Ion Battery Chemistries for Hybrid Electric Vehicles and Battery Electric Vehicles*, Elsevier Ltd., 2015. United Kingdom https://doi.org/10.1016/B978-1-78242-377-5.00007-8.

[43] H. Pan, Z. Cheng, P. He, H. Zhou, A review of solid-state lithium-sulfur battery: Ion transport and polysulfide chemistry, *Energy and Fuels.* 34, 2020, 11942–11961. https://doi.org/10.1021/acs.energyfuels.0c02647.

[44] K.M. Abraham, Z. Jiang, Electrochemical science and technology a polymer electrolyte-based rechargeable lithium/oxygen battery, *J. Electrochem. Soc.* 143, 1996, 1–5.

[45] G. Girishkumar, B. McCloskey, A.C. Luntz, S. Swanson, W. Wilcke, Lithium-air battery: Promise and challenges, *J. Phys. Chem. Lett.* 1, 2010, 2193–2203. https://doi.org/10.1021/jz1005384.

[46] T. Liu, J.P. Vivek, E.W. Zhao, J. Lei, N. Garcia-Araez, C.P. Grey, Current challenges and routes forward for nonaqueous lithium-air batteries, *Chem. Rev.* 120, 2020, 6558–6625. https://doi.org/10.1021/acs.chemrev.9b00545.

[47] K.N. Jung, J. Kim, Y. Yamauchi, M.S. Park, J.W. Lee, J.H. Kim, Rechargeable lithium-air batteries: A perspective on the development of oxygen electrodes, *J. Mater. Chem. A.* 4, 2016, 14050–14068. https://doi.org/10.1039/c6ta04510c.

[48] C. Wang, Z. Xie, Z. Zhou, Lithium-air batteries: Challenges coexist with opportunities, *APL Mater.* 7, 2019, 1–12. https://doi.org/10.1063/1.5091444.

[49] N. Imanishi, O. Yamamoto, Perspectives and challenges of rechargeable lithium–Air batteries, *Mater. Today Adv.* 4, 2019, 100031. https://doi.org/10.1016/j.mtadv.2019.100031.

[50] B. Scrosati, New approaches to developing lithium polymer batteries, *Chem. Rec.* 1, 2001, 173–181. https://doi.org/10.1002/tcr.7.

[51] J. Wang, S. Li, Q. Zhao, C. Song, Z. Xue, Structure code for advanced polymer electrolyte in lithium-ion batteries, *Adv. Funct. Mater.* 31, 2021, 1–35. https://doi.org/10.1002/adfm.202008208.

[52] B. Scrosati, F. Croce, S. Panero, Progress in lithium polymer battery R&D, *J. Power Sources.* 100, 2001, 93–100. https://doi.org/10.1016/S0378-7753(01)00886-2.

[53] J.Y. Song, Y.Y. Wang, C.C. Wan, Open Access LibrarySong, J.Y.; Wang, Y.Y.; Wan, C.C. Review of gel-type polymer electrolytes for lithium-ion batteries. *J. Power Sources* 77, 1999, 183–197, https://doi.org/10.1016/S0378-7753(98)00193-1., *J. Power Sources.* 77, 1999, 183–197. http://www.oalib.com/references/9023829.

[54] D. Sui, L. Xu, H. Zhang, Z. Sun, B. Kan, Y. Ma, Y. Chen, A 3D cross-linked graphene-based honeycomb carbon composite with excellent confinement effect of organic cathode material for lithium-ion batteries, *Carbon N. Y.* 157, 2020, 656–662. https://doi.org/10.1016/j.carbon.2019.10.106.

[55] O. Namsar, et al., Improved electrochemical performance of anode materials for high energy density lithium-ion batteries through $Sn(SnO_2)$-SiO_2/graphene-based nanocomposites prepared by a facile and low-cost approach, *Sustain. Energy Fuels.* 4, 2020, 4625–4636. https://doi.org/10.1039/d0se00597e.

[56] A. Bhargav, C.H. Chang, Y. Fu, A. Manthiram, Rationally designed high-sulfur-content polymeric cathode material for lithium-sulfur batteries, *ACS Appl. Mater. Interfaces.* 11, 2019, 6136–6142. https://doi.org/10.1021/acsami.8b21395.

[57] S. Huang, et al., Polypyrrole-S-coated MWCNT composites as cathode materials for lithium-sulfur batteries, *Ionics (Kiel).* 26, 2020, 5455–5462. https://doi.org/10.1007/s11581-020-03736-w.

[58] S. Pakseresht, T. Cetinkaya, A.W.M. Al-Ogaili, M. Halebi, H. Akbulut, Biologically synthesized TiO_2 nanoparticles and their application as lithium-air battery cathodes, *Ceram. Int.* 47, 2021, 3994–4005. https://doi.org/10.1016/j.ceramint.2020.09.264.

[59] S. Peng, T. Chen, C. Lee, H. Lu, S. Jessie, Optimal cobalt oxide (CO_3O_4): Graphene (GR) ratio in CO_3O_4 / GR as air cathode catalyst for air-breathing hybrid electrolyte lithium-air battery, 471, 2020, 1–9.

[60] A. Mohapatra, S. Tripathy, A critical review of the use of fuel cells towards sustainable management of resources, *IOP Conf. Ser. Mater. Sci. Eng.* 377, 2018, 0–5. https://doi.org/10.1088/1757-899X/377/1/012135.

[61] M.K. Mahapatra, P. Singh, Fuel Cells. *Energy Conversion Technology.* Elsevier B.V., 2013. United Kingdom. https://doi.org/10.1016/B978-0-08-099424-6.00024-7.

[62] R. O'Hayre, S.-W. Cha, W. Colella, F.B. Prinz, Chapter 1: Introduction, *Fuel Cell Fundamentals.* John Wiley & Sons, Hoboken, NJ, 2016, 1–24. https://doi.org/10.1002/9781119191766.CH1.

[63] O.Z. Sharaf, M.F. Orhan, An overview of fuel cell technology: Fundamentals and applications, *Renew. Sustain. Energy Rev.* 32, 2014, 810–853. https://doi.org/10.1016/j.rser.2014.01.012.

[64] G. Itskos, N. Nikolopoulos, D.S. Kourkoumpas, A. Koutsianos, I. Violidakis, P. Drosatos, P. Grammelis, *Energy and the Environment*, Elsevier B.V., 2016. Amsterdam. https://doi.org/10.1016/B978-0-444-62733-9.00006-X.

[65] Y. Xue, et al., A highly-active, stable and low-cost platinum-free anode catalyst based on RuNi for hydroxide exchange membrane fuel cells, *Nat. Commun.* 11, 2020, 1–8. https://doi.org/10.1038/s41467-020-19413-5.

[66] Y. Li, et al., Modifying the electrocatalyst-ionomer interface via sulfonated poly(ionic liquid) block copolymers to enable high-performance polymer electrolyte fuel cells, *ACS Energy Lett.* 5, 2020, 1726–1731. https://doi.org/10.1021/acsenergylett.0c00532.

[67] M. Hu, et al., Alkaline polymer electrolyte fuel cells without anode humidification and H_2 emission, *J. Power Sources.* 472, 2020, 228471. https://doi.org/10.1016/j.jpowsour.2020.228471.

[68] N.A.I.M. Ishak, S.K. Kamarudin, S.N. Timmiati, N.A. Karim, S. Basri, Biogenic platinum from agricultural wastes extract for improved methanol oxidation reaction in direct methanol fuel cell, *J. Adv. Res.* 28, 2021, 63–75. https://doi.org/10.1016/J.JARE.2020.06.025.

[69] S. Li, J. Wang, X. Lin, G. Xie, Y. Huang, X. Liu, H.J. Qiu, Flexible solid-state direct ethanol fuel cell catalyzed by nanoporous high-entropy Al–Pd–Ni–Cu–Mo anode and spinel $(AlMnCo)_3O_4$ cathode, *Adv. Funct. Mater.* 31, 2021, 1–9. https://doi.org/10.1002/adfm.202007129.
[70] F.R. Simões, M.G. Xavier, *6-Electrochemical Sensors*, Elsevier Inc., 2017, United Kingdom. https://doi.org/10.1016/B978-0-323-49780-0/00006-5.
[71] S.K. Chaulya, G.M. Prasad, Gas sensors for underground mines and hazardous areas, *Sensing and Monitoring Technologies for Mines and Hazardous Areas*. 2016, 161–212. https://doi.org/10.1016/b978-0-12-803194-0.00003-9.
[72] N.P. Shetti, D.S. Nayak, K.R. Reddy, T.M. Aminabhvi, *Graphene-Clay-Based Hybrid Nanostructures for Electrochemical Sensors and Biosensors*, Elsevier Inc., 2018. Amsterdam. https://doi.org/10.1016/B978-0-12-815394-9.00010-8.
[73] G. Manjavacas, B. Nieto, *Hydrogen Sensors and Detectors*, Elsevier Ltd., 2016, United Kingdom. https://doi.org/10.1016/b978-1-78242-364-5.00010-5.

9 Application of Nanomaterials for Electrochemical Devices

Rahul Singh

CONTENTS

9.1 INTRODUCTION

Nanomaterials have emerged as a significant class of materials that show great promise in solving the global future energy crises problem sustainably and effectively. Recently, enormous efforts have been made to enhance or develop high-performance electrochemical devices using different types of nanomaterials [1–4]. Nanomaterial offers the wide possibility to play with their nanostructure to achieve desired material properties. Nanostructure materials dominate the present electronic and portable

DOI: 10.1201/9781003216308-9

device market and play a vital role in transforming the future of electronic industries. Nanomaterials are easily distinguishable from the other bulk material in terms of size and shape. They offer remarkably new physical and chemical properties, plus they show enormous scope in the formation of the desired nanostructure. Typically, some of the prominent properties popularly achieved after the desired transformation are as follows: If semiconductor material is reduced to a specific size below few nanometers, enhancement in the bandgap due to quantum confinement is observed [5]. Similarly, when bulk material is transformed to nanomaterial through a specific protocol, extraordinary physicochemical properties appear, such as significant enhancement in the surface-to-volume ratio [6], surface chemistry of the material [7], magnetic material transforms to paramagnetic [8], and a few nanometer orders of magnitude can potentially make the nanoparticle feasible for drug delivery [9]. Nanomaterials can generate excellent electrical, optical, thermal, mechanical, surface, magnetic, and catalytic properties that are totally unique to their bulk counterparts. Moreover, it is possible to twist or turn the properties as per requirement by tuning the size, shape, and synthesis circumstances.

This chapter describes the importance of nanostructured materials in electrochemical devices. The main aim is to give a broad perspective of different electrochemical systems and how nanomaterial influences their output performance. Moreover, the basic principle of electrochemical systems such as lithium-ion batteries, polymer electrolyte membrane fuel cells (PEMFCs), photochemical solar cells, and electrochemical double-layer capacitors (EDLCs) are discussed in great detail. The prime objective is to provide fundamental concepts and equations to the readers. Apart from that, we also demonstrated each layer or component used in the fabrication of electrochemical cells. In particular, we explain and illustrate numerous terms relating to nanomaterials, which are feasible for developing energy systems.

9.2 SOLAR CELLS

Typically, a solar cell is a semiconductor-based technology that generates electricity in the form of direct current or electrical power (unit watts or kilowatts) when illuminated by incident photons. The power conversation efficiency η of any solar cell is as follows:

$$\eta = \frac{V_{OC} \cdot J_{SC} \cdot FF}{P_{incident}} \tag{9.1}$$

Here, V_{OC} is an open-circuit voltage of the cell, J_{SC} is the short circuit current density, FF is the fill factor, and $P_{incident}$ is the power obtained from incident radiation of the solar cell. To achieve efficient cell performance, the fundamental requirement is that the absorption spectra of the sensitizer layer must be well matched with absorbed light spectra with AM1.5G solar spectrum irradiated on earth [10]. However, photovoltaic technology has seen a series of technological advancements, breakthroughs from crystalline p–n junction silicon solar cells, and is categorized as first-generation solar cells. Whereas second-generation solar cell covers the amorphous silicon, polycrystalline silicon, cadmium, and indium-based thin-film solar cell. On the other hand, organic and inorganic nanomaterial-based solar cells are considered next-generation (or third-generation) solar cells.

9.3 BAND DIAGRAM AND OPERATIONAL PRINCIPLE OF NANOCRYSTALLINE SOLAR CELLS

The energy gap between the valence and conduction band of semiconductors is known as the bandgap. It is essential to provide efficient electron transfer from the sensitizer (dye or quantum dots (QDs) to the semiconductor oxides (ZnO or TiO_2) conduction bands. Still, it remains a topic of great interest due to the lack of standard metal oxides (semiconductors). Thus, the lowest unoccupied molecular orbital (LUMO) of the sensitizer should be adequately tuned or electronically coupled with the orbital of semiconductor oxides to produce efficient injection of electrons [11]. Typically, every sensitizer has its limitation to absorbs specific wavelengths only. For example, a ruthenium-based sensitizer can harvest almost full spectra of visible light and offering 920 nm wavelength as a cutoff. Under illumination conditions, the photoexcited electrons get sufficient energy to travel from the valence band to the conduction band (S^*). These photoexcited electrons are further injected into the conduction band of the nanocrystalline semiconductor, as illustrated in the schematic diagram in Figure 9.1.

Furthermore, electrons are diffused through the nanocrystalline semiconductor, which is connected to the external load. This injection process at the nanocrystalline

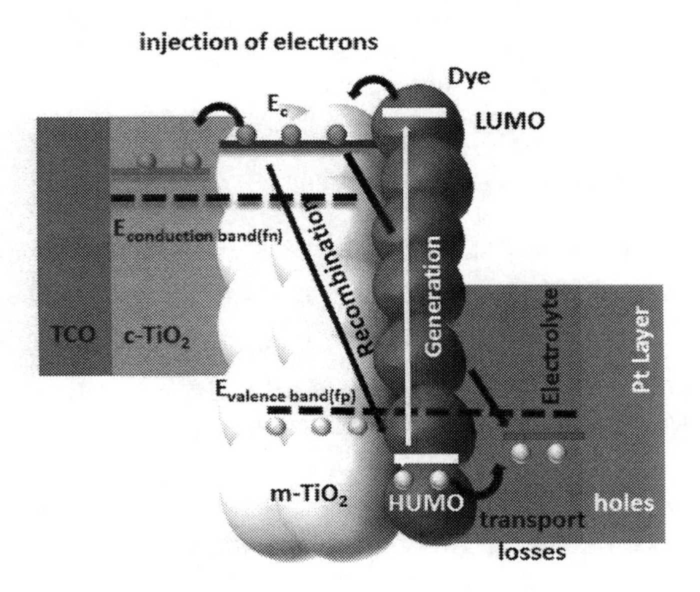

FIGURE 9.1 Schematic representation of the band diagram and different layers of a dye-sensitized solar cell.

side also indirectly depends on the electrolyte and other components of the cell. Fundamentally, in the case of iodide-based redox couple (I^-/I_3^-), the injection of electron takes place within Femto to picosecond time, i.e., 10^{-15}–10^{-12} seconds. This time is so efficient that after the injection of electrons, electrons' negligible possibility travels back to the oxidized sensitizer. Highly efficient electron injection is the main reason for the low diffusion coefficients of nanocrystalline semiconductors, i.e., $10^{-4}cm^2/s$ for TiO_2. Therefore, the possibility of the recombination of electrons may occur through the semiconductor oxides to I_3^- in electrolytes. Similarly, loss of electrons is also possible from transparent conducting oxide layer to I_3^- of electrolyte [12]. Thus, the efficient injection of electrons plays a significant role in designing reliable, high-performance nanocrystalline solar cells.

9.4 THE IMPORTANCE OF THE NANOSTRUCTURE

As discussed in the earlier section, the primary and efficient driving force for injecting electrons is the energy difference between sensitizer oxidation potential and the semiconductor's conduction band. Enormous efforts have been diverted toward the optimizations of the metal oxides semiconductor conduction bandgap with a sensitizer. The prime requirement for delivering highly efficient solar cells, one should develop metal oxide semiconductor high porous or a mesoscopic oxide thin film. High porosity will significantly enhance the surface area and provide an excess amount of photosensitizer to expose incident photons. Mesoporous TiO_2-based nanostructures are widely studied and used as electron transporting electrodes. It is prepared mainly from the sol-gel technique followed by sintering only for developing efficient cells. Although there is no standard semiconductor oxides material for comparison, the cell performance is primarily dominated by the material synthesis conditions, properties with pre- and post-treatments, which significantly controls the electronic properties of the nanostructure [13]. Thus, fabricating a more vibrant semiconductor oxide layer for reliable and ultrafast electron transfer is required to overcome the recombination losses of electrons in these cells to make them viable.

9.5 QUANTUM DOT SENSITIZER

Discrete levels in conduction and valence band of QDs make them unique and provide an enhanced electronic and optical properties compared to a number of bulk semiconductor properties [14–16]. The confinement effect in discrete bands put QDs apart from a few conventional semiconductors. Typically, QDs are defined as a nanomaterial with a definite size in a few nm, and their size must be smaller than the exciton radius. Interestingly, one can tune the QD bandgap by changing the shape and size of the QD semiconductor material at the nanoscale. QDs have the ability to form isolated nanoparticles (such as dispersion or colloidal) or crystalline solid nanostructure forms. Low-temperature QD synthesis and their thin-film forming capability using solution processes have open new ways for implementing QDs in various applications. Notably, in solar cell applications, QDs have already shown enormous potential in developing efficient quantum dot solar cells (QDSCs).

It is possible by utilizing excess energy in the form of thermal heat produced by incident photons when striking at much faster than the semiconductor bandgap. These high-speed incident photons lead to the formation of multiple excitons per photon absorbed by the semiconductor, resulting in the desolation of extra energy, certainly increasing the cell's temperature and becoming the reason for electron–hole formation. Thus, using QDs in solar cells as a sensitizer can significantly reduce excess energy dissipation by confining the charge carriers, as successfully achieved by lead selenide (PbSe) and lead sulfide (PbS)-based QDSCs.

9.6 ELECTROCHEMISTRY AND NANOSCALE MATERIALS

Electrochemistry is a vast field, and it has been studied for the past several decades and still offering enormous promises to modern electrochemical science. It mainly deals with the charge transfer process in electrode and solution interfaces, including solid-state, solid-liquid/gel, and liquid/gel-liquid/gel types. Electrochemistry in nanomaterials has made electron transport super-fast due to their unique physicochemical properties offered by metal-semiconductor-based nanomaterials. Nanomaterials are the material of great choice in various applications, including energy storage systems, photovoltaics, catalysis, sensors, drug delivery, etc. [17]. The prime reason is that it offers remarkably high surface area and controllable electronic well as optical properties. It is important to note that the shape and size of the nonmaterial play a crucial role in governing the extraordinary chemical properties. Therefore, to fundamentally understand the relationship between the nanostructure and its properties, an electrochemical measurement is required. The electrochemical information received from the nonmaterial helps to gain the technological edge in developing next-generation high-performance systems.

9.7 ELECTROCHEMISTRY AND SIZE EFFECTS

It is well articulated in various literature that variation material size at the nanoscale (<100 nm) regulates the output characteristic in terms of production, architecture, and application of the material compared to the bulk material. The electrochemical process provides an edge to understand the change in nanomaterial size in terms of electrical and optical. Electrochemical methods give a reliable foundation in developing advanced technology. Variations in size and shape of nanomaterial can be measured directly by electrochemical techniques, which identify the changes during reaction or on the surface. The main significance of the size effect on nanostructure material offers freedom to navigate or control the building block required for nanodevice and their engineering.

On the other hand, the in-situ electrochemical method provides insight into controlling the size and shape of nanomaterial for developing next-generation smart materials. Interestingly, the change in the electrochemical process due to the surface chemistry of nanomaterial inside any device can easily be studied through this technique in real time by interfacing the electrochemical system with different optical and spectroscopic instruments (such as Scanning Electron Microscope (SEM), Transmission Electron Microscopy (TEM), X-ray scatting, UV-vis, etc.). On the other hand, device performance, optimization of interface layers, the interaction of ions, efficient charge transport, surface charge, active site enhancement, and high surface area nanostructure are all standard requirements in most electrochemical systems and are greatly influenced by the nonmaterial design and its size. Thus, novel design strategies and innovative ways of controlling size (such as core-shell engineered nanostructure) can significantly impact the properties of the nanomaterials and will able to show the path for obtaining desirable reliable, high-performance material.

9.8 CHALLENGES OF CHARGE TRANSFER

Typically, in semiconductors, bandgap energy is required to excite the electron from the nucleus, creating a free electron and hole pair. In contrast, other electronic materials conduct electrons directly under the applied field. Therefore, the higher the electronic conductivity of the material, the better is the electron transfer. However, when the charge is required to transfer through the semiconductor-semiconductor interface or semiconductor-metal interface, several other factors are taken into consideration for efficient charge transfer. If the excess amount of charge concentrates at one end, resulting in exceeding the equilibrium value, thus extra charge carriers are forced to recombine and the process terms as recombination loss. Therefore, longer lifetimes are required for producing highly efficient cells. Fermi level equilibrium between semiconductor-metal is crucial and mainly governed by metal-based nanomaterial and required properly engineered for efficient charge transport. Efficient charge transfer between a semiconductor-metal-based nonmaterial can be easily achieved by surface passivation and by confining semiconductors. It will enable the fast transfer of charge from semiconductors to metal nonmaterial and enhance the charge transfer rate.

9.9 NANOMATERIALS AND NANOSTRUCTURED FILMS AS ELECTROACTIVE ELECTRODES

Organic and inorganic-based electroactive materials have been used in developing electrodes, but organic material has been widely studied because of its remarkable electronic and structural properties. Polypyrrole (PPy), poly (3,4-ethylenedioxythiophene) (PEDOT), polyaniline (PANi), carbon, carbon nanotubes, and so on are popularly used as electroactive materials for designing electrodes. Due to simple preparation techniques of polymer conducting layers such as electrodeposition or oxidative polymerization, these conducting materials are widely applied in various electrochemical device fabrications. Carbon-based materials have shown enormous potential in constructing electrodes, such as cathode and anode. Both electrodes utilize some form of carbon-based material in developing cells, such as rechargeable lithium-ion batteries, fuel cells, and EDLCs. However, for third-generation solar cells, carbon-based material is used in preparing counter electrodes. It is important to note that specific types of material, quantities, sizes, and forms are used for different electrochemical cells. The main purpose of utilizing electroactive materials in all these devices is straightforward, to enhance the charge transfer rate to work efficiently.

Electroactive materials are mainly used in the electrode in various applications, including solar cells, energy storage systems, sensors, light-emitting diodes, photoresists, etc. Electroactive materials are categorized as electronic and ionic conducting materials. Usually, nanomaterials are altered with different nanoparticles to achieve high surface area or high active sites on electrodes to deliver efficient electrochemical reactions between electrodes. It means the higher the electroactive area, the better the charge transfer. Therefore, one can determine the electroactive area by measuring the charge transfer.

9.10 NANOMATERIALS AS ELECTROLYTES

An electrolyte is an ion-conducting media (electronic insulators) composed of ionic liquid or molten salts, which exist in both liquid or solid forms. As electrolytes are placed between the cathode and anode electrodes in electrochemical cells, it is essential to be safe, efficient, and stable enough for long-lasting device performance. Conventional liquid electrolytes constitute ionic salts (LiTSFI and $LiPF_6$, etc.), a mixture of organic solvents polycarbonate (PC), ethylene carbonate (EC), dimethyl carbonate (DMC), ethyl methyl carbonate (EMC), and diethyl carbonate (DEC). As liquid electrolytes face enormous problems in handling, leaking, drying, and high inflammable issues dramatically reduce the cell's performance. Similarly, advanced polymer technology-based solid polymer electrolytes (SPEs) are macromolecules capable of ionic conduction in electrochemical devices. SPE works efficiently in electrochemical systems, provides device stability for the long run, and has shown promising outcomes.

On the other hand, several approaches have been adopted to overcome electrolyte stability by adding additives, nano filers, nano clay, nanorod, nanoparticles, nanowire, etc. The nanocomposite material shows enormous potential in enhancing device stability. And, it provides a series of other merits, including high ionic conductivity, wide electrochemical stability window, better thermal and mechanical strength. These nanocomposite materials are incorporated with the host polymer matrix and ionic salts to produce polymer nanocomposite-based SPEs.

9.11 NANOSCALE ELECTRONIC AND IONIC TRANSPORT

Both electronic and ion transport phenomena are required in electrodes and electrolytes, respectively. It enables power production from electrochemical systems. The fundamental mechanism of mass transfer and charge transfer depends primarily on materials properties plus several other associated factors. It is because most of the electrochemical methods do not follow linear current and voltage relations. Thus, it becomes a very complex system; therefore, it is initially required to consider several physicochemical parameters and material compositions, etc., for defining. Generally, in electronic transport, charge carriers, i.e., electrons and holes, are transported apart from one electrode to another electrode. Metals follow Ohm's law. However, electron transport in an electrochemical system's composed of semiconductors given by the drift-diffusion equation.

$$J_e = q\left(n\ \mu_e\ E + D_e\ \frac{dn}{dx}\right) \tag{9.2}$$

where J_e is the electron current density, q is the charge of an electron, n is the number of charge carriers (electrons), E is the applied electric field, D_e is the diffusion coefficient, and $\frac{dn}{dx}$ is the carrier gradient.

As described in the previous section, electrolytes conduct ionic charge, measured in terms of ionic conductivity in S/cm. Ionic charges coordinated with the host polymeric material through Coulombic interaction, resulting in the dissociation of ionic species in the medium. Ionic conductivity is expressed in the form of the following equation:

$$\sigma = n_i \; q_i \; \mu_i = \frac{1}{R_b}\left(\frac{d}{A}\right) \tag{9.3}$$

Here, n_i is the number of free ions, q_i is the number of charges, μ_i is the mobility, R_b is the bulk resistance of the electrolyte, d is the distance between electrodes, and A is the active area of the electrolyte.

9.12 ENERGY CONVERSION AND STORAGE IN ELECTROCHEMISTRY

The electrochemical process plays a vital role in designing and producing energy systems for generating power. It considers as an advanced future technology, which enables energy management and helps control pollution and greenhouse gases. In most electrochemical systems, the electrochemical reaction occurs between an electrode and electrolyte, making ions and electrons transport further, producing a sustainable form of energy. Interestingly, there is no limit for nonmaterial, which means that the same material can be utilized in energy conversion and energy storage processes by following specific protocols. For example, the nano form of carbon material helps energy conversion in the solar cell. The same material displays capacitive behaviors in the supercapacitor, collector in the fuel cell, and the same material acts as energy storage material in a battery. In all systems, the electrochemical process remains common with the stating material. Therefore, it is recommended that it is essential to select the appropriate material type or design them as state-of-the-art materials for other applications.

9.13 OVERVIEW OF THE PRINCIPLES OF OPERATION OF ENERGY CONVERSION AND STORAGE DEVICES

A series of fundamental concepts, phenomena, and mechanisms are associated with the operation of any energy conversion and storage devices [18]. To understand the electrochemical principle behind any electrochemical system, one needs to dissect the system into a number of parts. Then, it will be easy to understand the operational mechanism clearly. Typically, several processes occur in a cell, such as chemical, thermal, optical, mechanical, and electrical; then, it is transformed to the desired form, such as electrical energy or vice versa. Such processes are governed by physical and chemical laws and depend on material properties and their architecture design. Thus, it is recommended to separately study or analyze the fundamental principle for electrodes, electrolytes, and interfaces one by one. Below we have demonstrated each electrochemical system individually.

9.14 LITHIUM-ION BATTERIES

Lithium-ion batteries are energy storage devices, which are composed of complex chemical materials. Typically, a discharging lithium-ion battery contains a positive electrode called a cathode (aluminum current collector side), a negative electrode as an anode (copper current collector side), and a porous separator (filled with ion-conducting electrolyte), as illustrated in Figure 9.2. Ion-conducting lithium-based electrolytes are placed in between cathode and anode electrodes. The positive electrode is called an anode in the charging process, and the cathode works as a negative electrode, while, in discharging, it becomes vice versa. Throughout this chapter, charging case will be considered to avoid any confusion. The series of most promising lithium source materials used for developing cathode electrodes are LCO ($LiCoO_2$), LMO ($LiMn_2O_4$), NMC ($LiNiMnCoO_2$), LFP ($LiFePO_4$), NCA ($LiNiCoAlO_2$), and LTO ($Li_4Ti_5O_{12}$). The critical performance of the batteries is mainly evaluated in terms of specific energy, specific power, safety, performance, life span, and not the least, its cost. Out of all these batteries, electric vehicle industries have shown great interest in LFP, LMO, NCA, and NCM-based batteries due to their better stability and tunable chemistry [19,20].

Generally, deintercalation processes occur during charging, and lithium ions leave from Li metal. In contrast, during discharging, intercalation processes take place, where lithium ions return to their original positions. On the other hand, graphitic carbon-based negative electrodes are most usual in conventional battery systems. Here, the intercalation process takes place in the graphite electrode during charging; lithium-ion intercalates between graphite layers. However, deintercalation processes occur during discharging, and lithium ions leave the graphite layer. Reaction at graphite electrode is as follows:

$$\text{Li C}_6 \ \text{Li}^+ + e^- + 6\text{C} \tag{9.4}$$

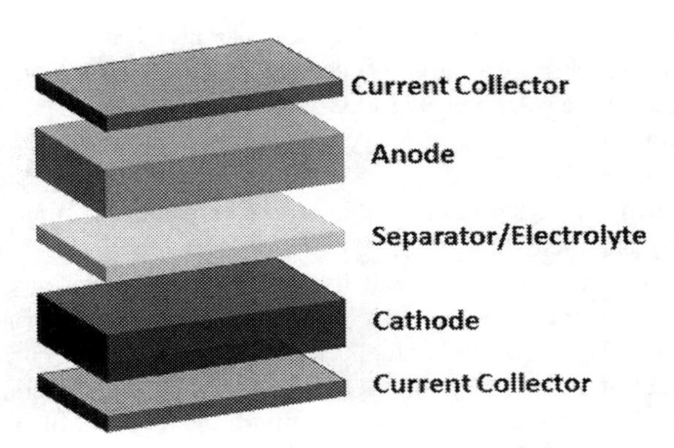

FIGURE 9.2 Schematic diagram of different layers in lithium-ion battery.

A mixture of active material, conducting additives, solvents, and binder is applied using a coating process to prepare electrode material. It is crucial to optimize the material composition, conductivity, and good interfacial contact for producing high efficient cell performance. Both positive and negative electrodes are separated by a separator filled with electrolytes. This electrolyte is also loaded in both electrodes' pores. Basically, the electrolyte is a lithium salt (LiTSFI, $LiPF_6$, etc.)-based solution prepared using different organic solvents. A series of other solvents are used in the electrolyte, such as EC, propylene carbonate, DMC, DEC, etc. These solvents are considered to produce high conducting electrolytes up to 10 mS/cm but are highly combustible, which causes substantial safety concerns. However, a separator is used between both electrodes to prevent the short-circuiting and direct contact of positive and negative electrodes. It is observed that a tiny quantity of metallic lithium on the electrode surface can cause violent reactions that lead to consuming electrolytes, which results in the formation of dendrites and might produce heat or thermal runaway. To make it commercially viable, one needs to overcome the fundamental challenges associated with lithium-ion batteries: safety, cost, and bulky size. State-of-the-art solid-state polymer electrolyte and gel polymer electrolyte-based Li-ion batteries have the potential to overcome the issues mentioned above significantly. These new electrolyte-based batteries still require a series of innovations but still are on the way for commercialization.

9.15 FUEL CELLS

Proton exchange membrane fuel cells are one of the clean, green, and noise-free power-generating technologies, as presented in the schematic diagram in Figure 9.3. In this technology, mainly hydrogen and oxygen are used as fuel, followed by two electrochemical half-reactions:

$$H_2 \ 2H^+ + 2e^- \tag{9.5}$$

$$\tfrac{1}{2} O_2 + 2H^+ + 2e^- \ H_2O \tag{9.6}$$

Both chemical equations force the fuel to transfer electrons to the external load, which results in useful power output. These reactions took place at the interface of the electrode and the electrolyte. An electrolyte is basically responsible for forming an ionic nanochannel, allowing specific ions to pass through while restricting electrons and oppositely charged ions. Thus, one can define a fuel cell as an electrochemical energy conversion system that works on electrochemistry. Therefore, it directly converts chemical energy into electrical energy as long as fuel is supplied. Majorly, fuel cells are categorized into five different types but work on the same electrochemical principles. They include PEMFC, phosphoric acid fuel cell (PAFC), alkaline fuel cell (AFC), solid-oxide fuel cell (SOFC), and molten carbonate fuel cell (MCFC) [21]. Out of all, PEMFC and SOFC are highly efficient and competent fuel cell technology because they can work not only on H_2 fuel but also on other kinds of fuels. For example, methanol, ethanol, and formic acids are alternative fuels for PEMFC, whereas methane, butane, gasoline, and carbon monoxides are alternate

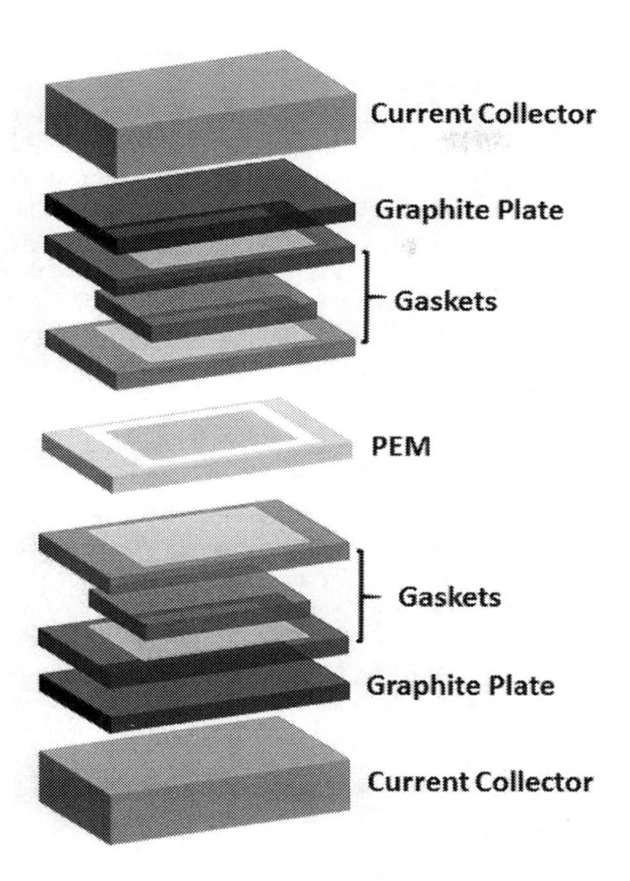

FIGURE 9.3 Schematic diagram of various layers present in developing fuel cell stack.

fuels for SOFC. Therefore, selecting an appropriate charge carrier may change the chemical reaction. In PEMFC, half-reaction is governed by proton (H^+), which leads to producing H_2O at the cathode side. On the other hand, half-reaction in SOFC is governed by the mobile oxygen ions (O^{2-}), which produces H_2O at the anode side. Similarly, other types of fuel cells work with different charge carrier types, such as OH^- or CO_3^{2-}. Thus, one side is filled with desired fuel (anode) and the other side with an oxidant (cathode). It is easy to understand that this process when electrons are released during a chemical reaction is said to be oxidation. While when electrons are consumed during a chemical reaction, it is called reduction. Notably, in PEMFC, both half-cell reactions are termed hydrogen oxidation reaction (HOR) at the anode side and oxygen reduction reaction (ORR) at the cathode side. It is important to note that both anode and cathode can be represented with any terminal positive or negative (it hardly matters).

To enhance the speed of the electrochemical reaction at the electrode and electrolyte interface, platinum-based catalysts are used in fuel cells. Optimized amounts of catalyst and their nanoengineering in designing the electrode are essential for delivering efficient power output. The fuel cell output performance is determined from the current and voltage curve, also known as the polarization curve. The polarization curve mainly depends on the catalytic behavior, fuel transport activity, ion-exchange capacity of the membrane, leakage, fuel crossover, etc. Thus, the power output of the fuel cell is given as the product of potential and current. Due to irreversible losses associated with the fuel cell system, it is challenging to achieve voltage output up to thermodynamic level voltage. Fundamentally, three significant losses are associated with a fuel cell system that is as follows: (a) activation loss mainly arises due to electrochemical reaction losses, (b) Ohmic losses arise from the ionic and electronic conduction, and (c) concentration losses because of mass transport loss. Thus, actual voltage is given by subtracting all three voltage drops from the predicted theoretical thermodynamic voltage.

9.16 PHOTOELECTROCHEMICAL SOLAR CELLS

Photochemical solar cells are also termed photovoltaic cells or electrochemical cells, as illustrated in the schematic diagram in Figure 9.4. Photochemical solar cells mainly consist of a sensitizer layer made up of dye, quantum dots, nanomaterials, organic and inorganic materials, etc. It is primarily responsible for converting incident photons (solar radiation) to electrical current. At present, photovoltaic is considered the most efficient and clean way to produce and store energy from solar energy. Semiconductor and electrolyte or hole transport layer help the generated electron and holes move apart in the cell. Typically, photoanode with electrolyte leads to band bending in the

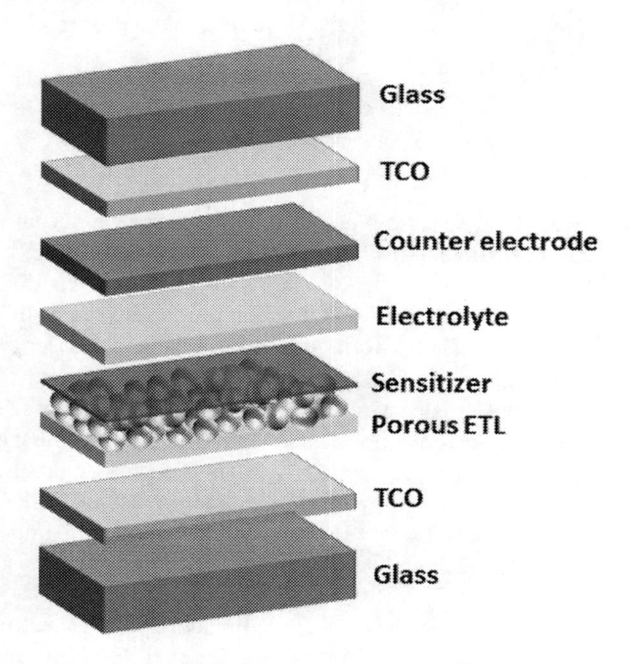

FIGURE 9.4 Schematic diagram of different layers of the photochemical solar cell.

depletion region, which governs the electrons into the semiconductor's conduction band, where holes in the valence band are moved to the electrolyte, followed by an oxidation process. At the counter electrode side, electrons led to take part in the reduction reaction. Under illumination conditions, photoanode exhibits the negative potential, and the band bending tends to zero; meanwhile, the Fermi level remains shifted toward the negative potential. At this point, one can extract maximum voltage from the photoanode under flat band conditions, which is up to barrier heights. When an electron passes through the electrolyte (oxidation and reduction process) to the external circuit, recycling the redox couple takes place, called a regenerating solar cell.

The following process describes the basic principle of the photoelectrochemical solar cell:

When the incident photon strikes the cell's surface, a series of reactions immediately starts taking place inside the cell, which leads to the output power. At the photoanode side, the first reaction occurs due to photon absorption by the sensitizer dye (denoted by S), which transforms into an excited state S^*. Process terms as photoexcitation and reaction are as follows:

$$S + \ hv\ S^* \ (\text{photoexcitation}) \tag{9.7}$$

Suddenly after this, two different processes get the possibility to occur or both at the same time. Either the excited state molecules decay/role back to the ground state directly or it may encounter oxidative quenching means transporting charged electron to semiconductor oxide (ZnO, TiO_2, etc.) conduction band. Mainly, this process is called as injection of electrons, and the reaction is as follows:

$$S^* \ S + hv' \,(\text{emission}) \tag{9.8}$$

$$S^* \ S + e^- \ (\text{injection}) \tag{9.9}$$

After electron injection, these electrons travel all the way from semiconductor oxides to transparent conducting oxides to external load to another side of electrode (counter) and finally reached the electrolyte region. At this point, the dye gets oxidized (hole is already generated at ground state) releases donor ions in the form of iodide inside electrolyte. This process is commonly known as regeneration of cells and is given by the following equation:

$$2S^+ + 3I^- \ 2S + (\text{regeneration}) \tag{9.10}$$

The recombination process takes place with the charged electron of the semiconductor oxide in the absence of the redox mediator, resulting in no photocurrent or output power at the external load. Loss of electron in a solar cell is popularly called recombination and is given by the following equation:

$$S^+ + e^- (\text{semiconductor oxide})\, S\, (\text{recombination}) \tag{9.11}$$

It is noted that the charged electron that traveled to the counter electrode through external load can promote oxidized redox couple (I^-). Thus, it leads the oxidation and reduction process at the redox mediator (I_2/I^-), resulting in cyclic process. This reaction is said to be the regeneration of iodide ions and is represented by the following equation:

$$I_3^- + 2e^- \; 3I^- \left(\text{regeneration}\right) \tag{9.12}$$

All the processes taking place inside the cell due to the incident photon produce an electric current.

9.17 ELECTROCHEMICAL DOUBLE-LAYER CAPACITORS

EDLCs are also termed as supercapacitors, ultracapacitors, symmetrical capacitors, asymmetrical capacitors, and electrochemical capacitors. The schematic diagram of EDLC is presented in Figure 9.5. The purpose and fundamental principle behind these names are the same, i.e., to store a large quantity of electric energy in between two different layers of electrodes. Because of its highly porous electrode, it has a very high storage capacity in terms of capacitance per unit volume compared to the other types of capacitors. High capacitance results from the very fine separation of the electronic and ionic charges at the electrode nanopores and electrolyte interface [22]. The fundamental equation that governs the capacitance value is as follows:

$$C = \varepsilon_0 \frac{A}{d} \tag{9.13}$$

Here, A is the porous electrode surface area and d is the distance between the porous electrodes. Hence, by determining the surface properties of the electrode material,

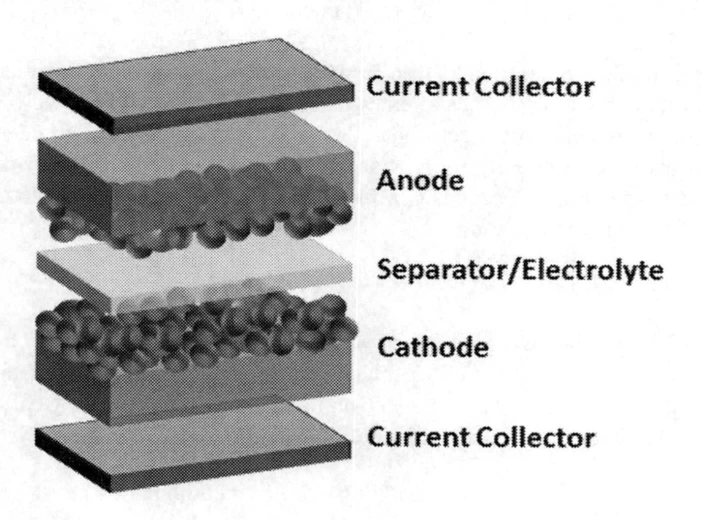

FIGURE 9.5 Schematic diagram of electrodes and electrolyte layers in electrochemical double-layer capacitors.

one can directly estimate the capacitances of the system. Popularly used materials for developing high-performance cells are carbon, conducting polymer, hybrid polymeric material, metal oxides, other organic-inorganic hybrid nanomaterials, etc. On the other hand, electrolytes are crucial for delivering high energy capacity. Therefore, breakdown voltage should be taken into consideration before selecting an appropriate solid-state or aqueous electrolytes. Last but not least, separators also play an essential role in maintaining the reliable performance of the EDLC. Materials such as paper, plastic, glass, and ceramic fibers are usually utilized for liquid or gel-type electrolyte-based EDLC.

9.18 WHAT RELEVANCE HAS NANOTECHNOLOGY FOR FUEL CELL SYSTEMS

Nanotechnology plays an essential role in designing every fuel cell component from electrodes to the gas diffusion layer to the catalyst to the binder to polymer electrolyte membrane. However, a fuel cell is a simple electrochemical system consisting of two electrodes with a polymer electrolyte membrane in-between, as illustrated in Figure 9.6. A fuel cell's output performance is governed by developing quality membrane electrode assembly (MEA). It can be achieved through MEA-interfacial nanoengineering and nanocatalysts amount. On the other hand, membranes are also greatly influenced by the ion nanochannel conducting species (anion or proton). Thus, nanotechnology in fuel cells provides novel material properties, improved and enhanced fundamental mechanism and principles, multifunctionality of materials, nano-structuring of the MEA, and helps in overcoming open-circuit potential loss. These features will undoubtedly stabilize the cell perforce and open up a new path for developing an ideal fuel cell system.

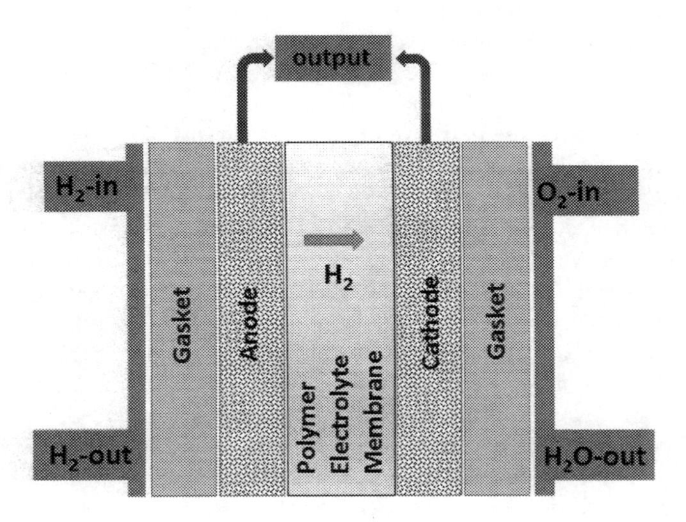

FIGURE 9.6 Schematic diagram of the fuel cell and their working.

9.19 FUEL CELL TECHNOLOGY AND NANOTECHNOLOGY

Nanotechnology has enormous potential for developing next-generation fuel cell technology. We have categorized this section into three-part, nanocatalyst, nanostructure in Polymer Electrolyte Membrane (PEM), and nanoengineered MEA.

Nanotechnology in electrode assembling: It is the costliest and essential part of the fuel cell containing platinum metal as catalyst. Researchers collaborate with companies for developing nanoparticles of platinum, thus reduction in cost as well as in quantity [23]. Other nanomaterial-based catalysts are also used successfully in fuel cell application, but performance is still not in level with platinum.

Nanotechnology in polymer electrolyte membrane preparation: The formation of numerous distinct varieties of nanostructure in PEM such as porous structure [24], layered structure [25], micellar structure interconnected by narrow pores [26], nanochannel network [27], polymer bundles [28], etc. Nano-level ordering in the polymer electrolyte membrane with the nonequilibrium composition has significant advancement in recently advanced polymer electrolyte materials.

Nanotechnology in MEA: It is one of the core components of the cell and is responsible for producing electrochemical reactions when fuel is supplied. The level of nanoengineering in the preparation of the catalyst layer, microporous layer, and gas diffusion layer significantly dominates the stability of cell performance [29]. Thus, nanotechnology in fuel cells gives an edge to improve the charge transport without any degradation of the cell performance and provides a path for standardizing the fuel cell technology for industrial uses.

9.20 OUTLOOK AND SUMMARY

The development of new technologies is overgrowing, whereas technological evolution promotes a focused purpose, commercial production, mainly governed by performance, cost, and environmental impact. Transformation of technology from laboratory-scale to the industrial level, many parameters have to be considered, such as reliability, efficient performance, large area, low cost, easy processability, high throughput, low toxicity, etc. Although nanomaterial has shown significant growth in device application from the past several decades in terms of efficient device performance, cost, and toxicity, there is still much to be done to achieve ideal stable devices. There is no evidence of developing novel architectures or fabrication techniques.

This chapter presents the fundamental concept and basic principles of electrochemical devices. Moreover, we show how nanomaterials are a new generation of promising material for enhancing the performance of electrochemical systems. Replacing conventional materials with nanomaterials offers ease of device processability, opening up the commercial path. Our emphasis mainly provides a versatile point of view in terms of energy conversion and energy storage devices. All the electrochemical system working principles were explicitly described, while the focus was nanomaterial and its implementation. Nanomaterial's physicochemical properties and charge transfer mechanism of each electrochemical cell were discussed in detail. The initial section covers the background and importance of nanomaterials in electrochemical devices.

COMPETING INTERESTS

The authors of this chapter declare that there are no competing interests.

REFERENCES

[1] M.I.A. Abdel Maksoud, et al., Advanced materials and technologies for supercapacitors used in energy conversion and storage: A review, *Environ. Chem. Lett.*, 19 (2020) 375–439.
[2] Y. Qiao, C.M. Li, Nanostructured catalysts in fuel cells, *J. Mater. Chem.*, 21 (2011) 4027–4036.
[3] G. Chen, J. Seo, C. Yang, P.N. Prasad, Nanochemistry and nanomaterials for photovoltaics, *Chem. Soc. Rev.*, 42 (2013) 8304–8338.
[4] K. Brainina, N. Stozhko, M. Bukharinova, E. Vikulova, Nanomaterials: Electrochemical properties and application in sensors, *Phys. Sci. Rev.*, 3 (2018) 165–222.
[5] A.P. Alivisatos, Semiconductor clusters, nanocrystals and quantum dots, *Science*, 271 (1996) 933–937.
[6] G. Graziano, Go with the flow, *Nat. Rev. Chem.*, 2 (2018) 47–47.
[7] Y.H. Lee, et al., Nanoscale surface chemistry directs the tunable assembly of silver octahedra into three two-dimensional plasmonic superlattices, *Nat. Commun.*, 6 (2015) 6990.
[8] A. Akbarzadeh, M. Samiei, S. Davaran, Magnetic nanoparticles: Preparation, physical properties, and applications in biomedicine, *Nano. Res. Lett.*, 7 (2012) 144.
[9] M.J. Mitchell, M.M. Billingsley, R.M. Haley, M.E. Wechsler, N.A. Peppas, R. Langer, Engineering precision nanoparticles for drug delivery, *Nat. Rev. Drug Discov.*, 20 (2021) 101–124.
[10] P. Würfel, U. Würfel, *Physics of Solar Cells: From Basic Principles to Advanced Concepts*, John Wiley & Sons, Weinheim (2016).
[11] M. Grätzel, Perspectives for dye-sensitized nanocrystalline solar cells, *Prog. Photo.: Res. Appl.*, 8 (2000) 171–185.
[12] A. Hagfeldt, G. Boschloo, L. Sun, L. Kloo, H. Pettersson, Dye-sensitized solar cells, *Chem Rev*, 110 (2010) 6595–6663.
[13] K. Kalyanasundaram, *Dye-sensitized Solar Cells*, CRC Press, Lausanne (2010).
[14] A.J. Nozik, Quantum dot solar cells, *Phys. E: Low-dimen. Syst. Nano.*, 14 (2002) 115–120.
[15] G.H. Carey, A.L. Abdelhady, Z. Ning, S.M. Thon, O.M. Bakr, E.H. Sargent, Colloidal quantum dot solar cells, *Chem Rev*, 115 (2015) 12732–12763.
[16] D. Bimberg, M. Grundmann, N.N. Ledentsov, *Quantum Dot Heterostructures*, John Wiley & Sons, West Sussex (1999).
[17] M.H. Sun, et al., Applications of hierarchically structured porous materials from energy storage and conversion, catalysis, photocatalysis, adsorption, separation, and sensing to biomedicine, *Chem Soc Rev*, 45 (2016) 3479–3563.
[18] V.S. Bagotsky, A.M. Skundin, Y.M. Volfkovich, *Electrochemical Power Sources: Batteries, Fuel Cells, and Supercapacitors*, John Wiley & Sons, Hoboken, NJ (2015).
[19] M. Yoshio, R.J. Brodd, A. Kozawa, *Lithium-Ion Batteries*, Springer, New York (2009).
[20] M. Li, J. Lu, Z. Chen, K. Amine, 30 years of lithium-ion batteries, *Adv Mater*, 30 (2018) e1800561.
[21] W. Vielstich, A. Lamm, H. Gasteiger, *Handbook of Fuel Cells: Fundamentals, Technology, Applications*, Volumes 3 and 4, p. 2690, John Wiley & Sons, Chichester (2003).
[22] P. Sharma, T. Bhatti, A review on electrochemical double-layer capacitors, *Energ. Con. Manag.*, 51 (2010) 2901–2912.
[23] K. Kodama, T. Nagai, A. Kuwaki, R. Jinnouchi, Y. Morimoto, Challenges in applying highly active Pt-based nanostructured catalysts for oxygen reduction reactions to fuel cell vehicles, *Nat. Nanotechnol.*, 16 (2021) 140–147.

[24] Y.Y. Jiang, et al., A novel porous sulfonated poly(ether ether ketone)-based multi-layer composite membrane for proton exchange membrane fuel cell application, *Sustain. Energ. Fuels*, 1 (2017) 1405–1413.
[25] X.B. Li, et al., Triple-layer sulfonated poly(ether ether ketone)/sulfonated polyimide membranes for fuel cell applications, *High Perform Polym*, 26 (2014) 106–113.
[26] T.D. Gierke, G.E. Munn, F.C. Wilson, The morphology in nafion perfluorinated membrane products, as determined by wide- and small-angle x-ray studies, *J. Polym. Sci.: Polym. Phys. Ed.*, 19 (1981) 1687–1704.
[27] K.D. Kreuer, S.J. Paddison, E. Spohr, M. Schuster, Transport in proton conductors for fuel-cell applications: simulations, elementary reactions, and phenomenology, *Chem. Rev.*, 104 (2004) 4637–4678.
[28] A.L. Rollet, O. Diat, G. Gebel, A new insight into Nafion structure, *J. Phys. Chem. B.*, 106 (2002) 3033–3036.
[29] L. Xing, et al., Membrane electrode assemblies for PEM fuel cells: A review of functional graded design and optimization, *Energy*, 177 (2019) 445–464.

10 Rechargeable Batteries with Nanotechnology

Meenal Gupta, Santosh J. Uke, Yogesh Kumar, Sweta Sharma, Ashwani Kumar, Pushpa Singh, Amit Saxena, Bhaskar Bhattacharya, and Yogesh Kumar

CONTENTS

10.1 INTRODUCTION

Electrochemistry is a discipline of chemistry that studies chemical processes that produced at electrodes. Electrochemistry has diver's applications in daily life; this includes electroplating, separations of salt, organic and inorganic synthesis, sensors, pollution control, energy, etc. Over the globe, the most of the electrochemist use the boon of electrochemistry especially conversion of chemical energy to electrical energy. The different batteries, supercapacitors and fuel cells are the most common gifts of electrochemistry to the mankind. The batteries are used to power a variety of devices, including electric vehicles, smartphones, electronic tablets, watches,

DOI: 10.1201/9781003216308-10

pacemakers and many others. Chemical reactions in batteries provide power that can be transformed into useful work. The transport of the electrons also occurs in the electrode material, electrode current collector interface and outer circuit in electrochemical applications. The processes take place in a system called the cell. In many systems, the reactions take place in a cell, where electrons are transferred between electrodes.

A rechargeable battery is an electrical device, which has many electrochemical cells together. Batteries involve electrically reversible electrochemical reactions and therefore known as secondary cell. While discharging of the battery, stored charges depleted slowly however during charging chemical reaction get reversed to restore new charges. With the advent of portable devices such as notebook computers, cell phones, MP3 players and cordless power tools, the demand for rechargeable batteries has increased dramatically in recent years. In 1859, the invention of lead-acid element by French physicist Gaston Plante opened a research filed toward new technology based on battery and still researches are going on battery system. Plante batteries with lead anodes, lead dioxide cathodes and sulfuric acid electrolyte are the pioneers of modern car batteries.

Both the primary batteries (nonrechargeable) and secondary batteries (rechargeable) work in exactly similar way. Electrochemical reactions developed between the both electrodes (anode and cathode) and electrolyte to generate electricity. However, for rechargeable batteries, the reaction is reversible. When external power is applied to the secondary battery, the electron flow generated during the discharge is reversed from negative to positive and the charge of the battery is restored. After nickel-metal hydride (Ni–MH) and nickel-cadmium (Ni–Cd) batteries, the most common rechargeable battery is Lithium-ion battery (LIB), which is available in market presently. The construction of a rechargeable battery is shown in Figure 10.1 [1].

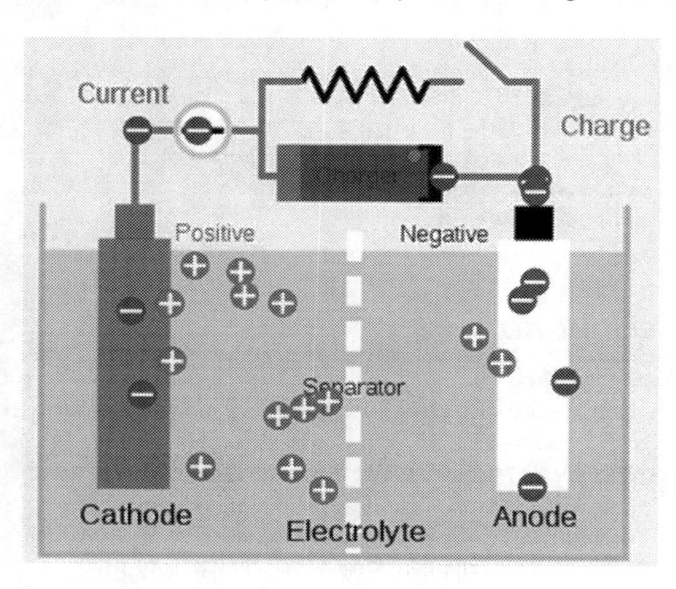

FIGURE 10.1 Construction of rechargeable battery [1].

10.2 NANOMATERIALS FOR RECHARGEABLE BATTERIES

The demand for energy materials is increasing at its fast pace; lots of new materials are been explored in order to meet the demand of the market. Nanomaterials of batteries in applications are gaining momentum day by day because of its versatile properties; it is meeting the demands of smart power enhanced and efficient batteries which can be used in electric vehicles, storage devices, plug in vehicles. In the present context, the climate change has given new dimensions for the search of those nanomaterials, which are environmentally friendly and can be easily recycled so that the dependence on fossil fuels can be minimized and the pollution can be checked at a large scale. Research on application about the nanomaterials in batteries was done by the global nanomaterials in batteries and supercapacitor market and the report predicted that the market size will grow for the period from 2021 to 2031. In Table 10.1, it has been shown the different parameters needed for the commercializations of nanomaterials in batteries. Moreover, the ongoing research in different companies of the countries for the exploration of different nanomaterials, depending on the various issues, are listed in Table 10.1.

The rechargeable batteries are becoming more and more important in our daily lives with their powerful ability to effectively store electrical energy in chemical form. Replacing traditional liquid electrolytes with polymer electrolytes (PEs) is considered to be one of the most feasible solutions for the development of higher energy density and safer electrochemical energy storage systems, which are eagerly used in electric vehicle applications. In recent years, to coordinate the advantages and to modify the material according to our needs using organic phase and the inorganic phase electrolyte, the introduction of organic–inorganic hybrid nanomaterials in PE has attracted more and more attention. Polyhedral oligomeric silsesquioxanes (POSSs) are one of the most attractive latest technologically important hybrid nanostructured

TABLE 10.1
Different Companies of the Countries for the Exploration of Different Nanomaterials Depending upon the Various Issues

Countries	Brands of the Companies	Scope and Segment	Analysis based on Following Factors
United States, Southern Asia, Canada, France, Germany, United Kingdom, South Korea, Taiwan, China, Japan, Brazil, Russia	Ampirius Inc.; BAK Power; Be-Dimensional; Bodi Energy; Dongxu Optoelectronic Technology Co. Ltd.; Nexeon; Ray Techniques Ltd.; Skeleton Technologies Group OA; HE3DA S. R. O	Company; Region (Country); Type; Application; Revenue; Forecast by region; Participants; Stakeholders	Market overview; Industry and applications; Prospects of growth in revenue; Competition in the market: Region wide consumptions and production; Upgradation in technologies; Analysis of supply chains; Landscapes

material having combined properties of organic–inorganic. In its structure, organic functional groups are attached to the inorganic nano-sized cores. POSS is known for its low density, adjustable surface properties, high thermal stability and good mechanical strength with polymers in the form of nanocomposites, which can be used as nano-filler in different systems. In recent years, the paradigm shift from solid polymer electrolyte (SPE) to nanocomposite hybrid polymer electrolyte (NCHPE) in rechargeable battery applications was highlighted mostly in many research reports.

The increases in conduction mechanism enhance the capacity of electrodes. The reduction in diffusion length in lithium ion at nanoscale is one most important application of nanotechnology in alkaline batteries. The thin films fabricated using the nanomaterials increase the transport properties of the electronic conduction. It helps in ion storage by increasing the surface area of the electrode materials. Mesoporous-ordered structure generally favors the kinetic of electrode. It also helps in enhancing the life cycle of batteries. The main area of research is to find those nanomaterials which can be used in electrodes having a high surface area as it will increase the energy density and capacity of the batteries; moreover, it will also increase the lifespan and efficiency of the batteries. Safety and cost-effectiveness are also the primary factors for the search of new nanomaterials.

The nano-batteries are also being manufactured using nanotechnology. These batteries can be combined together to form macro battery with increased efficiency. The nanomaterials can also be used as coating in order to separate the electrodes, thus causing low self-discharge. Although nanomaterials have lots of advantages, there are some shortcomings such as low density with large surface area which result in high resistance exists, thus reducing the conductivity and stability of the battery. Nevertheless, nanoparticles are also difficult to synthesize as a result the manufacturing cost increases.

There are huge potentials of nanomaterials to be used in batteries, as it can be used as a coating material for electrodes. With its use, the surface the charging time will also get reduced. The of nanomaterials is used in different battery, viz., nickel-cadmium battery, nickel metal hydride battery, LIB, sealed lead-acid battery, sodium-ion battery, etc. Nanomaterials and nanotechnology are the future of recent batteries, although there are many challenges related to its efficiency and cost, still the research studies are going to explore new materials for battery.

10.3 LEAD-ACID BATTERY

In 1859, Lead-acid batteries which are the oldest rechargeable batteries were invented by French physicist Gaston Plante. These are one of the most common secondary batteries used mostly for loading large cell potential. Lead-acid batteries are capable to supply high value of current, which requires maintaining high power to weight ratio by cells. Lead-acid batteries are low-cost batteries with high power to weight ratio; therefore, they become a suitable candidate for their use in automobiles, golf cars, forklifts and other vehicles, which require high currents. The hazards include heavy mass, incompetence under low temperatures and incapability to maintain its capacity for long intervals of time through disuse [2].

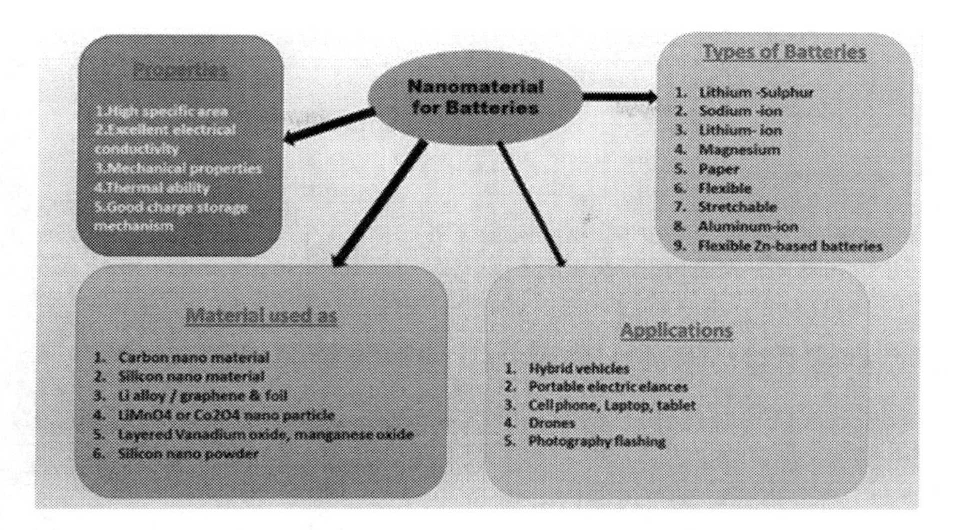

FIGURE 10.2 Nanomaterials for rechargeable batteries.

Usually, lead-acid battery consists of lead oxide (PbO_2) as cathode, lead (Pb) as anode and aqueous solution of concentrated sulfuric acid (H_2SO_4) as electrolyte. The chemistry behind lead-acid battery and its principal components are given in Figure 10.2.

The standard reversible electrochemical reactions in a lead-acid battery are shown as follows:

$$\text{At cathode}: PbO_2 + 3H^+ + HSO_4^- + 2e^- \rightleftharpoons 2H_2O + PbSO_4 \tag{10.1}$$

$$\text{At anode}: \ Pb + HSO_4^- \rightleftharpoons PbSO_4 + H^+ + 2e^- \tag{10.2}$$

$$\text{Overall reaction: } Pb + PbO_2 + 2H_2SO_4 \rightleftharpoons 2PbSO_4 + 2H_2O. \tag{10.3}$$

It can be seen that after complete discharge, anode and cathode both are converted into lead sulfate ($PbSO_4$); also, water produces after losing most of the dissolved H_2SO_4 in electrolyte. When fully charged, the cathode and anode are made of PbO_2 or Pb. The electrolyte converts back into concentrated H_2SO_4. The majority of the electrochemical energy is stored in such a fully charged state. Lead-acid batteries suffer with their low-energy density (~40 Wh/kg) [3] and have 85% Coulombic efficiency and 70% energy efficiency. They have the lowest storage capacity of any other rechargeable batteries and are typically large and heavy. Therefore, lead-acid batteries are unable to store a large amount of energy which can increase battery weight and practically limits their applicability in electric vehicles. The different components of a lead-acid battery are shown in Figure 10.3 [2].

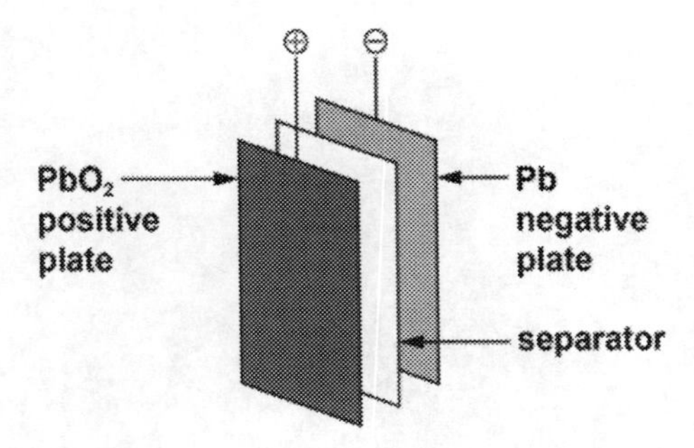

FIGURE 10.3 Chemistry and principal components of a lead-acid battery [2].

10.4 ALKALINE BATTERY

The alkaline electrolyte-based batteries were first developed in 1899 by Waldemar Jungner. Lewis Urry developed a button alkaline cell in 1949 and put it in the market by Ray-O-Vac Co, USA. In 1950, Lewis Urry invented zinc/manganese dioxide dry alkaline battery with high specific energy and relatively low cost. In 1957, Marsal, Larl and Urry filed a US patent (US2960558A) for the alkaline battery and granted it in 1960 [4]. Zinc alkaline batteries had leakage problems and decreased self-life in the late 1960s. To prevent the electrolytic action on impurity sites, a film of mercury amalgam was used on a zinc electrode [5]. A French company in the 1970s introduced a battery of better performance at a low cost with a new plastic-bonded negative electrode and sintered positive electrodes.

In the late 1980s, inspired by the increasing demand for high volumetric energy, Dr. Oshitani developed positive electrode foam technology which increased volumetric energy up to 30%. Mercury degrades the stability and purity of the electrode so the minimum use of mercury was required [6]. In 2005, after a century after the discovery of the Ni–Cd cell, Dr. Bernard et al. introduced a plastic-bonded positive electrode which exhibit high electrochemical performance at a lower cost using sintered electrode technologies. Modern alkaline batteries have manganese dioxide as a positive electrode and zinc negative one. This battery is called alkaline battery only because of an alkaline electrolyte used in it. Nickel-cobalt (Ni–Co) electrode materials give high capacitance, high abundance and good cycle stability [7]. Furthermore, nickel-based electrodes are redox active and good electrically conductive material for energy storage applications [7]. Moreover, the comparative studies between different types of batteries and commercially available nickel battery sizes are demonstrated in Tables 10.2 and 10.3, respectively.

10.4.1 Zinc Manganese Dioxide (Zn–MnO_2)

The zinc manganese dioxide (Zn–MnO_2) batteries are suitable for industrial applications where moderate amounts of electricity are needed. These batteries are commonly used worldwide in digital cameras, toys, flashlights, radios, compact disc players, etc. The low cost of material and high energy density application are other

TABLE 10.2
Comparative Studies between Different Types of Batteries

Types of Batteries	Cathode	Anode	Electrolyte
Alkaline	Manganese dioxide	Zinc	Aqueous alkaline
Lead acid	Lead dioxide	Lead	Sulfuric acid
Nickel	Nickel oxyhydroxide	Cadmium, hydrogen absorbing alloy	Potassium hydroxide
Lithiumion	Lithium nickel manganese cobalt oxide ($LiNiMnCoO_2$) Lithium nickel cobalt aluminum ($LiNiCoAlO_2$)	Carbon-based typically graphite	Lithium salt in an organic salt

TABLE 10.3
Commercially Available Nickel Battery Sizes

Sizes	D	C	AA	AAA	Sub C	Nine Volts	Button
Diameter in mm	34.2	26.2	14.5	10.5	22.2	26.5	Varies in Ni–MH
Length in mm	61.5	50	50.5	44.5	42.9	48.5	Variable size exists

advantages of $Zn–MnO_2$ battery, which enhance the scientists interest in commercialization and engineering. In $Zn–MnO_2$ battery, the negative and positive electrodes are Zn and MnO_2, respectively. In the discharging process of $Zn–MnO_2$ battery, only electrode materials (Zn & MnO_2) take part in the reaction, as seen in the reaction below:

$$Zn + 2MnO_2 \rightleftharpoons ZnO + Mn_2O_3 \tag{10.4}$$

During the reaction, the alkaline electrolyte KOH remains in equal amount of OH^- based on the contents of zinc in the KOH electrolyte and purity of manganese dioxide used; the nominal voltage of $Zn–MnO_2$ battery is 1.5 V; however, it varies from 1.50 V to 1.65 V. The $Zn–MnO_2$ battery has varying voltages, 1.3 V to 1.1 V, which depend on the current drawn and load level of discharge. After fully discharge, battery of 1 V potential still remains.

Recently, Zhang et al. reported a high capacitive rechargeable $Zn–MnO_2$ battery system with a mild-acidic zinc triflate electrolyte. The aqueous zinc/manganese triflate electrolyte formed the protective porous MnO_2 layer. In Figure 10.4a, $Zn–MnO_2$ batteries delivered 10% capacity depth of discharge and lower cycle stability. Recently, the rechargeability of $Zn–MnO_2$ battery has been improved by using a mild zinc-based acidic electrolyte. Figure 10.4b shows the reversible extraction/insertion of Zn^{2+} ions in the layered structure. A significant improvement finds in the cycling stability of zinc manganese dioxide battery employing $Mn(CF_3SO_3)_2$ and

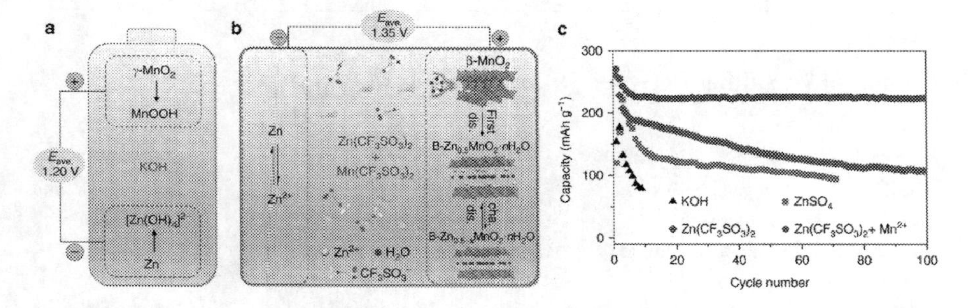

FIGURE 10.4 Zn–MnO_2 battery chemistry. Schematic illustration of (a) the primary alkaline Zn–MnO_2 battery using KOH electrolyte and (b) the rechargeable Zn–MnO_2 cell using CF_3SO_3-based electrolyte. (c) Comparison of the cycling performance of Zn–MnO_2 cells with electrolytes of 45 wt% KOH (at 0.32 C), 3 M $ZnSO_4$, 3 M $Zn(CF_3SO_3)_2$, and 3 M $Zn(CF_3SO_3)_2$ with 0.1 M $Mn(CF_3SO_3)_2$ additive at 0.65 C. nC equals the rate to charge/discharge the theoretical capacity (308 mA/hg) of MnO_2 in 1/n hours. (Reproduced from ref. [8] with permission from Springer Nature, © 2017.)

concentrated $Zn(CF_3SO_3)_2$ electrolyte which can be seen in Figure 10.4c. The cycle stability was found to be 94% after 2000 cycle with a high reversible capacity of 225 mA/h/g [8].

10.4.2 Nickel-Based Alkaline Batteries

With the growth of technology, the need for rechargeable batteries has increased manifold; lots of research studies are taking place to improve their efficiency so that they can be used in all types of electronic devices, vehicles, aeronautics industry, etc. In the last few years, the market has been commercialized in terms of hybridization topologies in order to increase the energy density and power density of the batteries. However, the electric double-layer capacitors and lead-acid batteries have solved many problems related to enhanced efficiency in terms of power and energy, longer life span, thermal stability. Still, the biggest challenge lies in the increased price as it is not cost-effective. In this context, nickel plays a significant role as it gives high energy density and better storage capacity at a very lower cost as compared with other batteries.

The nickel metal is the fifth most commonly found naturally occurring element on the earth surface. It has a silvery white shining appearance; it easily forms alloys because of its chemical properties. It has good catalytic and magnetic properties and can be easily recycled. From the mid-90s, its use is much more commercialized in battery technology and has a good market share. The most commonly used batteries like nickel cobalt aluminum use 80% nickel, whereas nickel manganese cobalt uses 33% nickel [9]. Most of the LIBs also rely on nickel. The nickel batteries are generally rechargeable and used in portable electronic devices, hybrid vehicles and stationary storage mechanisms. The main features include a discharge curve which is flat; it has a wide range of temperature operations. It is environmentally friendly as it can be easily recycled. Overall, it has a robust mechanism that is physically and chemically tolerant toward charging and discharging.

The nickel, along with manganese or aluminum, is used in LIBs to increase its longevity as well as proper capabilities. Also, nickel (tabs) strips are being made from ferromagnetic nickel alloy 201, which is highly conducting and noncorrosive in nature. Moreover, its resistance is also very low, making it viable to use as it is safe, heat resistant and there is no waste of energy. The nickel foam is used as it is porous with 75% to 95% void spaces, making it the perfect material to be used as a current collector due to its low density.

The nickel batteries consist of a cathode, anode and separator. The separator should be porous enough for the chemical reactions to take place. These three elements are mainly wound in Swiss roll or jelly roll type structure. At the top of this, the structure is a metal tab that connects the anode to a sealing plate, which seals the corrosive electrode and acts as a self-sealing vent that allows gases to escape if there is overcharging of the battery [10]. There are different types of nickel-based alkaline batteries, viz., nickel-iron (Ni–Fe) battery, nickel-zinc (Ni–Zn) battery, nickel-cadmium (Ni–Cd) battery and nickel-cadmium (Ni–Cd) battery. We have discussed in detail about these batteries in the following subsection.

10.4.2.1 Nickel–Iron (Ni–Fe) Batteries

Nickel-iron (Ni–Fe) batteries are rechargeable batteries and are much commercialized because they are much cheaper. It was the first commercialized nickel battery. The glory that adds to this battery is its long life and environmentally friendly. Nickel and iron as raw material are the most abundant material found on earth, which makes them cheaper [11]. It uses Fe as anode and Ni as cathode with KOH as an electrolyte. This is more resilient toward charging and discharging and maintaining its thermal stability and internal resistance [12]. Moreover, the energy density of nickel-iron batteries is in the range of 20–50 Wh/Kg, power density 65–90 W/Kg and cycle life 2000–5000 cycles. They have an excellent life cycle and are mainly used in locomotive applications, with the recent development of sintered electrodes where the nickel oxide cathode finds its applications in electric vehicles [13]. The main advantages of Ni–Fe battery are good resistance against vibrations. The operating life is long; hence, it is very much durable. It has good response toward charging and discharging of the battery. However, it is being observed that at low temperature its performance is very poor. Another setback is the rate of discharging as it is very high, nearly 40% every month. The specific energy is also very low, approximately 50 Wh/kg, which makes it less viable to get commercialized.

10.4.2.2 Nickel–Zinc (Ni–Zn) Battery

It is one of the rechargeable batteries which have good performance in high drain applications. As nickel and zinc are naturally occurring materials on the earth crusts, hence it is less expensive. The use of zinc electrodes has made it commercially competitive with other batteries existing in the market. This cell has got a voltage of 1.85 V in an open circuit when fully charged. A Ni–Zn battery has a similar curve in charging and discharging with a higher voltage (~1.7 V) as compared to Ni–Cd and Ni–MH batteries. It has got low internal resistance and high power density with excellent performance at low temperatures. Ni–Zn batteries use no toxic elements like Ni–Cd; hence, it is not a threat to our environment and can be easily recycled.

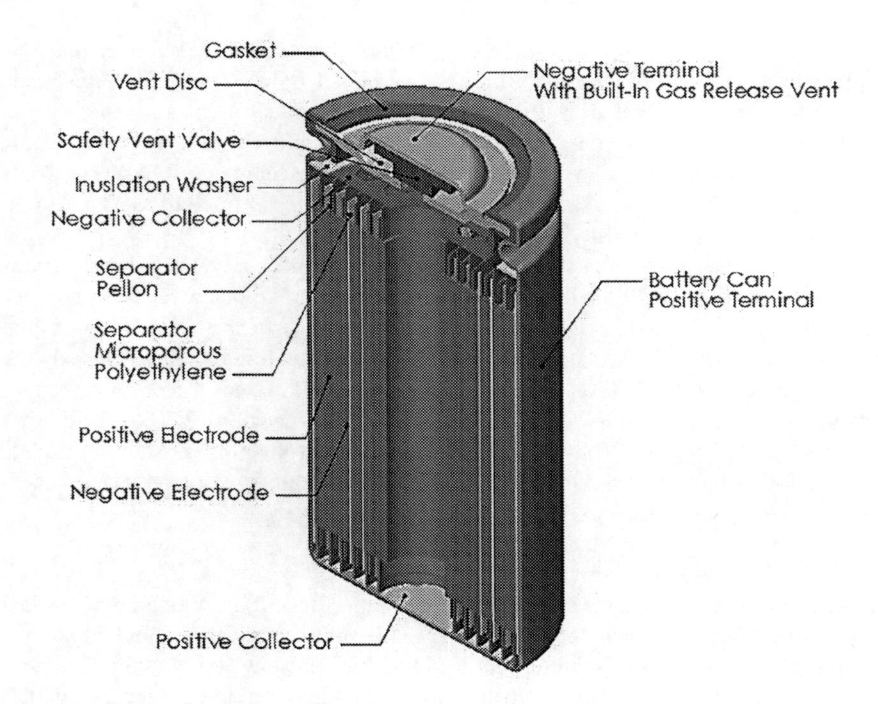

FIGURE 10.5 Internal structure of power Genix Ni–Zn Battery. (Reproduced from [15] with permission from IOP Publishing, © 2009.)

It is also inflammable as it uses on active materials, so it is safe to use [14,15]. Figure 10.5 demonstrates the internal structure of power Genix Ni–Zn battery [15].

In Ni–Zn batteries, nickel is also used as a positive electrode and zinc as a negative electrode. The chemical reaction at positive and negative electrodes is given in Equations (10.5) and (10.6), respectively.

$$2NiOOH + 2H_2O + 2e^- \rightarrow 2Ni(OH)_2 + 2OH^- \tag{10.5}$$

$$Zn + 2\ OH^- \rightarrow Zn + H_2O + 2e^- \tag{10.6}$$

Ni–Zn batteries are used for portable power, small-scale applications at high discharge rates. Ni–Zn batteries are relatively lower in cost than Li–ion batteries and can replace both Ni–Cd and Ni–MH batteries in most of the applications [6]. Ni–Zn batteries have high efficiency, high specific power and low cost. Both Ni and Zn can easily maintain their physical and chemical properties during cycling, and hence less effect on the environment. However, Zn is a self-corrosive material and soluble in KOH so Ni–Zn batteries show low discharge after a few cycles [16,17]. Although this battery was there for many years, the major challenges faced were the distribution of active materials on the electrode and the formation of dendrites during the charging and discharging process. Figure 10.6 shows the formation of dendrites in

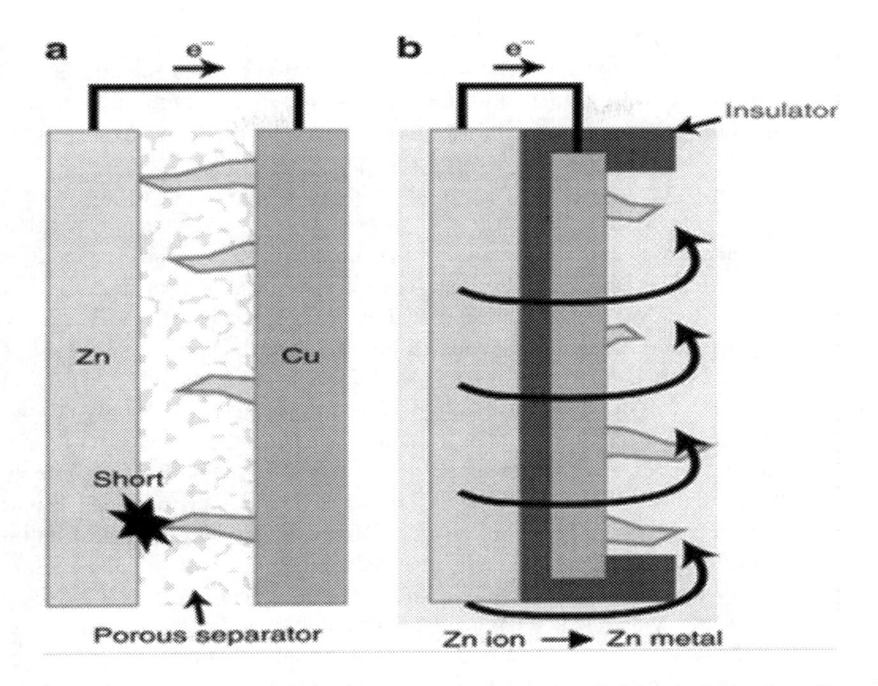

FIGURE 10.6 Formation of dendrites in Zn [18]. (Reproduced with permission from Nature, © 2016.)

TABLE 10.4
Features of Ni–Zn Battery

Energy Density (Wh/Kg)	Power Density (W/Kg)	Cycles Life Cycles	Anode	Cathode	Electrolyte
60–70	100–200	200–300	Nickel hydroxide	Zinc	Potassium hydroxide

Zn electrodes [18]. Table 10.4 demonstrates the important electrochemical properties of Ni–Zn battery.

The zinc electrode technology was based mainly on these principles, i.e., the requirement of good conductivity for the whole mass independent of the state of charge, a high level of porosity to be maintained at the anode of the cell, and treatment of soluble zincates were formed during the charging process. Moreover, to maintain the above principal and to other rectify the above-mentioned issues, a three-dimensional structure was made to support and collect current to reduce resistance and improve the electrical connections. The coating of a separator also helps in inhibiting the dendrite formations. Also, the copper foam was used to increase the porosity of the material at the zinc electrode, a composition of zinc oxide was mixed with additives like zinc alloys, conductive polymers having high adsorption capacity for zincates in order to trap zincates from the surface constituting a more conducting network and preventing dendrites formations. The use of calcium oxide as an additive helps in forming calcium zincates, which are less soluble than normal zinc hydroxides. Pasted powder technology or the pressed powder method is used to fabricate zinc electrode, with a substrate material like foil, foam mainly made of copper. This substrate maintains stability when the cathode undergoes polarization. This method also helps in changing electrode shape.

The advantages of Ni–Zn battery are good cycle life in terms of charging and discharging, high capability rate of nearly 25 C, very fast recharging, and cheaper than benign materials, which can be cycled to 100%, nontoxic, inflammable and safe to use as it is environmentally friendly in nature. However, the shortcomings of Ni–Zn batteries include heavy and bulky structure, low-energy density, discharge rate is very high and formation of dendrites which reduces final efficiency.

Ni–Zn batteries are used in consumer batteries and power tools. It is an excellent substitution for Ni–Cd, lithium-ion and Ni–MH batteries in terms of energy density, toxicity and price. Ni–Zn batteries can be used in hybrid electric vehicle (HEV) which provides similar power density and energy density to drive HEV vehicles at half the price of lithium ion. Ni–Zn batteries are used in micro hybrid (start/stop vehicles). In microhybrid vehicles, when the vehicles stop, then the engine shuts down, resulting in more consumption of fuel. A Ni–Zn battery has high power, half weight and more charge acceptance as compared to lead-acid battery at a competitive cost. The Ni–Zn battery at the utility scale meets the requirement of high power in terms of regulated frequency and provides an effective backup power system. Military requires a high level of sophisticated power supply which would be safe and powerful for use. In this regard Ni–Zn battery fulfills the entire criterion. Hence, Ni–Zn provides an excellent balance of power and energy density in order to meet the potential requirement of applications in various devices and vehicles in terms of safety, cycle stability and cost-effectiveness.

10.4.2.3 Nickel-Cadmium Battery

It is a rechargeable battery which can be recharged many times, and it maintains constant potential when discharged. It gives hassle-free service and finds applications in digital cameras, calculators, photoflash equipment, recorders, etc. Ni–Cd cell is available in all shapes and sizes ranging from AAA to D with a potential difference between 1.25 V and 1.35 V [19]. The potential of the Ni–Cd battery is 1.2 V, which is lower than the alkaline zinc-carbon cell (1.5 V), so the Ni–Cd battery cannot replace in all applications. However, many electronic appliances work at low voltages (0.90 V to 1.0 V per cell); the relatively steady 1.2 V of a Ni–Cd cell is enough to allow operation [20]. The battery is used safely in the temperature range from −20°C to 45°C. During charging, the battery temperature is around the same as the ambient temperature, but as the battery gets fully charged, the temperature will rise to 45°C–50°C. These batteries are capable of rapidly recharging to hundreds of instances and are tolerant including overcharging. However, compared to other batteries and even lead-acid batteries, nickel-cadmium batteries have limited power density and are also heavy in weight. They perform better if fully discharged every cycle before recharge. In any other case, cells may also exhibit a memory effect. Alkaline batteries show some application in backup power systems where low-temperature conditions, very high currents and high reliability are unique elements.

Large nickel-cadmium batteries are used to start aircraft engines and in emergency power structures. In addition, they are used in conjunction with solar-powered contemporary supplies to provide electrical power at night. Although they have some good characteristics, it has some drawbacks that nickel and cadmium both are toxic heavy metals that may cause health risk. Ni–Cd battery could not get commercial success because of its high cost (about USD 1000 kWh) which is ten times higher than lead-acid batteries.

Nickel-cadmium (Ni–Cd) structures are the most common small rechargeable battery for portable gadgets. Sealed cells are equipped with "jelly roll" electrodes, which deliver the high current. The Ni-Cd battery contains a cadmium negative electrode, a nickel (III) oxide-hydroxide positive plate and potassium hydroxide as an alkaline electrolyte. The chemical reactions during charging, discharging and overall are given in Equations (10.7)–(10.9), respectively.

$$2NiOOH + 2H_2O + 2e^- \rightarrow 2Ni(OH)_2 + 2OH^- \quad (10.7)$$

$$Cd + OH^- \rightarrow 2Cd(OH)_2 + 2e^- \quad (10.8)$$

$$2NiOOH + Cd + 2H_2O \rightarrow 2Ni(OH)_2 + Cd(OH)_2 \quad (10.9)$$

When the battery is discharged, then the cathode contains nickel hydroxide and the anode has cadmium hydroxide, whereas when the battery is charged, then reverse action takes place at anode, cadmium hydroxide changes into cadmium and at cathode nickel hydroxide changes into nickel. The electrolyte used is alkaline potassium hydroxide and a separator is also used.

This battery has metal cases with a safety valve; the anode and cathode isolated by the separator are rolled in the form of a jelly roll design or in a spiral shape. This design helps the battery to deliver maximum current comparable to an alkaline battery. Traditionally, Ni–Cd batteries were available in sealed types, where the release of gas occurs when it is overcharged. Vented cells have a low-pressure valve that generates oxygen and hydrogen when it undergoes a cycle of charging and discharging. This structure makes the battery safer, lighter and economical for use. Vented Ni–Cd batteries can operate for a wide range of temperature and has long life.

Ni–Cd batteries have a number of advantages as compared to other batteries available in market. It works extremely well in every robust condition and has long-term storage when fully discharged. As compared to other lead-acid batteries, Ni–Cd battery has got a high energy density and longer life in terms of charging and discharging. When compared to alkaline batteries, it has reversible chemical reactions which make them reusable and long-lasting. When compared to Ni–MH and lithium-ion batteries, this battery is much cheaper and has a lower self-discharge rate. Hence, it can be seen that Ni–Cd is a good choice for applications in photography and other portable devices as it has good specific energy and pulse power performance, and at the same time, it is relatively inexpensive as compared with other batteries

However, the disadvantages of Ni–Cd battery include as it is much costlier than lead-acid battery because cadmium and nickel cost more. The biggest threat is environmental hazards as cadmium is very toxic in nature. Under the battery directives,

TABLE 10.5
Characteristics of Ni–Cd Battery

Energy Density (Wh/L)	Specific Energy (Wh/Kg)	Specific Power (W/Kg)	Cycle Durability Cycles	Self-Discharge Rate	Nominal Voltage	Charge/Discharge Efficiency
50–150	40–60	150	2000	10% month	1.2 V	70%–90%

this battery has been banned in many countries. Recycling of this battery is also the biggest problem. It has the problem of thermal runaway in which the current keeps on rising until the battery destroys itself. It also exhibits a negative temperature resistance in which the internal resistance decreases as the temperature increases. Memory effect is another issue related to its charging; in this effect, it is seen that the battery retains the characteristics of previous charging and gives the false impression of charging. In Table 10.5, the range of various parameters of Ni–Cd battery has been shown. It has been observed that the range of cyclic durability is nearly 2000 cycles and charge/discharge efficiency is nearly 70% to 90%.

10.4.2.4 Nickel-Metal Hydride (N–MH) Battery

It's always a good choice to use rechargeable battery packs containing lithium ions cells as it gives excellent high energy density, storage and high voltage, for many applications such as mobiles, bio-medical instruments, electric hybrid vehicles, etc. Customized nickel-metal hydride is a very good substitute, as it is cost-effective in manufacturing and has no potential hazards linked with lithium products. The technology of Ni–MH batteries is not new as it was available in the early 70s, but when compared with Ni–Cd and lead-acid batteries, it is environmentally friendly and has excellent cycle life, with good safety and reliability performance. Moreover, it does not require the complexity of battery management technology like lithium batteries, still satisfying the customer need as lithium pack [21, 22].

Ni–MH has a positive electrode of nickel hydroxide and a negative electrode is interstitial hydrogen in the form of a metal hydride; the electrolyte used here is alkaline potassium hydroxide. In Ni–MH, the "*M*" represents intermetallic compound as a negative electrode; this "*M*" is of the formula AB_5, where "*A*" is a rare earth element (mainly used are cerium, lanthanum, neodymium, praseodymium), whereas "*B*" side comprises of cobalt, nickel, aluminum or manganese. In some cells, the negative electrode "*M*" has the formula AB_2; here, *A* is generally vanadium or titanium, whereas *B* is either zirconium or nickel.

Nickel-metal hydride (Ni–MH) batteries are replacing nickel-cadmium batteries in many applications because of the absence of toxic cadmium and have two–three times higher capacity than Ni-–Cd battery per unit volume. The chemical reaction on the positive electrode is similar to Ni–Cd battery, whereas at the negative electrode, hydrogen absorbing alloy is used instead of Cd. The charging and discharging voltage of Ni–MH battery is 1.6 V and 1.25 V, respectively.

Ni–MH has higher self-discharge, lower efficiency and costly as compared with lead-acid and nickel-cadmium batteries. The cost rises mainly over the constraints

TABLE 10.6
Comparison of Ni–MH with Li–ion Battery

Cell Type	Cell Voltage (V)	Specific Energy (Wh/Kg)	Specific Power (W/Kg)	a. Energy Density (kWh/m^3)	Power Density (MW/m^3)	Efficiency	Cost	Size
Li–Ion	3.6	3–100	100–1000	80–200	0.4–2	99	Costly	Small and lighter
Ni–MH	1.2	1–80	<200	70–100	1.5–4	81	Less costly	

on the manufacturing process and safe disposal because of the concerns on cadmium toxicity. It has been labeled "Eco friendly". It has a profitable margin when recycling is done. Lithium ion has the major problem of "Thermal Runaway"; this battery has a better design for this, but as compared to Ni–Cd, it is not that effective in solving the problem of thermal runaway. The chemistry of Ni–MH and Ni–Cd is moreover the same because of the same design; Ni–MH has a capacity of 30% more than Ni–Cd batteries but less in memory. As compared to others, this requires high maintenance as crystalline formation takes place if a full discharge is not done regularly. In Table 10.6, a comparative study of Ni–MH versus Li–ion battery has been given. Li–ion battery is more efficient and lighter than Ni–MH, but it is costlier. The comparative study of Ni–MH and alkaline battery has been shown in Table 10.7; it has been shown that the performance of Ni–MH is much better than alkaline batteries.

It is very convenient to store and its transportation is easy because it is not governed by any regulatory controls. However, it should be stored in a cool place as performance gets degraded at high temperatures. When compared with a primary battery, it has greater advantage as it can work at extreme low temperatures of about –20°C. It operates well over a wide range of temperatures, i.e., 0° to 50° and life expectancy is also more. It can be recharged as many times as possible and shows efficiency even at a high rate of self-discharge. It has 50% of self-discharge when compared to Ni–Cd batteries. It has a critical trickle charge and generates heat while charging; it also takes longer time to get charged as compared to Ni–Cd batteries. Ni–MH has 50% less final battery pack production price than a lithium battery, and it is less than 75% of a lithium battery in terms of the development of the product.

The chemical reactions at negative and positive electrodes are given in Equations (10.10) and (10.11), respectively.

$$H_2O + M + e^- \rightleftharpoons OH^- + MH \tag{10.10}$$

$$Ni(OH)_2 + OH^- \rightleftharpoons NiO(OH) + H_2O + e^- \tag{10.11}$$

Ni–MH battery has a higher self-discharge rate as compared to Ni–Cd battery, which varies greatly with temperature, i.e., is 5%–20% on the first day and stabilizes around 0.5%–4% per day at room temperature. The cylindrical structure of Ni–MH battery is shown in Figure 10.7 with a cap positive electrode, negative electrode and separator [21].

TABLE 10.7
Comparison of Ni–MH Battery with Alkaline Battery

Application Features	Voltage	Discharge Capacity	Recharge Capability	Discharge Voltage Profile	Self-discharge Rate	Low Temperature Performance	Weight	Environmental Issues
Ni–MH	1.25 V	Lasts longer	Several hundred cycles	Flat	50%–80 % @ 12 months	Better than alkaline	Less lighter	Recycling can be done. Environmentally friendly
Alkaline	1.5 V	Less	NA	Sloped	Retains 80% @ 10 years		Lighter	Less options available

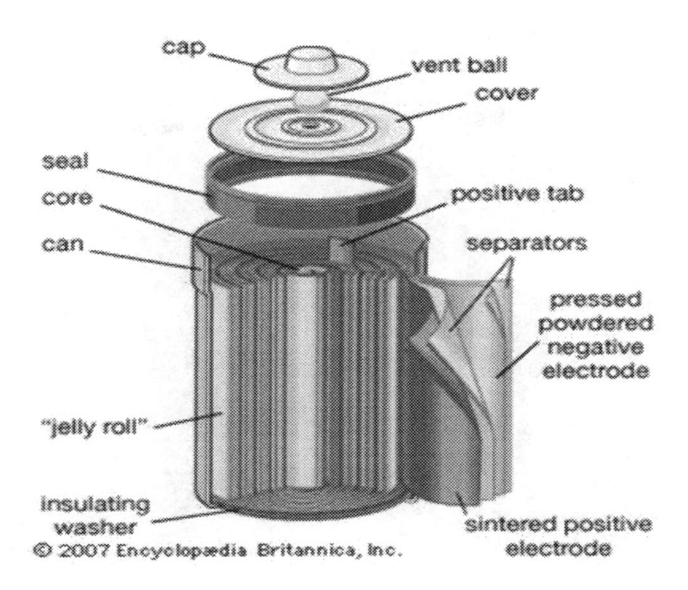

FIGURE 10.7 Nickel-cadmium batteries [21].

In order to have a hassle-free use of Ni–MH battery in terms of its maintenance and performance, basic knowledge of charging is very important as overcharging may damage the battery, which will result in loss of its capacity. Overcharging results in the formations of small crystals on the electrodes thus inhibiting the charging process. It is observed that the charging of Ni–MH is more complicated as compared to Ni–Cd battery. Designing of a charger is very important as in Ni–Cd a distinct bump is seen in output voltage when it is fully charged, but in Ni–MH, this bump is very small and difficult to detect. The charging efficiency of all forms of nickel-based batteries is about 70% of the full charge; initially, the temperature rise is less, but as the charge level increases, the efficiency level drops, thus raising the temperature of the battery [23,24]. Figure 10.8 shows the internal structure of the Ni–MH battery with a special reference to the separator and electrodes [23].

There are various methods for the charger to have constant current like timer charging by using an electronic timer which will help from getting overcharged; thermal detection from getting overheated; and negative delta voltage detection to detect the drop in voltage; compared to all, the step differential method is preferably used in advanced Ni–MH chargers in which initially fast charging of 1C is done followed by a cooling period, later the charge cycle is completed. In this process, the charger further applies reductions in current as the charging progresses.

Nickel-metal hydride (Ni–MH) batteries generally have self-discharge of 1% per day when they are used in a low-energy consummation mode or in a stand-by device, which affects their memory. Manufacturers are using BMS technology which is cost-effective in order to develop a system, which will trickle charges to reduce the negative effects of overcharging and to attain its maximum capacity. Mostly, a smart charger is used with a moderate rate of charging up to 2 hours to 3 hours.

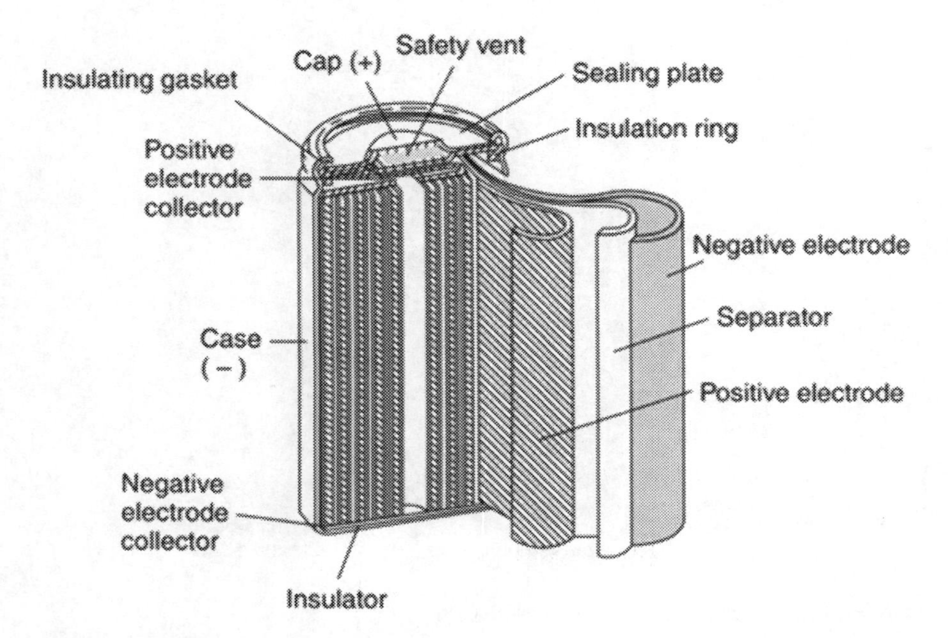

FIGURE 10.8 Structure of cylindrical Ni–MH battery [23]. (Reproduced from ref. [23] with permission from Elsevier, © 2001.)

The gases which are emitted due to overcharging are hydrogen and oxygen, so the battery enclosures of Ni–MH should be airtight and properly vented. Moreover, isolating the battery from the components which generate heat will also reduce the thermal stress. Figure 10.9 shows the charging and discharging of the battery in terms of the chemical reactions that take place inside the battery [25].

10.4.3 Nickel-Hydrogen ($Ni–H_2$) Battery

The nickel-hydrogen ($Ni–H_2$) battery was first patented in 1971 by Alexander et al. in US [26]. $Ni–H_2$ battery is a rechargeable power source based on nickel and hydrogen; here, hydrogen is used as fuel with conventional nickel electrode. $Ni–H_2$ is used in over 800 satellites with good reliability. The overall reaction of the cell is given in Equation (10.12).

$$2NiOOH + H_2 \rightleftharpoons 2Ni(OH)_2 \tag{10.12}$$

It can be observed by the reaction that nickel hydroxide is formed at the positive electrode and hydrogen is produced at the negative electrode. $Ni–H_2$ battery can handle up to ~20,000 charge cycles with 85% efficiency. $Ni–H_2$ battery using KOH as an electrolyte performed an energy density of 75 Wh/kg and a specific power of 220 W/Kg. $Ni–H_2$ battery has a discharge voltage of 1.25 V. In the recent few years, nickel-metal hydride batteries have been used in several commercial hybrid vehicles such as Toyota Prius and Honda Insight. Nickel electrode-based batteries are compared in Table 10.8.

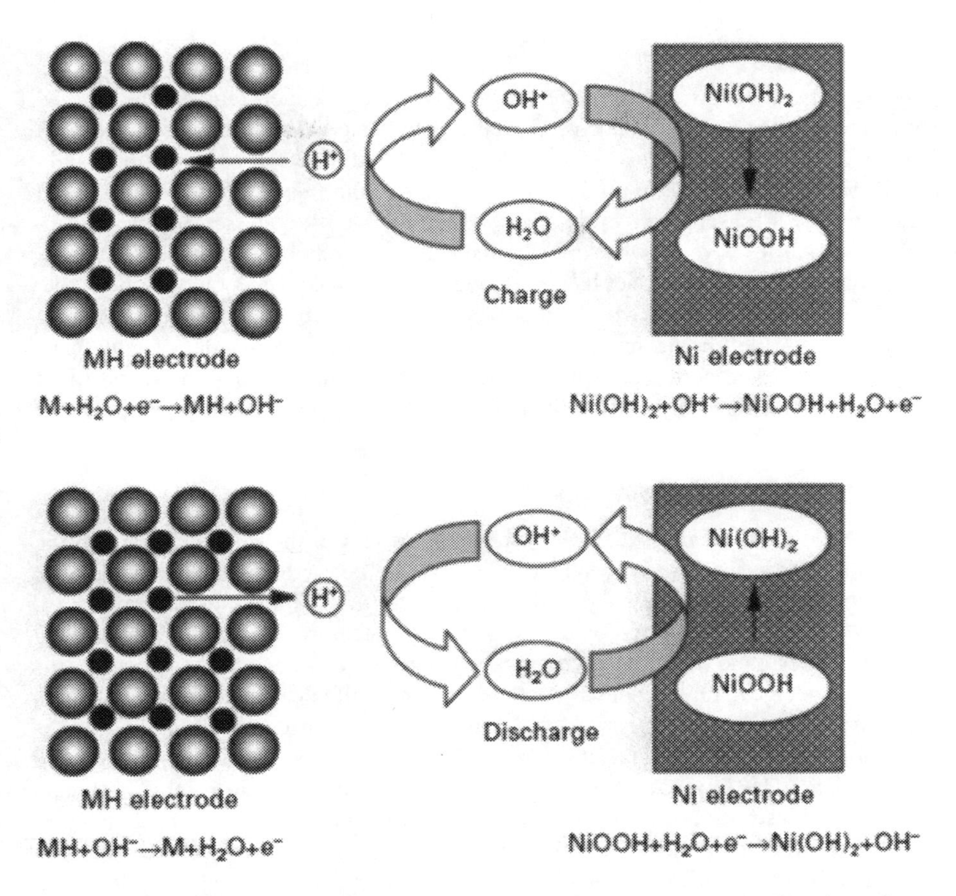

FIGURE 10.9 The charging–discharging process of the hydrogen atom dissociates from $Ni(OH)_2$ and is absorbed by the MH alloy and the hydrogen atom dissociates from the MH alloy and joins with NiOH to form $Ni(OH)_2$ [25]. (Reproduced from ref. [25] with permission from Elsevier, © 2001.)

TABLE 10.8
Nickel-based Batteries Comparative Parameter

Battery	Cell Voltage (V)	Specific Energy (Wh/kg)	Specific Power (W/kg)	Energy Efficiency (%)	Cycle Life
Nickel-Cadmium	1.2	50–60	200	70–75	>1500
Nickel–Iron	1.2	30–60	100	60–70	1500
Nickel–Metal Hydride	1.2	60–70	170–1000	70–80	>1000
Nickel–Zinc	1.2	80–100	170–1000	70–80	<500

10.4.4 Advantages of Alkaline Battery

The performance of an alkaline battery is good at low and ambient temperatures. These have a high energy density, a fairly long self-life, low internal resistance, better dimensional stability and very less leakage problem. Alkaline batteries perform equally well in both intermittent and continuous use. These are also well-performing high and low rates of discharge. An alkaline battery has a fairly long self-life. Moreover, alkaline batteries have drawbacks. Alkaline batteries are bulky as compared to lead-acid battery. Alkaline batteries have high internal resistance which reduces the output power. Alkaline batteries have different types of leakage from other batteries or cells; they can leak if left in the appliance for too long, and this corrosive leak can damage a device. However, with these disadvantages in mind, alkaline batteries are still an excellent choice for many uses, including developing a battery pack.

10.5 SODIUM-ION BATTERIES

The commercialization of different types of batteries in the market has given the option to search for new materials for electrode, which will fulfill all the criterion prerequisites for high efficiency of the batteries. In this context, sodium-ion batteries have received great attention due to its versatile features like better safety, good power delivery and it can be used for many purposes. The main advantage of this battery is a huge natural resource of sodium being available. The principle on which LIBs and sodium-ion battery works is almost the same; instead of Li^+ ion, Na^+ ion is used for charging and discharging process through intercalation and deintercalation mechanism. The lower tap density, lower cost, elevated operating potentials and high capacities have attracted lot of attention for cathode-based sodium transition-metal oxides. This battery uses electrolytes that are aqueous as well as nonaqueous like dimethyl carbonate, diethyl carbonate, propylene carbonate, etc. It can be used for electric vehicles and other power tools if the energy density is increased [27]. Figure 10.10 shows the use of electrolytes and binders for sodium-ion battery; in the figure, the layered oxide is seen at the cathode and metal oxides at anode. Table 10.9 shows the comparative study of sodium-ion battery with lead-acid battery and LIB. In terms of efficiency, sodium-ion battery is much more efficient as compared to the other two.

The cathode of sodium ion stores charges through the reaction mechanism. The advantages of sodium-ion battery over other batteries are their low costs and corrosion-free reactions. It is durable, and it does not get damaged if it is charged for a long time. It has got excellent power delivery with less energy density. From performance wise, it is not applicable to portable electric vehicles and electronic devices. As it is three times heavier than lithium, hence it is not lighter and when used with organic solvents; it is not at all safe as it may cause fire. Figure 10.11 shows the working potential versus a specific capacity of the various materials at anode and cathode. The use of various electrolytes and binders has also been shown in the diagram for sodium-ion battery.

The two-dimensional layer of transition-metal oxides is based on the stacking sequence of metal oxide between the layers. The arrangement is made by

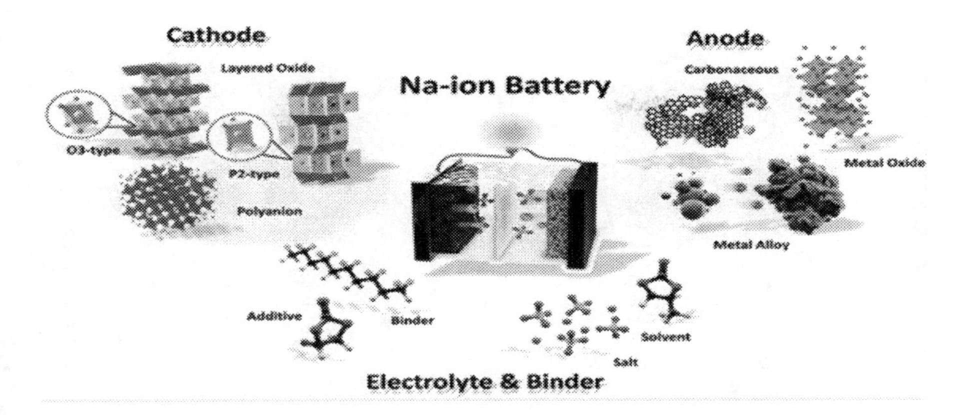

FIGURE 10.10 Cross section of sodium-ion battery [27]. (Reproduced with permission from Royal Society of Chemistry, © 2017.)

TABLE 10.9
Comparison of Sodium-ion Battery with Lead Acid Battery and Lithium-ion Battery

Battery Types	Lead Acid	Lithium-ion	Sodium-ion
Volumetric energy density (Wh/L)	80–90	200–683	250–375
Gravimetric energy density (Wh/Kg)	35–40	120–260	75–150
Cyclic stability	Moderate	High	High
Efficiency	70%–90%	85%–95%	Up to 92%
Range of temperature	−20°C–60°C	15°C–35°C	−20°C–60°C
Availability of resources	Availability limited	Not easily available scarce	Available in abundance
Safety	Moderate as it is toxic	Low	It is very safe to use
Price	Economical	Costly	Economical

sandwiching sodium ion layers between octahedral or prismatic structures. Based on the phase transition of the O_3 type and P_2 type, there is a structure variation, and at low temperature, the synthesis of these compounds takes place, breaking the M–O bonds through thermal analysis. There are lots of possibilities for upgrading the electrodes utilizing the sodiated anode material along with transition-metal oxides. In this, the two- or three-dimensional layer also uses fluorides for better efficiency of the electrodes. Polyanion materials have shown considerable thermal stability due to the covalent bonds in the deeply charged oxide state. For Sodium-ion batteries (SIBs), depending upon insertion reaction, Ti-based oxides as well as carbonaceous oxides are being used as anode material. Carbon materials, especially hard carbons,

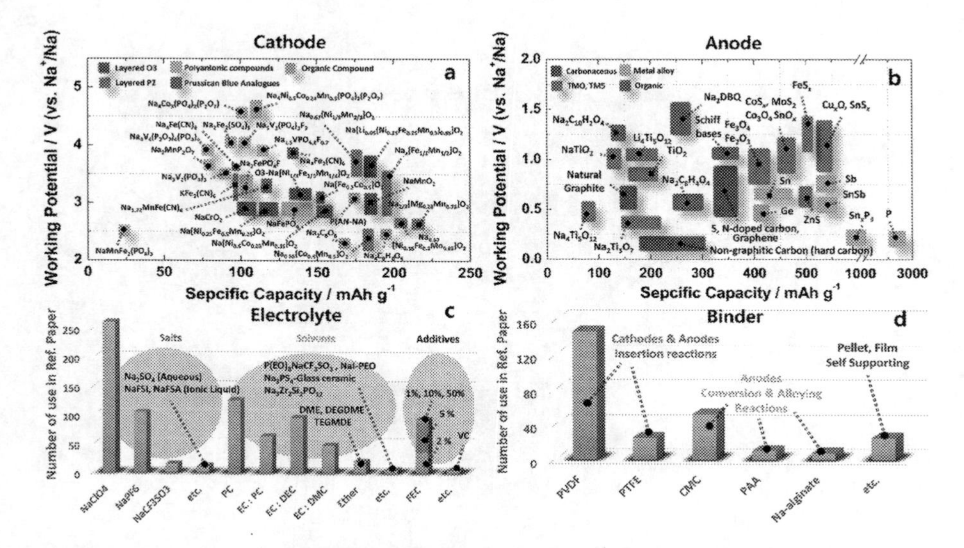

FIGURE 10.11 Working potential for Na-ion battery [27]. (Reproduced with permission from Royal Society of Chemistry, © 2017.)

are widely used because they have the potential to accommodate the sodium ions in its structure. They have a low operating potential. Companies of different countries depend a lot on many parameters such as types of electrodes used, with different manufacturing techniques. In Table 10.10, a comparative study has been made based on the companies of different countries.

The sodium-ion batteries have received lots of attention due to their abundance availability; SIB cathode material consists of metal oxide, metal sulfides, compounds of oxoanionic, polymers, Prussian blue analogue, etc. Lots of research are still in progress in order to make this battery more commercialized. Tables 10.11 and 10.12 summarize the advancements in anode and cathode material for sodium-ion batteries, respectively.

10.6 MG-ION BATTERY (MIB)

For the next-generation electrochemical power sources, large-scale portable and stationary electrical device applications, in addition to the LIB, the secondary MIB is one of the most hopeful alternative solutions. The MIBs have safer, richer abundances than LIBs. The magnesium metal has diverse characteristics with high natural abundance (approximately 10^4 times that of lithium, which helps to reduce the cost of electrode material). It is further incorporated into low-cost energy storage devices. The magnesium metal is more environmentally stable in nature, excellent thermodynamic properties, high volumetric capacity (3833 mA/h/cm^3), dendrite-free deposition, higher melting point than lithium, low reduction potential up to –2.4 vs. SHE, high Columbic efficiency, etc. [45,46]. Such remarkable inherent characteristics of Mg metal prove that the Mg metal is an ideal and promising electrode for MIBs. In addition, compared to nickel-cadmium and lead-acid batteries, MIBs

TABLE 10.10
Comparative Studies between Different Types of Batteries

Company Name	Country	Electrode	Characteristics	Uses
Faradion limited	United Kingdom 2011	Cathodes–oxides Anode–carbon and liquid electrolyte	Carbothermal reduction method of synthesis is used for electrode of $Na_3M_2(PO_4)_2\ F_3$	It is used in E–bike and E–Scooter
Tiamat	France 2017	Polyanionic	Cylindrical cells	Mainly used in power market
Hi–Na battery technology Co. Ltd.	China 2017	Cathodes of Na–Fe–Mn–Cu-based oxide Anode is of carbon	Energy density of 120 Wh/Kg	Used as a power bank
Natron	Stanford University, United States	Prussian blue electrode with an aqueous electrolyte	—	—
Altris AB	Uppsala University, Sweden 2017	Iron-based Prussian and carbon as an electrode	Low-energy process	Stationary energy storage
CATL Co. Ltd.	China 2021	Porous carbon as cathode and Prussian blue as anode	Specific energy density 160 Wh/Kg. Sodium ion has more volume and stable with structure	Electric vehicles and stationary storage battery

provide a significantly higher energy density. Also, in contrast to lead and cadmium, magnesium and lithium are less expensive and environmentally friendly in nature. The use of Mg metal as electrode material in MIBs reduces the final cost of battery and results in a more stable alternating power source than LIBs.

Nevertheless, the development of the magnesium-ion battery technology is not as fast as LIBs. This is because there are certain challenges among the development of magnesium-based cathode material for batteries. This includes high reversible capacity, high operating voltage, and nonavailability of appropriate electrolyte, which can allow the reversible release of Mg^{2+} ions from a Mg metal anode [47–50]. The sluggish nature of Mg^{2+} ions in solid electrolytes makes large voltage hysteresis and low magnetization degrees for most of the material [50]. Most of the electrolyte used in MIBs allows the development of passivating surface films, which impede the electrochemical performance during the charging–discharging process [46,51,52]. Another important problem associated with MIBs is the unavailability of a suitable electrolyte. The electrolyte which can neither accept nor donate proton can be the most suitable electrolyte for MIBs [46]. The second most significant difficulty in the development of MIBs is the limited choice of cathode material by the inability to intercalate Mg ions in many hosts [48]. To enlighten the structures and chemistries of the materials developed for magnesium-ion cathodes, in this section, we discuss the

TABLE 10.11
Recent Research Advancements in Anode Material for Sodium-ion Battery

S. No.	Anode Material	Charge Storage Mechanism	Potential	Initial Reversible Capacity	Current Density (mA/g)	Capacity After Cycle	References
1	Carbon microspheres	Intercalation and pore filling	0.005–3	202 at 30 mA/g	30	183 @50	[28]
2	3D porous Fe_3O_4–C	Conversion reaction	0.005–3.0	321 at 50 mA/g	100	277 @200	[29]
3	*N*-doped expended graphite	Intercalation	0.01–3.00	373.37 at 100 mA/g	1000	200 @1000	[30]
4	Mo–S_2 Carbon fiber	Conversion and interaction mechanism	0–3	529 @200 mA/h	50	452 @50	[31]
5	Amorphous Carbon	Adsorption at defect sits and	–0.01–2	284 at 100 mA/g	30	266.96 @100	[32]
6	MnO_2–Nanaoflowers	Conversion reaction and some extant of sodium	0.01–3.0	349.0 at 50 mA/g	50	177.1 @100	[33]
7	MoS_2–C	Conversion reaction	0.01–2.5	510 at 100 mA/g	100	390 @100	[34]
8	Graphene nanosheets	Adsorption and interlayer intercalation	0.01–2	220 at 30 mA/g	100	176 @300	[35]
9	Se_4P_4	Conversion alloying mechanism	0.01–3	1048 at 50 mA/g	50	904 @60	[36]
10	Disodium terephthalate	Intercalation and redox	0–2	295 at 30 mA/g	30	265 @90	[37]

TABLE 10.12
Recent Research Advancements in Cathode Material for Sodium-ion Battery

S. No.	Cathode Materials	Voltage Range	Current Density (mA/g)	Initial Reversible Capacity	Capacity After Cycles	References
1	P_3–$Na_{0.9}Ni_{0.5}Mn_{0.5}O_2$	1.5–4.5	100	141 at 10 mA/g	79 @100	[38]
2	O_3–$Na_{1.0}Mn_{1/3}Fe_{1/3}Cu_{1/6}Mg_{1/6}O_2$	2.3–3.6	20	135 at 10 mA/g	108 @100	[39]
3	Sodium copper hexacyanoferrate	2–4.2	50	89 at 50 mA/g	71.7 @1000	[40]
4	P_3–$Na_{2/3}Ni_{1/4}Mg_{1/12}Mn_{2/3}O_2$	0–3.5	133	125 at 133 mA/g	97.5 @100	[41]
5	$Na_4Fe_3(PO_4)_2P_2O_7/C$	2.04–2.22	1000	128.5 at 20 mA/g	71.6 @4000	[42]
6	O_3–$NaCrO_2$	1.5–4.2	200	123 at 10 mA/g	94 @500	[43]
7	$Na_4MnV(PO_4)_3/C/GA$	3.3–3.4	550	109.8 at 55 mA/g	87.7 @500	[44]

summary of the recent progress of cathode material and electrolyte for MIBs with special emphasis on strategies for future research initiatives. To solve these issues, in recent years, several strategies have been reported in literature; this includes the use of mesoporous and high specific surface area nanostructured material with high divalent ion mobility as cathode for MIBs [53,54]. The cathode fabricated from such material decreases the diffusion length for Mg ion into the cathode during the charging–discharging process [55–57].

10.6.1 Cathode Material for MIBs

Till date, Chevrel phase oxides, chalcogenides, carbon-based nanomaterials and polyanions have been investigated as a potential cathode material for MIBs. The mostly used cathode material for MIBs includes the Chevrel phase molybdenum sulfide (Mo_6S_8) [58], molybdenum disulfide (MoS_2) [59], orthorhombic molybdenum oxide MoO_3 [60,61], vanadium bronze (V_3O_8) [62], vanadium pentoxide (V_2O_5) [63], magnesiated vanadium pentoxide ($Mg_xV_2O_5$) [64], nickel hexacyanoferrate ($NiFe(CN)_6$) [65], copper hexacyanoferrate ($CuFe(CN)_6$) [66], and various manganese-based oxides such as hollandite MnO_3 [67], todorokite MnO_3 [68], birnessiteMnO_3 [67], Mg_6MnO_8 [69], manganese silicate ($Mg_{1.03}Mn_{0.97}SiO_4$) [70], iron silicate ($MgFeSiO_4$) [71], cobalt silicate($MgCoSiO_4$) [72], etc.

Considering the sluggish nature and unsatisfactory cycling life of cathode material used in MIBs, recently, various strategies and structural modifications, viz., mesoporous, hierarchical two- and three-dimensional materials as cathode for MIBs have been implemented. Chevrel-phase Mo_6S_8 is one of the important materials which is facilitated by the high mobility and fast interfacial charge transfer. Nowadays, Mo_6S_8 has gained considerable attention toward the application of cathode material at room in MIBs. Mo_6S_8 has superb intercalation kinetics and best reversibility. Also, at potential 1.2 V, the initial capacity for Mo_6S_8 is reported in literature to be up to 120 mA/hg [73,74]. To study the effect of leaching chemistry on the performance of Mo_6S_8-based cathodes of MIBs, Lancy et al. have reported interesting results: as they accommodate two Mg atoms per formula unit when $Cu_2Mo_6S_8$ leached in I_2/AN (acetonitrile) or in HCl/H_2O. Also, the capacity fading of the MIBs is reported because of the inability to extract Mg ions during charging at room temperature. The $Cu_2Mo_6S_8$ cathode in MIBs resulted in a specific capacitance of 90–100 mA/hg with excellent stability over large numbers of charge–discharge cycles [75]. Further, Choi et al. modify Mo_6S_8 by Cu metal and reported the increase in rate capability discharge of (99.1 mA/hg) than pristine with (85.6 mA/hg) at 1.0 V vs. Mg/Mg^{2+}. Moreover, in this study, the in-situ formation of $Cu_xMo_6S_8$ is studied further. Also, the electrochemical insertion of Cu from Cu nanoparticle/graphene composite to the Mo_6S_8 host and the schematic of Mg^{2+} insertion and extraction during the replacement reaction of Cu in the Mo_6S_8 is demonstrated in Figure 10.12 [76].

Wu et al. demonstrated the use of Lattice engineering to settle the issue of sluggish kinetics, which is one of the important reasons for dissatisfactory Mg-storage capabilities in two-dimensional layered materials. For this study, using heterogeneous monolayers of MoS_2 and graphene, they have fabricated the van der Waals' heterostructures. Further, they reported that the Mg-diffusion barrier was reduced

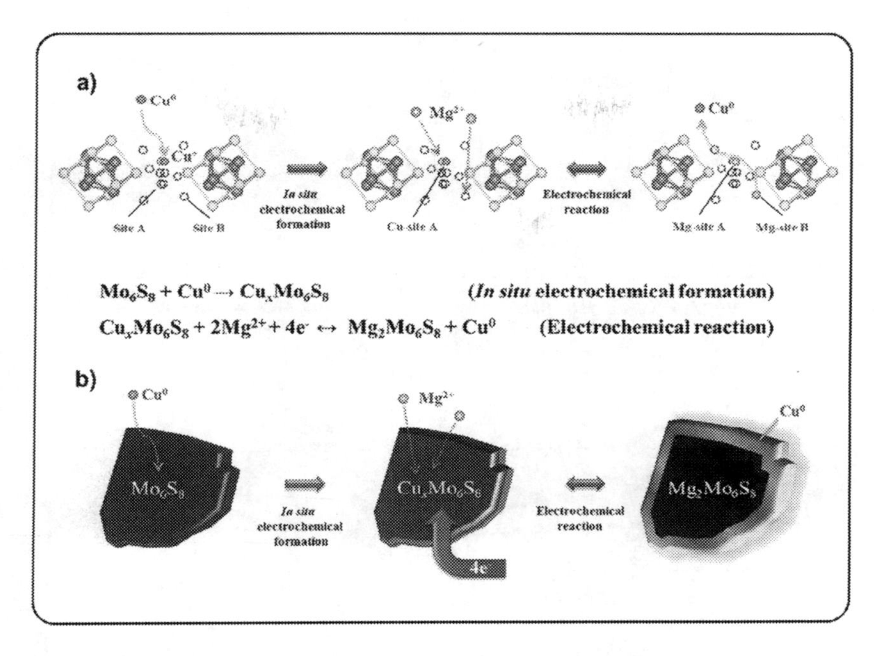

FIGURE 10.12 Schematics of (a) Cu replacement reaction in the Mo_6S_8 and (b) the proposed solid-state Cu replacement reaction structure during Mg^{2+} insertion and extraction. (Reproduced with permission with ACS, publication [76]. (Reproduced with permission from ACS Publication, © 2017.)

by 0.4 eV, and in consequence, the diffusion rate enhances 11 times higher than the diffusion rate of original MoS_2. As a result of the assisted diffusion kinetics, the Mg-storage capacity for MoS_2/GR and its rate performance are reported to be 210 mA/hg and 90 mA/hg at 500 mA/g, respectively [77]. To enhance the electrochemical performance and enhances the Mg-ion kineticsm, Liu et al. synthesized and studied the MoS_2/graphene hybrid material as cathode for MIBs. The facile lithium-assisted sonication method was used for the synthesis of cathode material. In this study, graphene is inserted in MoS_2, this influences the Mg^{2+}-ion insertion–deinsertion in the host MoS_2/graphene hybrid cathode, as a result, the electrochemical behavior of MoS_2/graphene hybrid cathode in MIBs enhances. The initial capacity and cyclability after 50 cycles for MoS_2/graphene cathode are reported to be 115.9 mA/hg and 82.5 mA/hg, respectively. Furthermore, the cyclic voltammetry curves at various scan rates, typical GCD profiles at various current densities and electrochemical impedance spectra of different MoS_2/graphene-15 composites are demonstrated in Figure 10.13 [78]. The reported MoS_2/graphene-15 hybrid fascicled MIBs demonstrated excellent electrochemical behavior as compared to bare MoS_2 or graphene [78]. Furthermore, the issues related to Mg-ion adsorption at host material and diffusion, Yang et al., using the Density Functional Theory, investigated the MoS_2 nanoribbon as cathode material for MIBs and reported the maximum theoretical of 223.2 mA/hg [79].

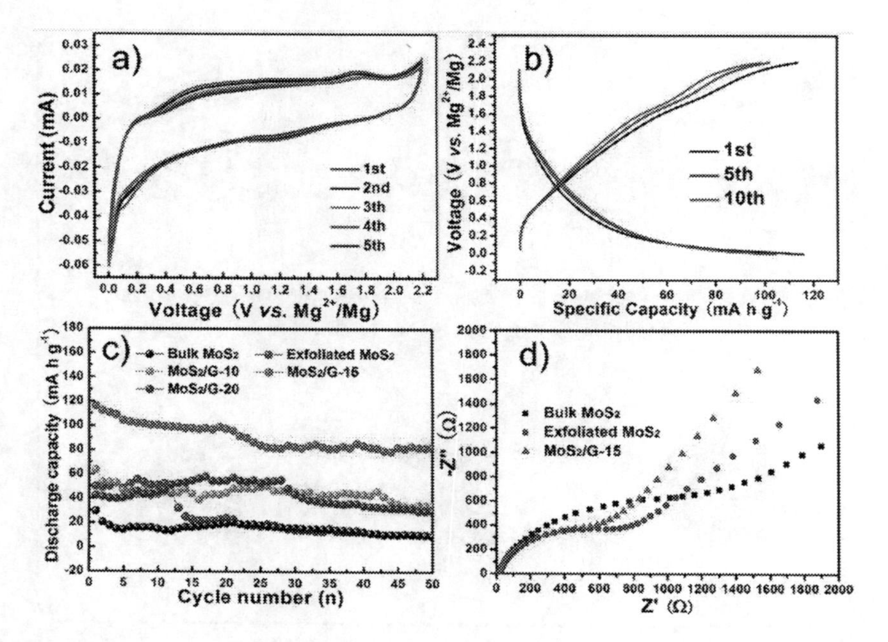

FIGURE 10.13 (a) The cyclic voltammetry curves of MoS_2/graphene-15 electrode scanned at a rate of 0.5 mV/s in the voltage window of 0 V–2.2 V vs. Mg^{2+}/Mg. (b) Typical GCD profiles of MoS_2/graphene at a current density of 20 mA/g. (c) Cycling performance of bulk MoS_2, exfoliated MoS_2, and the MoS_2/graphene-15 composites at 20 mA/g. (d) Electrochemical impedance spectra of bulk MoS_2, exfoliated MoS_2, and MoS_2/graphene-15 electrodes [78]. (Reproduced with permission from Elsevier, © 2017.)

10.6.2 Transition-Based Cathode Material for MIBs

In recent years, over the transition-metal sulfide, transition-metal oxides have gained enormous attention as the cathode material for MIBs. Owing to the many attractive properties such as good electrochemical characteristics, higher working potentials, better thermal stability etc., transition-metal oxides reported to be superior cathode material than the transition-metal sulfide for MIBs. Many transition-metal oxides such as MnO_2 [80], V_2O_5 [81], $VOPO_4$ [82], TiO_2 [83], etc. also offers the similar advantages. Zhang et al. [84] reported the nano-sized hollandite phase α-MnO_2 cathode material for MIBs. The reported cathode material demonstrated excellent electrochemical performance in electrolyte Mg_2 $(\mu\text{-}Cl)_3{\cdot}_6(OC_4H_8)((N(Si(CH_3)_3)_2)_nAlCl_{4-n})$ (n=1, 2) with a specific capacitance of 280 mA/hg within the potential range of 0.3 V–1.5 V. Ju et al. reported the MoS_2 nano flowers as the cathode material for hybrid ion Li^+/Mg^{2+} battery and reported the remarkable enhancement of the electrochemical performance due to Mg stripping/plating at the anode side and Li^+ intercalation at the cathode side with a small contribution from Mg^{2+} adsorption at MoS_2 cathode. The MoS_2 nano flower-based hybrid Li^+/Mg^{2+} hybrid ion battery demonstrated high capacity, excellent rate capability, and long cycle life of 243 mA/hg at the 0.1 C rate and 108 mA/hg at the 5 C rate, respectively. Also, the capacity retention over 2300 cycles is reported to be 87.2 % [85]. Zhu et al. have used the introduction of defects in cathode material strategy for the enhancement of energy density and stability of MIBs and accelerated the diffusion of Mg^{2+} in cathode materials. For demonstration, they have reported the synthesis of defective 2D MoS_2 nanosheets via the hydrothermal method and used as a cathode for MIBs. The discharge specific capacity of MIBs is found to be 152 mA/hg [86]. Liu et al. synthesized the ε-MnO_2 via facile potentiostatic electrodeposition and utilized for cathode material for MIBs. The interconnected nanoflakes of ε-MnO_2 and direct contact with the electrolyte are responsible for high electrochemical performance. The specific capacity and energy density of binder-free ε-MnO_2 were reported to be 259.3 mA/hg at 0.5 Ag^{-1} and 98.6 Wh/kg, respectively. The as-prepared cathode shows high stability and 94.3% retention in capacitance over 400 cycles [80].

10.7 MAGNESIUM-SODIUM (MG–NA) HYBRID ION BATTERIES

With the advance in the use of renewable energy, various nanomaterials have been used to make the batteries very efficient; LIBs satisfy all the requisite criterion, but the major challenge is the limitation of energy density. It also suffers the problems of dendrites formation, thermal runaway and most important part is that it is costly. Lithium was replaced by MIB, a magnesium-ion battery which has lots of advantages as an electrode material such as an excellent volumetric capacity of 3833 mA/h/cm^3 and gravimetric capacities of 2205 mA/hg, which is double as compared to lithium, lower flammability and safe to use; it has lower negative redox potential and forms non-dendrite morphology [3,87,88]. Magnesium-ion batteries require cathode of high energy density, enhanced rate capability and good cyclability. The major challenge in the selection of cathode of MIB is the undesirable effects of the host materials that prevent the insertion of magnesium. Some of the cathodes used are Chevrel phase MO_6S_8 and spinel TiS_2, but the energy densities were very less and a lower voltage was achieved less than 1.5 V. As compared to MIB, sodium-ion batteries are also gaining lots of momentum in the field of energy materials. SIB batteries [89] use Na as a cathode because there is a strict prohibition of Na as an anode because it is highly reactive and has dendrite formation issues. However, its plenty abundance has made it an attractive electrode material as it will lower the cost of manufacturing. Figure 10.14 [89] shows the charging and discharging of Mg/Na hybrid battery in the charging process, the reduction process takes place at magnesium anode and oxidation takes place at cathode. The converse effect takes place during the discharging process.

There are many interesting features of sodium-magnesium hybrid ion batteries. Choosing magnesium as an electrode material has many advantages such as high capacity, availability, low cost and its safety. Sodium ion as a cathode material has lots of importance like generating high voltage, high energy density, found in abundance, low cost and good cyclability. The electrolyte should allow the transport of Mg ion and Na-ion both in charging and discharging of the battery. Its oxidation and reduction process should match with the operating voltage of the electrodes. This battery mainly works on the principle of sodium-ion insertion or extraction in the cathode and dissolution or deposition of magnesium ion at the anode during the cycles of charging and discharging. The stability of the oxidation state of the magnesium electrolyte [89] should not exceed the Na negative electrode. Moreover, the operating range of the cathode should be in the range the discharge voltage should not be very low. The electrolyte should be free from any materials which will affect

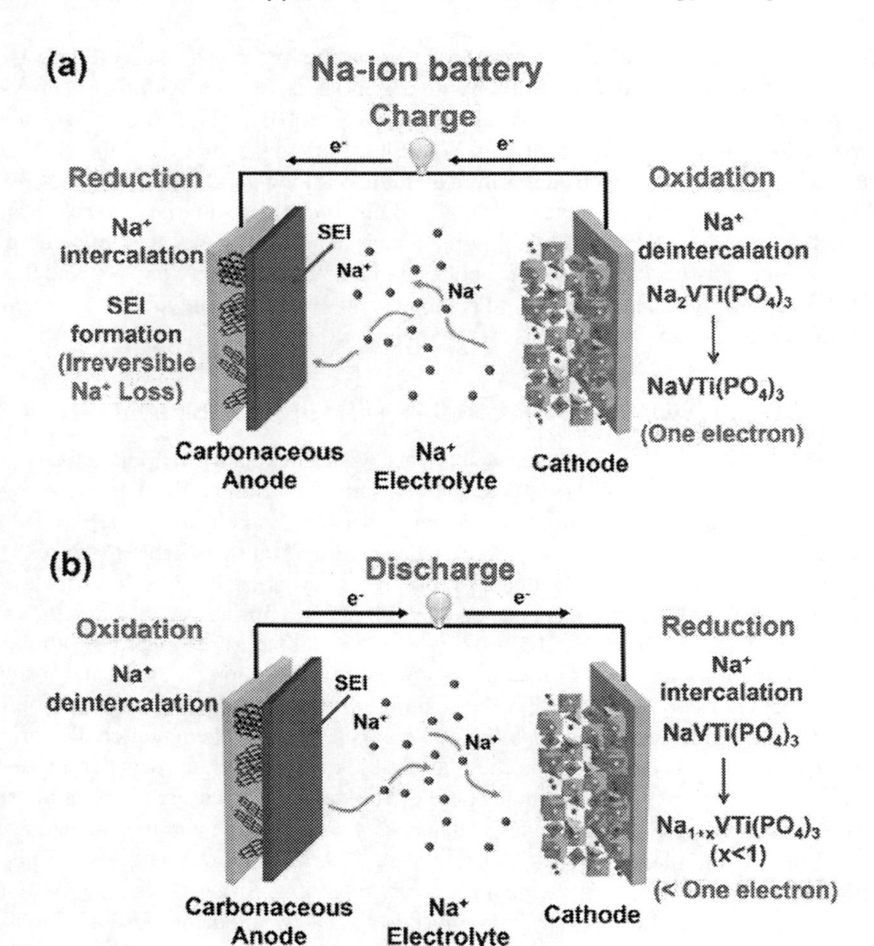

FIGURE 10.14 Charging and discharging phenomenon of Mg/Na hybrid battery [89]. (Reproduced with permission Elsevier, © 2020.)

the deposition or dissolution of Mg. Moreover, the recent research highlights that sodium magnesium hybrid ion battery is demonstrated in Table 10.13.

The future of sodium and magnesium hybrid ion batteries is truly bright. Because this hybrid battery utilizes the dendrite-free deposition of magnesium at anode and has the fast intercalation process of sodium ion at the cathode, which is very much required for the charge storage mechanism of the battery. This also increases the energy density of the battery. It can be commercialized by improving the materials of the electrolytes.

10.8 CONCLUSIONS AND FUTURE PROSPECTIVE

The recent advances in the energy storage system alternative to the lithium batteries/LIBs have been reviewed in detail. The cathode, anode and electrolyte materials

TABLE 10.13
Recent Research Advancements in Anode and Cathode Material for Sodium–Magnesium Hybrid Ion Battery

Types of Battery	Anode	Cathode	Electrolyte	Characteristics	References
$MgNaCrO_2$	Mg	$NaCrO_2$	$PhMgCl-AlCl_3$, MgAPC, $NaCB_{11}H_{12}$, dual salt electrolyte	Energy density –183 Wh Kg^{-1} at voltage 2.3 V average 50 cycles	[90]
Sodium–Magnesium hybrid battery	Mg metal	FeS_2 (nanocrystal)	Dual salt containing Mg^{+2} and Na^{+1}	Good cathodic capacity; excellent Coulombic efficiency; rate capability	[91]
Mg Berlin green hybrid cell	Mg	Na (open frame work Berlin Green cathode)		Average discharge voltage is 2.2 V Stable cyclability 50 cycles	[92]
Mg–Na hybrid	Mg	$Na_3V_2(PO_4)_3$		Voltage 2.6 V Energy density 150 Wh Kg^{-1} Good capacity	[93]
NVTP	Mg	$Na_2VTi(PO_4)_3$ NASICON structure	Dual ion electrolyte Mg^{+2} and Na^{+1}	High capacity of 168 mA hg^{-1} Cyclability of 1000 cycles	[89]
Mg/Na hybrid aqueous battery	$NaTi_2(PO_4)_3$	Mn_3O_4	Mg^{+2} and Na^{+1} hybrid electrolyte	Discharge potential at 1.2 V Energy density of 23.6 Wh Kg^{-1} Excellent cyclability	[94]

used in the different batteries have been discussed thoroughly. The operating principle of different batteries such as alkaline batteries, sodium-ion battery, magnesium-ion battery and sodium-magnesium hybrid ion battery with the key advancement in their architecture and energy efficiency is discussed in detail. The electrochemical technologies related to electrode materials and electrolytes have been discussed with the recent advancements including adopted methods of synthesis, use of nanostructured material for cathode and anode and applicability of solid electrolyte. While reviewing the status of different battery systems, it has been observed that there is a lot of scope for the development of nickel-based batteries, sodium-ion batteries and magnesium-based batteries. These batteries can replace the monopoly of lithium batteries in the future.

REFERENCES

[1] Rechargeable Batteries, Chemistry LibreTexts. (2013). https://chem.libretexts.org/Bookshelves/Analytical_Chemistry/Supplemental_Modules_(Analytical_Chemistry)/Electrochemistry/Exemplars/Rechargeable_Batteries. (Accessed: August 24, 2021).

[2] G.J. May, A. Davidson, B. Monahov, Lead batteries for utility energy storage: A review, *J. Energy Stor.* 15 (2018) 145–157.

[3] Y. Liang et al., A review of rechargeable batteries for portable electronic devices, *InfoMat.* 1 (2019) 6–32.

[4] P.A. Marsal, K. Karl, L.F. Urry, Dry cell, US2960558A, (1960). https://patents.google.com/patent/US2960558A/en. (Accessed: August 11, 2021).

[5] A.R. Mainar, L.C. Colmenares, J.A. Blázquez, I. Urdampilleta, A brief overview of secondary zinc anode development: The key of improving zinc-based energy storage systems, *Int. J. Energy Res.* 42 (2018) 903–918. https://doi.org/10.1002/er.3822.

[6] D. Linden, T.B. Reddy, eds., *Handbook of Batteries*, 3rd ed., McGraw-Hill: New York, 2002.

[7] W. Li et al., Nickel cobalt sulfide nanostructure for alkaline battery electrodes, *Adv. Func. Mater.* 28 (2018) 1705937. https://doi.org/10.1002/adfm.201705937.

[8] N. Zhang et al., Rechargeable aqueous zinc–manganese dioxide batteries with high energy and power densities, *Nat. Commun.* 8 (2017) 405. https://doi.org/10.1038/s41467-017-00467-x.

[9] Nickel in batteries | Nickel Institute, (n.d.). https://nickelinstitute.org/about-nickel/nickel-in-batteries/. (Accessed: August 22, 2021).

[10] P.-J. Tsais, L.I. Chan, [11]Nickel-based batteries: Materials and chemistry, in: Z. Melhem (Ed.), *Electricity Transmission, Distribution and Storage Systems*, Woodhead Publishing, Swaston, 2013, pp. 309–397. https://doi.org/10.1533/9780857097378.3.309.

[11] J.P. Barton, R.J.L. Gammon, A. Rahil, Characterisation of a nickel–iron battolyser, an integrated battery and electrolyser, *Fron. Energy Res.* 8 (2020) 318. https://doi.org/10.3389/fenrg.2020.509052.

[12] Y. Zeng, X. Zhang, X. Mao, P.K. Shen, D.R. MacFarlane, High-capacity and high-rate Ni–Fe batteries based on mesostructured quaternary Carbon/Fe/FeO/Fe_3O_4 hybrid material, *Science.* 24 (2021) 102547. https://doi.org/10.1016/j.isci.2021.102547.

[13] H. Wang et al., An ultrafast Nickel–Iron battery from strongly coupled inorganic nanoparticle/nanocarbon hybrid materials, *Nat. Commun.* 3 (2012) 917. https://doi.org/10.1038/ncomms1921.

[14] Z. Lu, X. Wu, X. Lei, Y. Li, X. Sun, Hierarchical nanoarray materials for advanced nickel–zinc batteries, *Inorg. Chem. Front.* 2 (2015) 184–187. https://doi.org/10.1039/C4QI00143E.

[15] J. Phillips, S. Mohanta, M. Geng, J. Barton, B. McKinney, J. Wu, Environmentally friendly nickel–zinc battery for high rate application with higher specific energy, *ECS Trans.* 16 (2009) 11. https://doi.org/10.1149/1.3087437.

[16] M.A. Nazri et al., Screen-printed nickel–zinc batteries: A review of additive manufacturing and evaluation methods, *3D Print. Add. Manu.* 8 (2021) 176–192. https://doi.org/10.1089/3dp.2020.0095.

[17] Y. Liu, Z. Yang, J. Yan, Zinc hydroxystannate as high cycle performance negative electrode material for zn/ni secondary battery, *J. Electrochem. Soc.* 163 (2016) A3146. https://doi.org/10.1149/2.1411614jes.

[18] Avoiding short circuits from zinc metal dendrites in anode by backside-plating configuration - PubMed, (n.d.). https://pubmed.ncbi.nlm.nih.gov/27263471/. (Accessed: August 29, 2021).

[19] U. Koehler, Chapter 2–general overview of non-lithium battery systems and their safety issues, in: J. Garche, K. Brandt (Eds.), *Electrochemical Power Sources: Fundamentals, Systems, and Applications*, Elsevier, 2019: pp. 21–46. https://doi.org/10.1016/B978-0-444-63777-2.00002-5. Netherland.

[20] N. Omar et al., Analysis of nickel-based battery technologies for hybrid and electric vehicles, in: reference module in chemistry, In *Reference Module in Chemistry, Molecular Sciences and Chemical Engineering*, Elsevier, 2014. https://doi.org/10.1016/B978-0-12-409547-2.10740-1.

[21] P. Krivik, P. Baca, Electrochemical energy storage, in: Energy Storage - Technologies and Applications. London: IntechOpen, 2013. Available: https://www.intechopen.com/chapters/42271; doi: 10.5772/52222 (Accessed on 16 Aug 2022)..

[22] K. Scott, Recycling | Nickel–Metal hydride batteries, in: J. Garche (Ed.), *Encyclopedia of Electrochemical Power Sources*, Elsevier, Amsterdam, 2009: pp. 199–208. https://doi.org/10.1016/B978-044452745-5.00401-9.

[23] A. Taniguchi, N. Fujioka, M. Ikoma, A. Ohta, Development of nickel/metal-hydride batteries for EVs and HEVs, *J. Power Sources*. 100 (2001) 117–124. https://doi.org/10.1016/S0378-7753(01)00889-8.

[24] E. Lemaire-Potteau, M. Perrin, S. Genies, Batteries | charging methods, in: J. Garche (Ed.), *Encyclopedia of Electrochemical Power Sources*, Elsevier, Amsterdam, 2009: pp. 413–423. https://doi.org/10.1016/B978-044452745-5.00885-6.

[25] F. Feng, M. Geng, D.O. Northwood, Electrochemical behaviour of intermetallic-based metal hydrides used in Ni/metal hydride (MH) batteries: a review, *Int. J. Hydro. Energy*. 26 (2001) 725–734. https://doi.org/10.1016/S0360-3199(00)00127-0.

[26] B.I. Tsenter, V.M. Sergeev, A.I. Kloss, Hermetically sealed nickel-hydrogen storage cell, US3669744A, 1972. https://patents.google.com/patent/US3669744/en (Accessed: August 15, 2021).

[27] J.-Y. Hwang, S.-T. Myung, Y.-K. Sun, Sodium-ion batteries: present and future, *Chem. Soc. Rev.* 46 (2017) 3529–3614. https://doi.org/10.1039/C6CS00776G.

[28] M. Fan, H. Yu, Y. Chen, High-capacity sodium ion battery anodes based on CuO nanosheets and carboxymethyl cellulose binder, *Mater. Technol.* 32 (2017) 598–605. https://doi.org/10.1080/10667857.2017.1295628.

[29] N. Wang, Q. Liu, Y. Li, J. Chen, J. Gu, W. Zhang, D. Zhang, Self-crosslink assisted synthesis of 3D porous branch-like Fe3O4/C hybrids for high-performance lithium/sodium-ion batteries, *RSC Adv.* 7 (2017) 50307–50316. https://doi.org/10.1039/C7RA09348A.

[30] M. Hu, H. Zhou, X. Gan, L. Yang, Z.-H. Huang, D.-W. Wang, F. Kang, R. Lv, Ultrahigh rate sodium ion storage with nitrogen-doped expanded graphite oxide in ether-based electrolyte, *J. Mater. Chem. A*. 6 (2018) 1582–1589. https://doi.org/10.1039/C7TA09631C.

[31] Q. Shen, P. Jiang, H. He, C. Chen, Y. Liu, M. Zhang, Encapsulation of $MoSe_2$ in carbon fibers as anodes for potassium ion batteries and nonaqueous battery–supercapacitor hybrid devices, *Nanoscale*. 11 (2019) 13511–13520. https://doi.org/10.1039/C9NR03480C.

[32] Y. Li, L. Mu, Y.-S. Hu, H. Li, L. Chen, X. Huang, Pitch-derived amorphous carbon as high performance anode for sodium-ion batteries, *Energy Stor. Mater.* 2 (2016) 139–145. https://doi.org/10.1016/j.ensm.2015.10.003.

[33] Z. Zhang, X. Zhao, J. Li, Facile synthesis of nanostructured MnO_2 as anode materials for sodium-ion batteries, *Chem. Nano. Mat.* 2 (2016) 196–200. https://doi.org/10.1002/cnma.201500194.

[34] Y.-X. Wang, et al., Reversible sodium storage via conversion reaction of a MoS_2–C composite, *Chem. Commun.* 50 (2014) 10730–10733. https://doi.org/10.1039/C4CC00294F.

[35] X.-F. Luo, et al., Graphene nanosheets, carbon nanotubes, graphite, and activated carbon as anode materials for sodium-ion batteries, *J. Mater. Chem. A.* 3 (2015) 10320–10326. https://doi.org/10.1039/C5TA00727E.

[36] Y. Lu, P. Zhou, K. Lei, Q. Zhao, Z. Tao, J. Chen, Selenium Phosphide (Se_4P_4) as a new and promising anode material for sodium-ion batteries, *Adv. Energy Mater.* 7 (2017) 1601973. https://doi.org/10.1002/aenm.201601973.

[37] Y. Park, et al., Sodium terephthalate as an organic anode material for sodium ion batteries, *Adv. Mater.* 24 (2012) 3562–3567. https://doi.org/10.1002/adma.201201205.

[38] T. Risthaus, et al., $P_3Na_{0.9}Ni_{0.5}Mn_{0.5}O_2$ Cathode material for sodium ion batteries, *Chem. Mater.* 31 (2019) 5376–5383. https://doi.org/10.1021/acs.chemmater.8b03270.

[39] Z.-Y. Li, et al., Modulating the electrochemical performances of layered cathode materials for sodium ion batteries through tuning coulombic repulsion between negatively charged TMO_2 slabs, *ACS Appl. Mater. Interfaces.* 10 (2018) 1707–1718. https://doi.org/10.1021/acsami.7b15590.

[40] Z. Wang, et al., Ion-exchange synthesis of high-energy-density prussian blue analogues for sodium ion battery cathodes with fast kinetics and long durability, *J. Power Sources.* 436 (2019) 226868. https://doi.org/10.1016/j.jpowsour.2019.226868.

[41] Y.-N. Zhou, et al., Air-stable and high-voltage layered p3-type cathode for sodium-ion full battery, *ACS Appl. Mater. Interfaces.* 11 (2019) 24184–24191. https://doi.org/10.1021/acsami.9b07299.

[42] X. Pu, et al., $Na_4Fe_3(PO_4)_2P_2O_7$/C nanospheres as low-cost, high-performance cathode material for sodium-ion batteries, *Energy Storage Mater.* 22 (2019) 330–336. https://doi.org/10.1016/j.ensm.2019.02.017.

[43] Y. Wang, et al., Electrochemical performance of large-grained $NaCrO_2$ cathode materials for Na–Ion batteries synthesized by decomposition of $Na_2Cr_2O_7{\cdot}2H_2O$, *Chem. Mater.* 31 (2019) 5214–5223. https://doi.org/10.1021/acs.chemmater.9b01456.

[44] H. Li, et al., Rational architecture design enables superior Na storage in greener NASICON-$Na_4MnV(PO_4)_3$ Cathode, *Adv. Energy Mater.* 8 (2018) 1801418. https://doi.org/10.1002/aenm.201801418.

[45] M. Mao, T. Gao, S. Hou, C. Wang, A critical review of cathodes for rechargeable Mg batteries, *Chem. Soc. Rev.* 47 (2018) 8804–8841.

[46] D. Aurbach, Z. Lu, A. Schechter, Y. Gofer, H. Gizbar, R. Turgeman, Y. Cohen, M. Moshkovich, E. Levi, Prototype systems for rechargeable magnesium batteries, *Nature.* 407 (2000) 724–727.

[47] J. Giraudet, D. Claves, K. Guérin, M. Dubois, A. Houdayer, F. Masin, A. Hamwi, Magnesium batteries: Towards a first use of graphite fluorides, *J. Power Sources.* 173 (2007) 592–598.

[48] P. Novák, R. Imhof, O. Haas, Magnesium insertion electrodes for rechargeable nonaqueous batteries—a competitive alternative to lithium?, *Electro. Acta.* 45 (1999) 351–367.

[49] M.S. Whittingham, Lithium batteries and cathode materials, *Chem. Rev.* 104 (2004) 4271–4302.

[50] J.B. Goodenough, Y. Kim, Challenges for rechargeable Li batteries, *Chem. Mater.* 22 (2010) 587–603.

[51] J.D. Genders, D. Pletcher, Studies using microelectrodes of the Mg (II)/Mg couple in tetrahydrofuran and propylene carbonate, *J. Electro. Chem. Inter. Electro.* 199 (1986) 93–100.

[52] O.R. Brown, R. McIntyre, The magnesium and magnesium amalgam electrodes in aprotic organic solvents a kinetic study, *Electro. Acta.* 30 (1985) 627–633.

[53] Y. Zhu, et al., Hydrated $Mg_xV_5O_{12}$ cathode with improved Mg^{2+} storage performance, *Adv. Energy Mater.* 10 (2020) 2002128.

[54] M.M. Huie, D.C. Bock, E.S. Takeuchi, A.C. Marschilok, K.J. Takeuchi, Cathode materials for magnesium and magnesium-ion based batteries, *Coordi. Chem. Rev.* 287 (2015) 15–27.

[55] M. Rastgoo-Deylami, M.S. Chae, S.-T. Hong, $H_2V_3O_8$ as a high energy cathode material for nonaqueous magnesium-ion batteries, *Chem. Mater.* 30 (2018) 7464–7472.

[56] C. Pei, F. Xiong, J. Sheng, Y. Yin, S. Tan, D. Wang, C. Han, Q. An, L. Mai, VO_2 nanoflakes as the cathode material of hybrid magnesium–lithium-ion batteries with high energy density, *ACS Appl. Mater. Interfaces.* 9 (2017) 17060–17066.

[57] L. Blanc, et al., Toward the development of a high-voltage Mg cathode using a chromium sulfide host, *ACS Mater. Lett.* 3 (2021) 1213–1220.

[58] J. Bae, H. Park, X. Guo, X. Zhang, J. Warner, G. Yu, High-performance magnesium metal battery via switching passivation film into solid electrolyte interphase, *Energy Environ. Sci.* 14(8) (2021) 4391–4399.

[59] Y. Shao, et al., 3D crumpled ultrathin 1T MoS_2 for inkjet printing of Mg-ion asymmetric micro-supercapacitors, *ACS Nano.* 14 (2020) 7308–7318.

[60] L.F. Wan, D. Prendergast, Ion-pair dissociation on α-MoO_3 surfaces: Focus on the electrolyte–cathode compatibility issue in Mg batteries, *J. Phys. Chem. C.* 122 (2018) 398–405.

[61] P.G. Bruce, F. Krok, J. Nowinski, V.C. Gibson, K. Tavakkoli, Chemical intercalation of magnesium into solid hosts, *J. Mater. Chem.* 1 (1991) 705–706.

[62] D. Wu, J. Zeng, H. Hua, J. Wu, Y. Yang, J. Zhao, $NaV_6\,O_{15}$: A promising cathode material for insertion/extraction of Mg^{2+} with excellent cycling performance, *Nano Res.* 13 (2020) 335–343.

[63] A. Mukherjee, S. Taragin, H. Aviv, I. Perelshtein, M. Noked, Rationally designed vanadium pentoxide as high capacity insertion material for Mg-Ion, *Adv. Func. Mater.* 30 (2020) 2003518.

[64] A. Das, et al., Electrolytes for magnesium-ion batteries next generation energy storage solutions for powering electric vehicles, in: *Ceramic and Specialty Electrolytes for Energy Storage Devices*, CRC Press, 2021. pp. 177–191. Florida, US.

[65] Y. Li, Q. Dang, W. Chen, L. Tang, M. Hu, Recent advances in rechargeable batteries with prussian blue analogs nanoarchitectonics, *J. Inorganic and Organ. Polym. Mater.* (2021) 1–17.

[66] P. Marzak, M. Kosiahn, J. Yun, A.S. Bandarenka, Intercalation of Mg^{2+} into electrodeposited prussian blue analogue thin films from aqueous electrolytes, *Electro. Acta.* 307 (2019) 157–163.

[67] S. Rasul, S. Suzuki, S. Yamaguchi, M. Miyayama, Synthesis and electrochemical behavior of hollandite MnO_2/acetylene black composite cathode for secondary Mg-ion batteries, *Solid State Ionics.* 225 (2012) 542–546.

[68] N. Kumagai, S. Komaba, H. Sakai, N. Kumagai, Preparation of todorokite-type manganese-based oxide and its application as lithium and magnesium rechargeable battery cathode, *J. Power Sources.* 97 (2001) 515–517.

[69] N. Kuganathan, E.I. Gkanas, A. Chroneos, Mg_6MnO_8 as a magnesium-ion battery material: Defects, dopants and Mg-ion transport, *Energies.* 12 (2019) 3213.

[70] Y. Tuhudahong, Y. Nuli, Q. Chen, J. Yang, J. Wang, Vanadium doped $Mg_{1.03}Mn_{0.97}SiO_4$ cathode materials for rechargeable magnesium batteries, *Mat. Sci.* 2 (2012) 139–144.

[71] J. Heath, H. Chen, M.S. Islam, $MgFeSiO_4$ as a potential cathode material for magnesium batteries: iIon diffusion rates and voltage trends, *J. Mater. Chem. A.* 5 (2017) 13161–13167.

[72] Q.D. Truong, M.K. Devaraju, I. Honma, Nanocrystalline $MgMnSiO_4$ and $MgCoSiO_4$ particles for rechargeable Mg-ion batteries, *J. Power Sources.* 361 (2017) 195–202.

[73] L.F. Wan, B.R. Perdue, C.A. Apblett, D. Prendergast, Mg desolvation and intercalation mechanism at the MO_6S_8 chevrel phase surface, *Chem. Mater.* 27 (2015) 5932–5940.

[74] H.D. Yoo, I. Shterenberg, Y. Gofer, G. Gershinsky, N. Pour, D. Aurbach, Mg rechargeable batteries: An on-going challenge, *Energy Environ. Sci.* 6 (2013) 2265–2279.
[75] E. Lancry, E. Levi, Y. Gofer, M. Levi, G. Salitra, D. Aurbach, Leaching chemistry and the performance of the MO_6S_8 cathodes in rechargeable Mg batteries, *Chem. Mater.* 16 (2004) 2832–2838.
[76] S.-H. Choi, et al., Role of Cu in MO_6S_8 and Cu mixture cathodes for magnesium ion batteries, *ACS Appl. Mater. Interfaces*. 7 (2015) 7016–7024.
[77] C. Wu, et al., MoS_2/graphene heterostructure with facilitated Mg-diffusion kinetics for high-performance rechargeable magnesium batteries, *Chem. Eng. J.* 412 (2021) 128736.
[78] Y. Liu, L.-Z. Fan, L. Jiao, Graphene intercalated in graphene-like MoS_2: A promising cathode for rechargeable Mg batteries, *J. Power Sources*. 340 (2017) 104–110.
[79] S. Yang, D. Li, T. Zhang, Z. Tao, J. Chen, First-principles study of zigzag MoS_2 nanoribbon as a promising cathode material for rechargeable Mg batteries, *J. Phys. Chem. C.* 116 (2012) 1307–1312.
[80] Z. Liu, et al., Binder-free MnO_2 as a high rate capability cathode for aqueous magnesium ion battery, *J. Alloys Comp.* 869 (2021) 159279. https://doi.org/10.1016/j.jallcom.2021.159279.
[81] Reversible Mg-Ion Insertion in a Metastable One-Dimensional Polymorph of V_2O_5-ScienceDirect, (n.d.). https://www.sciencedirect.com/science/article/pii/S245192941730520X. (Accessed: August 21, 2021).
[82] L. Zhou, et al., Interlayer-spacing-regulated $VOPO_4$ nanosheets with fast kinetics for high-capacity and durable rechargeable magnesium batteries, *Adv. Mater.* 30 (2018) 1801984.
[83] X. Cai, et al., MOF derived TiO_2 with reversible magnesium pseudocapacitance for ultralong-life Mg metal batteries, *Chem. Eng. J.* 418 (2021) 128491. https://doi.org/10.1016/j.cej.2021.128491.
[84] R. Zhang, et al., α-MnO_2 as a cathode material for rechargeable Mg batteries, *Electro. Commun.* 23 (2012) 110–113. https://doi.org/10.1016/j.elecom.2012.07.021.
[85] Y. Ju, et al., Li^+/Mg^{2+} hybrid-ion batteries with long cycle life and high rate capability employing MOS_2 nano flowers as the cathode material, *Chem.–A European J.* 22 (2016) 18073–18079.
[86] F. Zhu, H. Zhang, Z. Lu, D. Kang, L. Han, Controlled defective engineering of MOS_2 nanosheets for rechargeable Mg batteries, *J. Energy Stor.* 42 (2021) 103046. https://doi.org/10.1016/j.est.2021.103046.
[87] M. Matsui, Study on electrochemically deposited Mg metal, *J.Power Sources*. 196 (2011) 7048–7055.
[88] C. Ling, K. Suto, Thermodynamic origin of irreversible magnesium trapping in chevrel phase Mo6S8: importance of magnesium and vacancy ordering, *Chem. Mater.* 29 (2017) 3731–3739.
[89] Y. Zhang, J. Gui, T. Li, Z. Chen, S. Cao, F. Xu, A novel Mg/Na hybrid battery based on $Na_2VTi\ (PO_4)_3$ cathode: Enlightening the Na-intercalation cathodes by a metallic Mg anode and a dual-ion Mg^{2+}/Na^+ electrolyte, *Chem. Eng. J.* 399 (2020) 125689.
[90] R. Zhang, O. Tutusaus, R. Mohtadi, C. Ling, Magnesium-sodium hybrid battery with high voltage, capacity and cyclability, *Front. Chem.* 6 (2018) 611.
[91] M. Walter, K.V. Kravchyk, M. Ibanez, M.V. Kovalenko, Efficient and inexpensive sodium–magnesium hybrid battery, *Chem. Mater.* 27 (2015) 7452–7458.
[92] H. Dong, et al., A magnesium–sodium hybrid battery with high operating voltage, *Chem. Commun.* 52 (2016) 8263–8266.
[93] Y. Li, et al., A high-voltage rechargeable magnesium-sodium hybrid battery, *Nano Energy*. 34 (2017) 188–194.
[94] X. Cao, L. Wang, J. Chen, J. Zheng, A low-cost Mg^{2+}/Na^+ hybrid aqueous battery, *J. Mater. Chem. A.* 6 (2018) 15762–15770.

Index

Printed in the United States
by Baker & Taylor Publisher Services